Haemaphysalis Ticks of India

Haemaphysalis Ticks of India

G. Geevarghese and A.C. Mishra
National Institute of Virology,
Pune, Maharashtra, India

This publication forms a part of the centenary year celebration of the Indian Council of Medical Research (ICMR), New Delhi

ELSEVIER

AMSTERDAM • BOSTON • HEIDELBERG • LONDON • NEW YORK • OXFORD
PARIS • SAN DIEGO • SAN FRANCISCO • SINGAPORE • SYDNEY • TOKYO

Elsevier
32 Jamestown Road London NW1 7BY
225 Wyman Street, Waltham, MA 02451, USA

First edition 2011

Notices
Knowledge and best practice in this field are constantly changing. As new research and experience broaden our understanding, changes in research methods, professional practices, or medical treatment may become necessary.

Practitioners and researchers must always rely on their own experience and knowledge in evaluating and using any information, methods, compounds, or experiments described herein. In using such information or methods they should be mindful of their own safety and the safety of others, including parties for whom they have a professional responsibility.

To the fullest extent of the law, neither the Publisher nor the authors, contributors, or editors, assume any liability for any injury and/or damage to persons or property as a matter of products liability, negligence or otherwise, or from any use or operation of any methods, products, instructions, or ideas contained in the material herein.

British Library Cataloguing-in-Publication Data
A catalogue record for this book is available from the British Library

Library of Congress Cataloging-in-Publication Data
A catalog record for this book is available from the Library of Congress

ISBN: 978-0-323-16527-3

This book has been manufactured using Print On Demand technology. Each copy is produced to order and is limited to black ink. The online version of this book will show color figures where appropriate.

Transferred to Digital Printing in 2016

Contents

Preface

Out of the nine genera of ixodid ticks known in India, species of the genus *Haemaphysalis* Koch, 1,844 are among the most important as transmitters of animal and human diseases. The basic work on haemaphysaline ticks is the monographic work by Nuttall and Warburton in 1915 and for the Indian region by Sharif in 1928. However, illustrations/descriptions of haemaphysaline ticks available then in different collections were only partial or were not in detail. Original descriptions of most of these ticks have been usually based on adults and frequently only on one sex or one specimen.

According to Hoogstraal and Kim, "These ticks constitute the most useful assemblage in the superfamily Ixodoidea for displaying numerous interrelated structural-biological clues to affinities between tick species and groups and to historical and contemporary host associations. No other tick species shows this range of lucid clues and this variety of data for species of biological and evolutionary significance." It indicates the taxanomical importance of this group, which warranted a study on this group. Though a number of publications on individual species have been published, all these publications are scattered in different scientific journals, and there is no compiled information on the ticks of this genus with regard to taxonomy, ecology, biology, etc., in India. Therefore, in an effort to fill this lacuna, a monographic work has been undertaken with an objective to gather information on this group of ticks. The work presented in this monograph is discussed in three parts: Part 1 is an introduction; Part 2 includes taxonomy, biology, and ecology of individual species; and Part 3 gives a summary of the whole work.

The present monograph provides a compilation of taxonomy and biologic descriptions of the 42 haemaphysaline ticks that are found in the Indian subcontinent. In addition, taxonomic keys for individual species as well as for each subgenus have been provided. These species have been grouped into different subgenera using Hoogstraal and Kim's keys for different subgenera. Notes on ecology and biology of many of these species have been added. The phylogenetic relationships of these ticks have been discussed in relation to hosts. In addition, immatures of a few of these species (more than 10 species) have been described for the first time, including a new species. Keys provided for subgeneric and individual identification will be useful for easy identification of Indian haemaphysaline ticks.

Several of our colleagues in the division of Medical Entomology and Zoology, National Institute of Virology (NIV), gave valuable advice and liberal assistance during the course of preparation of this manuscript. We are indeed grateful to each of them. Special mention may be made here to Mr. Omkar Manadke who spent several days preparing the manuscript. We also acknowledge and remember the names of the late Dr. Harry Hoogstraal and the late Dr. V. Dhanda, for enabling us to use some of their illustrations of Indian haemaphysaline ticks. We are also thankful to the editors of the *Journal of Parasitology* who gave permission to reproduce certain illustrations. The editorial help received from Prof. K.M. Thomas and S.M. Jadhav is also gratefully acknowledged.

Part 1

General

1 Introduction

Ticks are obligate bloodsucking arthropods found in almost every region of the world.[1–4,6,37,49] They belong to the subfamily Ixodoidea of the order Acarina of class Arachnida.[1,105,110] They are very important vectors of human and animal diseases.[133] They surpass all other arthropods in the variety of pathogenic organisms transmitted to man and animals, which includes viruses, rickettsia, protozoa, and bacteria. Many of these diseases are zoonotic in nature, causing infection to the human population closely associated with cattle and forest. Ticks rank second only to mosquitoes as vectors of human and animal diseases. Some of the tick-borne human diseases such as Lyme disease and Crimean Congo hemorrhagic fever (CCHF) have caused panic recently in public health authorities in many countries including Europe and America. Tick-borne protozoan, rickettsial, viral, and bacterial diseases continue to affect man and livestock production in India and abroad. Tick-borne protozoan diseases such as theilereasis and babesiosis are burning veterinary problems in many countries including India.

It is essential to develop strategies to control these arthropods in order to contain the tick-borne diseases affecting man and animals. Usually, persons become infected with tick-borne diseases when they intrude into an established natural focus (biocenose) where the disease agent is quietly cycling between vertebrate reservoir hosts and their tick vectors. This intersection is increasingly frequent as human populations grow in number and mobility; as more persons reside, work, or recreate in tick-infested habitats; as changing environments favor ticks and increase their range, density, and likelihood of human interaction. It is upon an adequate understanding of taxonomy, vector biology, and ecology in various geographic regions of the country that a sound program of control measures can be instituted. Further, a basic knowledge of taxonomy is important for the proper recognition of the vector species and its variants. The strategies to be used need to be adapted to the local species and their behavioral characteristics. Knowledge of the habits and habitats of the vectors responsible for transmission of tick-borne diseases in different ecologic settings in the country is therefore an upshot for the successful pursuit of any control program. Presently, information on the systematics and ecology of Indian ticks is scattered in numerous scientific journals and magazines as research papers, published from India and abroad. There is no compiled work or monograph providing health research workers in India with accurate and up-to-date information on the distribution, ecology, disease relationship, and control of the ticks of India. It is hoped that this monograph on *Haemaphysalis*, one among a few to appear shortly on Indian ticks, will serve this purpose.

Haemaphysalis Ticks of India. DOI: 10.1016/B978-0-12-387811-3.00001-2

1.1 Tick Studies: Historical Background

Ticks, which have been around in much the same form for approximately 200 million years, are among the oldest and most successful groups of arthropods.[6] They were described as pests by the ancient Greeks and ever since have been treated in literature mostly with revulsion. These primitive, obligate, bloodsucking parasites prey on every class of vertebrate in all parts of the world. Ticks were the first arthropods to be clearly established as vectors of infectious disease, owing to the discovery in 1893 by Smith and Kilbourne of the role of *Boophilus annulatus* as the vector of *Babesia bigemina*, the protozoal cause of Texas cattle fever.[133] A decade or so after this discovery, mosquitoes were established as the vectors for malaria, yellow fever, and filariasis. In 1903, ticks were first proved as vectors of human disease, when J.E. Dutton, working in the Congo, discovered the principal cause of endemic relapsing fever (named *Babesia duttonii*) and its argasid (soft-tick) vector, *Ornithodoros moubata*.[48]

A perusal of literature reveals that the history of our knowledge of the Indian ticks dates from the time of Linnaeus who described *Acarus elephantinus* in 1758 and *Acarus indus* in 1767 from India.[104,126,127] Both these species were invalid according to Neumann (1911), who published a monograph on ticks, "Ixodidae" wherein he described a few tick species belonging to seven genera, *Haemaphysalis*, *Ixodes*, *Amblyomma*, *Aponomma*, *Dermacentor*, *Hyalomma*, and *Rhipicephalus* from India. In an earlier series of papers published between 1897 and 1910, he had described a number of new species belonging to the above genera. In 1907, Warburton published a short account of certain Indian species. In 1910 Warburton published a report on the ticks belonging to the Indian Museum wherein he added a number of additional tick species belonging to the above-mentioned genera. Nuttall and Warburton along with Robinson and Cooper published Part II and Part III of a comprehensive work entitled "Ticks: A Monograph of Ixodidae" in 1911 and 1915. This was followed by Part IV in 1926 by Robinson.

The next major work in Indian tick research was by Sharif (1928) who reported a checklist of 45 species from the collections in the Indian Museum, Calcutta.[127] The next checklist was by Sen (1938) who reported 50 species. The discovery of Kyasanur Forest Disease (KFD) from ticks in March 1957 marked the next milestone in the history of Indian tick studies and stimulated extensive studies on different aspects of studies on Indian ticks, which included taxonomy, distribution, biology and ecology, disease relationships, and control.[46,144] Subsequently the tick fauna of the KFD, which presented a perplexing array of nomenclature and identification problems, involving larvae, nymphs, and adults of 14 *Haemaphysalis* species, were solved by the classic work by Trapido et al. (1964) who prepared a key for the identification of these ticks. Detailed studies on the ecology of ticks in India are mostly confined to the ticks of KFD area. Great contributions by the late Dr. Harry Hoogstraal in the study of Indian ticks especially systematics is worth mentioning here. He has been instrumental in getting a number of new species described from India. The extensive field studies on *Haemaphysalis spinigera*, the main vector of KFD, by Bhat (1968) will remain classic in the field of ecology of Indian ticks. The checklist published by Jagannath et al. (1973) and Miranpuri and Naithani (1978) who reported 99 and 160 species, respectively, have helped tick studies in India. The latest checklist is by Geevarghese et al. (1997), who have reported 106 valid ticks after segregating the invalid species and synonyms from the list provided by earlier workers. Other workers who have contributed to the taxonomic and biologic studies of ticks in India are Drs V. Dhanda, K.R.P. Singh, Alwar, Heregoudar, Jagannath, Nagar, Gill, B.D. Sharma, Vazirani, etc. Most of these workers have carried out studies on tick fauna in certain Indian regions where they have been working.[91–94]

1.2 External Morphology of Ixodid Ticks[4,133]

The body of a tick is divisible into an anterior region, **gnathosoma** (capitulum), and a posterior region called the **idiosoma** or the main body part. In the Ixodidae, a sclerotized shield called the **scutum** covers the entire dorsal surface. However, in the female, larvae, and nymph it covers only the anterior region. Scutum is absent in Argasidae. There are a number of grooves on the scutum and these are named after their location on the scutum, namely cervical, lateral, marginal grooves. The scutal surface also may bear a number of small pits called **punctations**, which vary in size and number. There are various other structures like ridges, **mammillae**, and **discs** on the dorsum of Argasidae. The posterior end of the idiosoma is divided into a number of rectangular areas called **festoons**. These are absent in one genus of Ixodidae, namely *Boophilus* and in all genera of Argasidae. Eyes when present are situated on the lateral margins of the scutum, anteriorly in males or at the greatest width of scutum in the female. In Argasidae if the eyes are present, they are situated on the lateral on the supracoxal folds. The adults of certain genera such as *Dermacentor* and *Amblyomma* possess ornamental bodies, that is, some parts of the scutum may have patches of enameled colors. Important structures placed on the ventral side of the ticks are (1) v**entral plates**; (2) **genital aperture** placed anteriorly covered by a delicate sclerotized flap called the genital operculum; (3) **anus** on the posterior side; and (4) **spiracle**. The gnathosoma or capitulum consists of a posterior basis capitulum to which is attached the mouthparts. On the dorsal surface of basis capitulum in female ixodids, there is a pair of areas called ***porose areas*** in certain species of hard ticks. Two spur-like structures called **corona** are situated dorsally on either side of the posterolateral side of the basis capitulum. The mouthparts consist of a tonguelike projection called the **hypostome** over which lie dorsally a pair of chelicerae, which are used for piercing. Chelicerae consist of long movable shafts and cutting digits attached dorsoanteriorly. Hypostomes possess a number of rows of backwardly directed curved teeth, which vary in shape, size, and number. The number of rows of teeth in the hypostome is a diagnostic feature for ticks, for example, *Amblyomma*. There are two palps situated on each side of the mouthparts. Each palp consists of four segments, that is, segment nos. 1, 2, 3, and 4. Segment 2 is usually the longest one and may project basolaterally beyond the basis capitulum. Segments 2 and 3 may have basodorsal and/or basoventral retroverted spurs. These palpal segments are important from the identification point of ticks especially in *Haemaphysalis*.

The adults are provided with four pairs of legs, which are attached to idiosoma ventrally by immovable coxae. Each leg has six segments, namely coxa, trochanter, femur, tibia, metatarsum, and tarsus. A pair of claws and median pad or pulvillus is attached to the posterior side of the tarsus. Larvae have only three pairs of legs. A sensory organ called **Haller's organ** is situated dorsally on the tarsus of leg no. 1. The Haller organ consists of the posterior capsule and an anterior pit. This organ is unique and is believed to be olfactory in nature.

Body parts of typical male and female hard ticks, including various parts of capitulum are given in the illustrations below. Terminology used for these illustrations has been used throughout this book.

1.3 Tick Genera in India[3,86]

The family Ixodidae includes **13 genera** comprising approximately **650 species** known from all over the world. In India a total of **88 species** belonging to the following **seven genera** have been recorded.

Genus *Ixodes*

This is the largest genus in the family Ixodidae containing **245 species**, of which only 11 are known to occur in India. *Ixodes* are inornate ticks. The capitulum of the female is considerably longer than that of male. There are no eyes or festoons. The anal groove is placed anteriorly to the anus, whereas in all other genera the anal groove is posterior to the anus. In the male there are seven ventral plates. In India *Ixodes* species are known to occur in forests with heavy rainfall. They are known to mate off the hosts. They are generally three-host ticks.

Genus *Haemaphysalis*

At present 155 *Haemaphysalis* species are known throughout the world, of which 41 have been recorded in India and one new species is proposed in this book. These are small inornate ticks with short mouthparts, that is, brevirostrate. The basis capitulum is rectangular and the base of the second palpal segment is expanded, projecting laterally beyond the basis capitulum. The second and third palpal segments taper anteriorly so that the capitulum anterior to the basis capitulum appears to be triangular. There are no eyes in either sex. Festoons are present. They are generally three-host ticks.

Genus *Dermacentor*

These are usually ornate, brevirostrate ticks. There are 30 species known from all over the world of which three occur in India. The basis capitulum is rectangular dorsally. Coxa IV is greatly enlarged in the male, which has no ventral plates. They are generally three-host ticks.

Genus *Nosomma*

These are brevirostrate ticks with short palps. This genus comprises a single species *N. monstrosum*, which has been recorded from India and Southeast Asia. This is a three-host tick.

Genus *Hyalomma*

These are medium-sized inornate ticks with long mouthparts, that is, longirostrate. The basis capitulum is subtriangular dorsally and eyes are situated in orbit-like pits. The genus comprises 30 known species found all over the world, out of which nine have been recorded from India. The male has one pair each of adanal and accessory adanal plates. Coxa I is bifid. *Hyalomma* are tough, hardy ticks that survive where humidity is low and climatic conditions are extreme. *Hyalomma* species are either two- or three-host ticks.

Genus *Rhipicephalus*

These are small brevirostrate, reddish or blackish brown ticks, and are inornate. This genus contains 75 known species, eight of which occur in India, including those belonging to ***Boophilus*** group which has been merged recently with this genus. The basis capitulum is hexagonal dorsally and coxa I is bifid in both sexes. The male has adanal and accessory adanal shields or plates, the shapes of which are of diagnosable value. In ticks belonging to *Boophilus* group, the palpi are compressed and articles II and III have transverse sides on the dorsal and lateral surfaces. Festoons are absent. Males bear elongated posteriorly pointed adanal shields. *Boophilus* ticks are generally one-host ticks, whereas ***Rhipicephalus*** are mostly three-host ticks.

Genus *Amblyomma*

These are large ornate longirostrate ticks with festoons, but no ventral plates in the male. The genus is represented by 12 species in India, including those belonging to the ***Aponomma*** group which has been merged recently with this genus. There are 126 known species of this genus distributed all over the world. Species of the genus are characterized by the usually long mouthparts. The palpal article 2 is about 2 or 3 times as long as article 3. Almost all species belonging to *Amblomma* are three-host ticks, whereas *Aponomma* species are usually one-host and feed exclusively on amphibians.

1.4 Keys to Genera of Indian Ixodidae

1. Anal grooves surrounding the anus in front .... *Ixodes*
Anal groove distinct and contouring the anus behind or anal groove faint or indistinct 2

2. Without eyes 3
With eyes 4

3. Short conical palpi, article II projects laterally beyond the basis capituli, about twice as broad as long .... *Haemaphysalis*
Long narrow palpi, article II about twice as long as broad .... *Aponomma*

4. Palpi short and broad, that is, as broad as or broader than the length 5
Palpi longer than broad 6

5. Large ticks usually ornate, basis capituli rectangular dorsally 6
Small inornate ticks, basis capituli usually hexagonal dorsally 7

6. Large ticks usually ornate, basis capituli rectangular dorsally, in some species coxae I–IV of the male increases progressively in size; coxa IV largest, spiracular plate is oval or comma shaped .... *Dermacentor*
Scutum ornate, male with adanal; accessory or subanal plates; each subanal plate with three lobes .... *Nosomma*

7. Anal grooves distinct, with eyes and festoons, coxa I bifid, male possesses a pair of adanal shields and usually a pair of accessory adanal shields, spiracles bluntly or elongately comma shaped .... *Rhipicephalus*
Anal grooves are obsolete in female, weakly indicated in male, eyes present but festoons absent, slightly chitinized, spiracles are rounded or oval in both sexes, slightly chitinized palpi ridged dorsally and laterally, coxae bifid, small male with adanal and accessory shields ....*Boophilus*

8. Capitulum long and scutum inornate, male with adanal and subanal shields, palpal segment II is at least twice as long as segment III, festoons may or may not be present or they may be partially coalesced, eyes are spherical..........*Hyalomma*
Capitulum long and scutum inornate, male with adanal and subanal shields, palpal segment II is at least twice as long as segment III, festoons may or may not be present or they may be partially coalesced, eyes are generally flat and nonorbital, scutum ornate, male without ventral shields..............................*Amblyomma*

1.4.1 Family Argasidae

General Characteristics

1. Integument leathery, without a scutum.
2. Capitulum ventral or subterminal, not visible from the dorsal aspect.
3. Spiracular plates inconspicuous, located on supra cozal fluid between coxa III and IV.
4. Palpal articles of more or less equal length, article IV not recessed or subterminal.

With a definite sutural line separating the dorsal and ventral surfaces, lateral body margins with distinct rectangular discs (marginal discs). Body periphery flattened and usually distinct from dorsum with a definite sutural line separating the dorsal and ventral surfaces, lateral body margins with distinct rectangular discs (marginal discs).

Eyes absent, integument of adults and nymphs wrinkled, with numerous tiny folds and discs .... ***Argas***
Without a definite, sutural line between dorsal and ventral surfaces.
Adults with integument granular, with hypostome vestigial.
Nymphs with integument beset with spines and hypostome well developed.
Hypostomes vestigial .... ***Otobius***
Body periphery undifferentiated. Integument mammillated. Hypostome well developed with denticles in all stages .... ***Ornithodoros***
On the basis of recent molecular taxonomic studies, *Aponomma* has been merged with *Amblomma*, and *Boophilus* with *Rhipicephalus*.[6]

1.5 Biology and Life Cycle[82,90,133]

Hard Ticks

Ticks possess many unusual features that contribute to their remarkable success and vector potential. Hard ticks have three active developmental stages: larva, nymph, and adults. Soft ticks have two or more nymphal stages. The duration of each developmental stage varies depending upon the temperature, humidity conditions, and from species to species. The fed females of Indian species after a period of preoviposition continue oviposition for days together, sometimes ranging up to 30 days under the ambient temperature. In ticks there is a unique accessory genital structure called Gene's organs which facilitate oviposition (Figures 1.1 and 1.2). This organ has a balloon-shaped base provided with two anterolateral horns on each side. During oviposition period this organ is extended from the genital opening. During oviposition the eggs pass through the horns and get a wax coating. In the hard ticks, there is only one gonotrophic cycle and the females that have undergone oviposition once cannot feed and oviposit for a second time. Following oviposition, hatching begins and the emerging larvae disperse to seek hosts. Attached larvae take at least 2–3 days before they feed and drop from their hosts, find a sheltered microenvironment, and undergo ecolysis (or molting). The unfed nymphs seek hosts again. The engorged nymphs drop off their hosts, under molting in

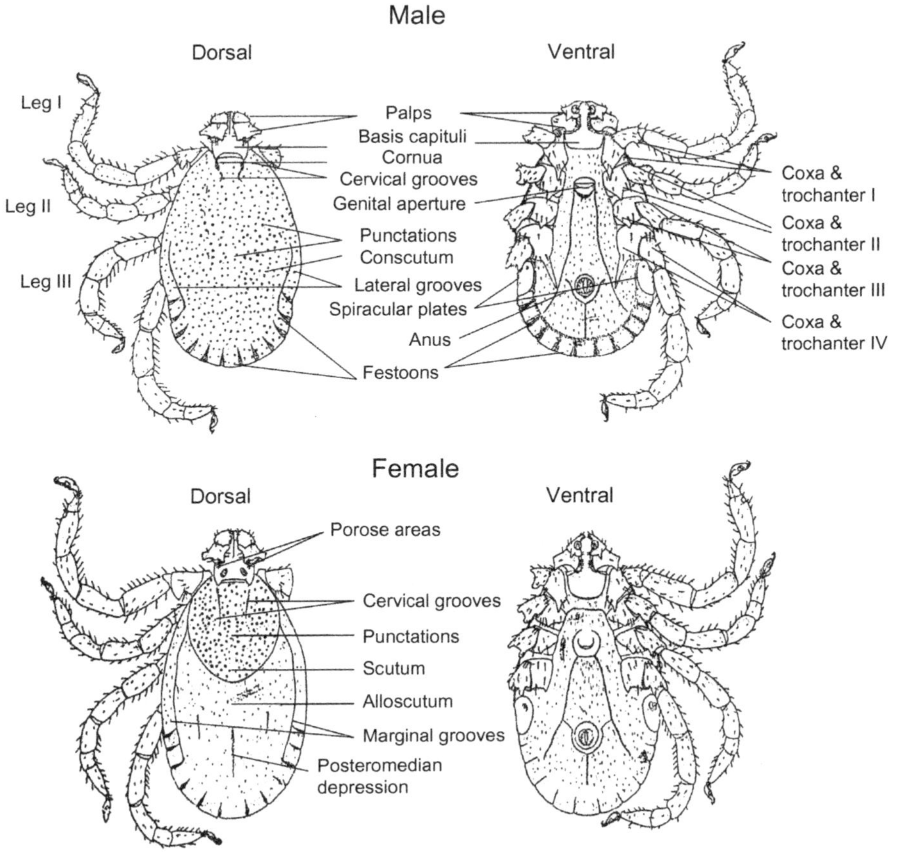

Figure 1.1 External morphological characters of ioxid ticks.

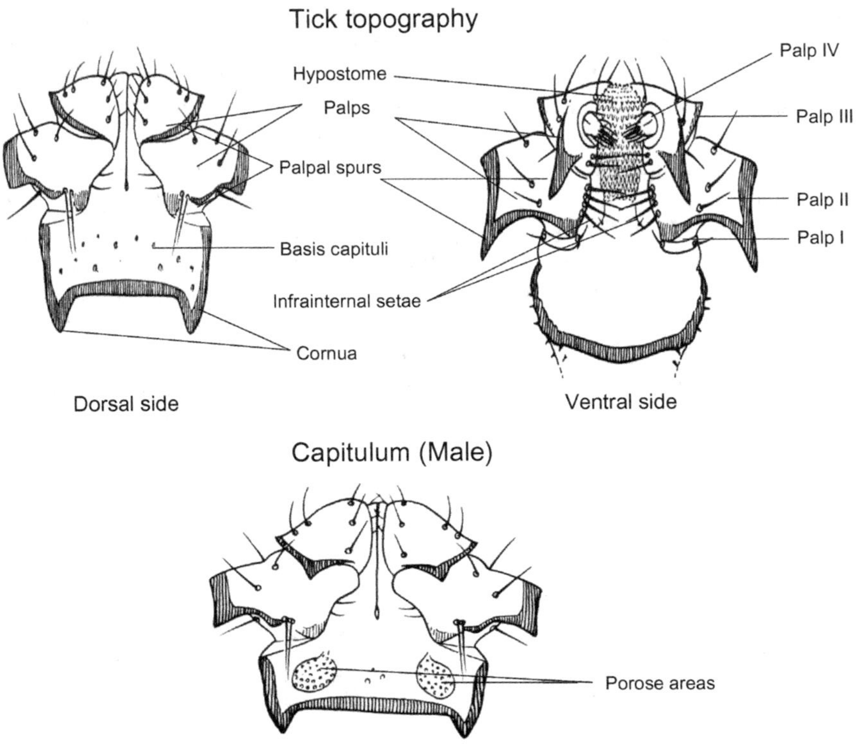

Source: Oriental Insects 1974; 8(4):505–520.

Figure 1.2 External morphological characters of capitulum.

appropriate niches in the natural environment. The adults that emerge from these fed nymphs attach to some hosts, feed, and mate, and the fed females drop to oviposit and complete the life cycle. In most of the hard ticks, mating takes place when the adults are feeding on the hosts. The eggs laid by the unmated females are not viable or fertile. In some of the *Haemaphysalis* ticks, the nymphs feed for 3–4 days, whereas adults feed for 6–7 days. Feeding periods vary in different species belonging to different genera. On the basis of the number of hosts and feeding habits, ticks are divided into three types, namely (1) one-host ticks which remain on a single host from the time of their first attachment as larvae to the time when they drop as fully fed adults, for example, *Boophilus microplus*; (2) two-host ticks in which the larvae attach themselves to the host, feed, and molt into nymphs which feed on the same host and then drop to the ground. The adults emerging from these nymphs attach to another host and drop down for egg laying, for example, *Hyalomma marginatum isaaci*. In this case, the fed larvae undergo molting in the host without detachment from the host; and (3) three-host ticks in which larvae, nymphs, and adults feed on separate hosts. The vertebrate hosts may be of the same species and or of different species, for example, all species of *Haemaphysalis*. Some of the tick species such as *Hya. a. anatolicum* behave as both two- and three-host ticks. Compared to hard ticks, soft ticks have a number of gonotrophic cycles and there is more than one nymphal instar in the life cycle. The duration of feeding is restricted to a few minutes as in the bedbug. It has been observed that in most of the hard ticks there is a rhythm with regard to the dropping of fed stages of the ticks which is synchronized with the daytime when the host is actively grazing in the field. Two species, namely *Hya. a. anatolicum* and *H. bispinosa* are exceptions to this rule. They drop only during night when their hosts, that is, cattle, are taking rest inside the cattle shed.

Soft Ticks

Argasid ticks exhibit remarkable diversity in their patterns of feeding behavior, in contrast to the much more uniform pattern found in the Ixodidae. Argasid ticks feed rapidly (except in larvae of certain **Argas** and **Ornithodoros** species). In contrast to Ixodidae, argasid nymphs and adults swell enormously during engorgement. They increase only 3–5 times of their body weight. The females feed and oviposit frequently (multiple gonotrophic cycles). The mated females deposit small egg masses (i.e., 500 eggs/cycle). There are often many nymphal molts in the life cycle. Almost all argasid ticks have a multi-host feeding pattern. In the majority of species that feed on mammals and birds, larvae seek hosts, feed rapidly (15–30 min), and drop to molt. Fed larvae molt to first stage nymphs (N1) followed by other nymphal instars, numbering usually 3–4. An important factor enabling the nymphs to feed rapidly is their ability to eliminate excess blood meal water in the form of coxal fluid which is a clear colorless liquid excreted from the coxal glands during or shortly after feeding. In the Argasidae, the passage of so many nymphal stages contributes to a much longer life cycle than in Ixodidae.

In Argasidae, adults become sexually active after emergence from the last nymphal molt and they do require a blood meal to initiate gametogenesis. Mating occurs before as well as after blood feeding and rarely on the host itself. However, the gonotrophic cycle is completely dependent upon blood meal. When oviposition is completed, the ticks remain vigorous, the hungry females then seek hosts, feed, and oviposit again. Usually these cycles never exceed six. This pattern of repeated gonotrophic cycles, often followed by long periods of waiting between blood meals, enables argasid ticks to disperse their progeny gradually over time. Thus, argasid strategy is different from that of Ixodid, where all of the progeny of a fecund female are produced in one massive surge of ovipositional activity.

Ticks are able to parasitize a variety of host species. The general host pattern of Indian ticks is given in Section 1.61.

1.6 Hosts

Hyalomma	Adults—Domestic animals Immatures—Domestic animals, small mammals, birds
Rhipicephalus	Adults—Domestic animals Immatures—Domestic animals and rodents
Amblyomma	Adults—Domestic and wild animals Immatures—Domestic animals
Haemaphysalis	Adults—Domestic and wild animals; birds Immatures—Domestic and wild animals, birds, small mammals, etc.
Ixodes	Adults—Domestic and wild animals; carnivores, flying squirrel, mongoose. Immatures—Domestic animals, small mammals
Boophilus	Adults and Immature stages—Domestic animals
Nosomma	Adults—Domestic animals, Immatures—small mammals
Dermacentor	Adults—Domestic animals, wild boar, bear, etc., Immatures—Rodents
Aponomma	Adults and Immatures—Reptiles
Otobius	Adults and Immatures—Horses
Argas	Adults and Immatures—Birds, herons, egrets, storks and spoonbills, hawks, vultures, bats
Ornithodors	Adults and Immatures—Domestic animals, cattle

1.7 List of *Haemaphysalis* Ticks from All Over the World

Genus *Haemaphysalis*

1. ***H. aborensis* Warburton, 1913**
2. *H. aciculifer* Warburton, 1913
3. ***H. aculeata* Lavarra, 1904**
4. *H. adleri* Feldman-Muhsam, 1951
5. ***H. anomala* Warburton, 1913**
6. *H. anomaloceraea* Teng and Cui, 1984
7. *H. anoplos* Hoogstraal, Uilenberg and Klein, 1967
8. ***H. aponommoides* Warburton, 1913**
9. *H. asiatica* Supino, 1897
10. *H. atheruri* Hoogstraal, Trapido and Kohls, 1965
11. *H. bancrofti* Nuttall and Warburton, 1915
12. *H. bandicota* Hoogstraal and Kohls, 1965
13. *H. bartelsi* Schulze, 1938
14. *H. bequaerti* Hoogstraal, 1956
15. ***H. birmaniae* Supino, 1897**
16. ***H. bispinosa* Neumann, 1897**
17. *H. borneata* Hoogstraal, 1971
18. *H. bremneri* Roberts, 1963
19. *H. calcarata* Neumann, 1902
20. *H. calva* Nuttall and Warburton, 1915

21. ***H. campanulata* Warburton, 1908**
22. ***H. canestrinii* Supino, 1897**
23. *H. capricornis* Hoogstraal, 1966
24. *H. caucasica* Olenev, 1928
25. *H. celebensis* Hoogstraal, Trapido and Kohls, 1965
26. *H. chordeilis* Packard, 1869
27. *H. colasbelcouri* Santos Dias, 1958
28. *H. concinna* Koch, 1844
29. *H. cooleyi* Bedford, 1929
30. *H. cornigera* Neumann, 1897
31. ***H. cornupunctata* Hoogstraal and Varma, 1962**
32. ***H. cuspidata* Warburton, 1910**
33. *H. dangi* Phan Trong, 1977
34. *H. danieli* Cerny and Hoogstraal, 1977
35. ***H. darjeeling* Hoogstraal and Dhanda, 1970**
36. ***H. davisi* Hoogstraal, Dhanda and Bhat, 1970**
37. *H. demidovae* Emel'yanova, 1978
38. ***H. doenitzi* Warburton and Nuttall, 1909**
39. *H. elliptica* Koch, 1844
40. *H. elongata* Neumann, 1897
41. *H. erinacei erinacei* Pavesi, 1884
42. *H. erinacei ornata* Feldman-Muhsam, 1956
43. *H. erinacei taurica* Pospelova-Shtrom, 1940
44. *H. erinacei turanica* Pospelova-Shtrom, 1940
45. *H. eupleres* Hoogstraal, Kohls and Trapido, 1965
46. *H. filippovae* Bolotin, 1979
47. *H. flava* Neumann, 1897
48. *H. formosensis* Neumann, 1913
49. *H. fossae* Hoogstraal, 1953
50. *H. fujisana* Kitaoka, 1970
51. ***H. garhwalensis* Dhanda and Bhat, 1968**
52. *H. goral* Hoogstraal, 1970
53. *H. grochovskajae* Kolonin, 1992
54. *H. heinrichi* Schulze, 1939
55. ***H. himalaya* Hoogstraal, 1966**
56. *H. hirsuta* Hoogstraal, Trapido and Kohls, 1966
57. *H. hispanica* Gil Collado, 1938
58. *H. hoodi* Warburton and Nuttall, 1909
59. *H. hoogstraali* Kohls, 1950
60. *H. houyi* Nuttall and Warburton, 1915
61. ***H. howletti* Warburton, 1913**
62. *H. humerosa* Warburton and Nuttall, 1909
63. *H. hylobatis* Schulze, 1933
64. *H. hyracophila* Hoogstraal, Walker and Neitz, 1971
65. ***H. hystricis* Supino, 1897**
66. *H. ias* Nakamura and Yajima, 1937
67. ***H. indica* Warburton, 1910**
68. ***H. indoflava* Dhanda and Bhat, 1968**

69. *H. inermis* Birula, 1895
70. ***H. intermedia* Warburton and Nuttall, 1909**
71. *H. japonica douglasi* Nuttall and Warburton, 1915
72. *H. japonica japonica* Warburton, 1908
73. *H. juxtakochi* Cooley, 1946
74. *H. kadarsani* Hoogstraal and Wassef, 1977
75. ***H. kashmirensis* Hoogstraal and Varma, 1962**
76. ***H. kinneari* Warburton, 1913**
77. *H. kitaokai* Hoogstraal, 1969
78. *H. koningsbergeri* Warburton and Nuttall, 1909
79. *H. kopetdaghica* Kerbabaev, 1962
80. ***H. kutchensis* Hoogstraal and Trapido, 1963**
81. ***H. kyasanurensis* Trapido, Hoogstraal and Rajagopalan, 1964**
82. *H. lagostrophi* Roberts, 1963
83. *H. lagrangei* Larrousse, 1925
84. *H. laocayensis* Phan Trong, 1977
85. *H. leachi* Audouin, 1826
86. *H. lemuris* Hoogstraal, 1953
87. *H. leporispalustris* Packard, 1869
88. *H. lobachovi* Kolonin, 1995
89. *H. longicornis* Neumann, 1901
90. *H. luzonensis* Hoogstraal and Parrish, 1968
91. *H. madagascariensis* Colas-Belcour and Millot, 1948
92. *H. mageshimaensis* Saito and Hoogstraal, 1973
93. ***H. megalaimae* Rajagopalan, 1963**
94. *H. megaspinosa* Saito, 1969
95. *H. menglaensis* Pang, Chen and Xiang, 1982
96. ***H. minuta* Kohls, 1950**
97. Hoogstraal and El Kammah, 1972
98. *H. moschisuga* Teng, 1980
99. *H. muhsamae* Santos Dias, 1954
100. *H. nadchatrami* Hoogstraal, Trapido and Kohls, 1965
101. ***H. nepalensis* Hoogstraal, 1962**
102. *H. nesomys* Hoogstraal, Uilenberg and Klein, 1966
103. *H. norvali* Hoogstraal and Wassef, 1983
104. *H. novaeguineae* Hirst, 1914
105. ***H. obesa* Larrousse, 1925**
106. *H. obtusa Donitz*, 1910
107. *H. orientalis* Nuttall and Warburton, 1915
108. ***H. ornithophila* Hoogstraal and Kohls, 1959**
109. *H. palawanensis* Kohls, 1950
110. *H. papuana* Thorell, 1883
111. *H. paraleachi* Camicas, Hoogstraal and El Kammah, 1983
112. ***H. paraturturis* Hoogstraal, Trapido and Rebello, 1963**
113. *H. parmata* Neumann, 1905
114. *H. parva* Neumann, 1897
115. *H. pavlovskyi* Pospelova-Shtrom, 1935
116. *H. pedetes* Hoogstraal, 1972
117. *H. pentalagi* Pospelova-Shtrom, 1935

118. *H. petrogalis* Roberts, 1970
119. *H. phasiana* Saito, Hoogstraal and Wassef, 1974
120. *H. pospelovashtromae* Hoogstraal, 1966
121. *H. primitiva* Teng, 1982
122. *H. psalistos* Hoogstraal, Kohls and Parrish, 1967
123. *H. punctaleachi* Camicas, Hoogstraal and El Kammah, 1973
124. *H. punctata* Canestrini and Fanzago, 1878
125. *H. qinghaiensis* Teng, 1980
126. *H. quadriaculeata* Kolonin, 1992
127. ***H. ramachandrai* Dhanda, Hoogstraal and Bhat, 1970**
128. *H. ratti* Kohls, 1948
129. *H. renschi* Schulze, 1933
130. *H. roubaudi* Toumanoff, 1940
131. *H. rugosa* Santos Dias, 1956
132. *H. rusae* Kohls, 1950
133. ***H. sambar* Hoogstraal, 1971**
134. *H. sciuri* Kohls, 1950
135. *H. semermis* Neumann, 1901
136. ***H. shimoga* Trapido and Hoogstraal, 1964**
137. *H. silacea* Robinson, 1912
138. ***H. silvafelis* Hoogstraal and Trapido, 1963**
139. *H. simplex* Neumann, 1897
140. *H. simplicima* Hoogstraal and Wassef, 1979
141. *H. sinensis* Zhang, 1981
142. ***H. spinigera* Neumann, 1897**
143. *H. spinulosa* Neumann, 1906
144. *H. subelongata* Hoogstraal, 1953
145. *H. subterra* Hoogstraal, El Kammah and Camicas, 1992
146. ***H. sulcata* Canestrini and Fanzago, 1878**
147. *H. sumatraensis* Hoogstraal, El Kammah, Kadarsan and Anastos, 1971
148. ***H. sundrai* Sharif, 1928**
149. *H. suntzovi* Kolonin, 1993
150. *H. susphilippensis* Hoogstraal, Kohls and Parrish, 1968
151. *H. taiwana* Sugimoto, 1936
152. *H. tauffliebi* Morel, 1965
153. *H. theilerae* Hoogstraal, 1953
154. *H. tibetensis* Hoogstraal, 1965
155. *H. tiptoni* Hoogstraal, 1953
156. *H. toxopei* Warburton, 1927
157. *H. traguli* Oudemans, 1928
158. *H. traubi* Kohls, 1955
159. ***H. turturis* Nuttall and Warburton, 1915**
160. *H. verticalis* Itagaki, Noda and Yamaguchi, 1944
161. *H. vidua* Warburton and Nuttall, 1909
162. *H. vietnamensis* Hoogstraal and Wilson, 1966
163. ***H. warburtoni* Nuttall, 1912**
164. ***H. wellingtoni* Nuttall and Warburton, 1908**
165. *H. xinjiangensis* Teng, 1980
166. *H. yeni* Toumanoff, 1944

167. *H. zumpti* Hoogstraal and El Kammah, 1974

(The names in **bold** indicate the tick species present in India.)

1.8 Keys to Subgenera[60,61,77,78,85]

1. Palpi are elongate without broadening of palpal segment 2
   Palpal segment broadened either in adults or immature stages...... 5
   Palpal segment broadened in both adults and immatures, capitular spurs are less developed 3
   Posterodorsal margin of palpal segment III with distinct or indistinct spur in one or both sexes, or with platelike projections; posteroventral spur moderate to prominent, triangular or hooklike, overlapping of anterior or whole part of palpal segment II 4
2. Basis capitulum short and laterally convex with no cornua, dental formula 3/3 in female while 2/2 in male and immatures, coxal spurs are small or obsolete, genital aperture is slitlike .... *Alloceraea*
   Adult basis capitulum is variously shaped and armed with variable cornua, some nymphs and larvae bear large coxal spurs, the adult palpal bears small ventral spurs ..... *Allophysalis*
3. Absence or mild development of spurs on capitulum, coxa, and trochanters .... *Ornithophysalis*
   Coxal and trochantal spurs are well developed, the basis capitulum has small to large cornua.................. *Rhipistoma*
4. Posteroventral spur on palpal segment III is moderate to extraordinary, posterodorsal margin of palpal segment III with distinct or indistinct, median or internal spur, cornua and coxal spur broadly or elongately triangular, coxa IV sometimes with extraordinarily long spurs in one sex .... *Kaiseriana*
   Posteroventral spurs of palpal segment III broadly triangular, overlapping anterior 1/3rd of palpal segment II, posterodorsal spur on palpal segment III is absent, coxal spurs are short, broadly triangular .... *Haemaphysalis*
5. Palpal segment broadened in adults only........... 6
   Palpal segment broadened in immature stages only 7
6. Adult palpi are campanulate in male and broadly campanulate in female, capitular and coxal spurs are moderately large and well developed............ *Segalia*
   Capitular spurs are much reduced........... *Aborphysalis*
7. No lateral expansion of basis capituli in immature stages, palps are compact but are slightly salient in nymphs of some species, adults and immatures have rectangular basis capituli, adult cornua are moderately large to large in males and obsolete or small in females................. *Herpetobia*
   In certain species adults show a tendency to broad palpi, lateral expansion of basis capituli of immature stages, immatures stages palps compact............... *Aboimisalis*

1.9 List of Indian *Haemaphysalis* Ticks According to Subgenera

Subgenus: *Aboimisalis*

- *Haemaphysalis cornupunctata*

Subgenus: *Aborphysalis*

- *Haemaphysalis aborensis*
- *Haemaphysalis kyasanurensis*
- *Haemaphysalis tezuensis*

Subgenus: *Alloceraea*

- *Haemaphysalis aponommoides*

Subgenus: *Allophysalis*

- *Haemaphysalis garhwalensis*
- *Haemaphysalis warburtoni*

Subgenus: *Haemaphysalis*

- *Haemaphysalis campanulata*
- *Haemaphysalis birmaniae*
- *Haemaphysalis darjeeling*
- *Haemaphysalis indoflava*

Subgenus: *Herpetobia*

- *Haemaphysalis sulcata*
- *Haemaphysalis sundrai*
- *Haemaphysalis kashmirensis*
- *Haemaphysalis himalaya*
- *Haemaphysalis nepalensis*
- *Haemaphysalis kumaonensis* (new species)

Subgenus: *Kaiseriana*

- *Haemaphysalis aculeata*
- *Haemaphysalis anomala*
- *Haemaphysalis bispinosa*
- *Haemaphysalis cuspidata*
- *Haemaphysalis davisi*
- *Haemaphysalis hystricis*
- *Haemaphysalis kinneari*
- *Haemaphysalis ramachandrai*
- *Haemaphysalis spinigera*
- *Haemaphysalis shimoga*
- *Haemaphysalis intermedia*
- *Haemaphysalis sambar*
- *Haemaphysalis turturis*
- *Haemaphysalis wellingtoni*

- *Haemaphysalis kutchensis*
- *Haemaphysalis paraturturis*
- *Haemaphysalis silvafelis*

Subgenus: *Rhipistoma*

- *Haemaphysalis canestrinii*
- *Haemaphysalis indica*

Subgenus: *Ornithophysalis*

- *Haemaphysalis doenitzi*
- *Haemaphysalis minuta*
- *Haemaphysalis howletti*
- *Haemaphysalis ornithophila*
- *Haemaphysalis megalaimae*

Subgenus: *Segalia, Santos Dias*

- *Haemaphysalis montgomeryi*
- *Haemaphysalis obese*

Part 2

Biology, Distribution, Taxonomic Description, and Disease Relationship of Individual *Haemaphysalis* Ticks

2 Structurally Primitive *Haemaphysalines*

A. Subgenus *Alloceraea*
B. Subgenus *Allophysalis*
C. Subgenus *Aboimisalis*

There are 17 structurally primitive haemaphysaline species included in four subgenera from all over the world. In India, only four species representing three subgenera among the structurally primitive (SP), i.e., *Alloceraea*, *Allophysalis*, and *Aboimisalis*, are present.

2.1 Subgenus *Alloceraea*

General Characteristics[85,110]

Only one species, *H. aponommoides*, is found in India, representing the subgenus *Alloceraea*. Its structure apparently represents an *Aponomma*-like primitive prototype haemaphysaline, which was presumed to have evolved in the late Paleozoic or early Mesozoic era. The basis capitulum is short and laterally convex, has posterior–lateral projections, and lacks cornua; the palpi are narrowly elongate or clavate; segment III lacks a ventral spur and segment IV is in apical pit; dental formula is 3/3; coxal spurs are small or obsolete; the body integument is leathery; and the genital aperture is slit-like. The apron of the male is short and rough. *Alloceraea* males, nymphs, and larvae each also have a laterally convex or otherwise laterally projecting basis capituli lacking posterior cornua and elongate palpi lacking a ventral spur, but their dental formulae are 2/2. These are the only males with a 2/2 dental formula in the entire genus *Haemaphysalis*.

Keys to Identify Species of Subgenus Alloceraea

Male

Basis capitulum is laterally convex or otherwise laterally projecting; lacking posterior cornua, and elongate palpi lacking a ventral spur, but their dental formulae are 2/2; punctations are numerous, superficial, noncontiguous, closely spaced, and

Haemaphysalis Ticks of India. DOI: 10.1016/B978-0-12-387811-3.00002-4

irregular in size and distribution; lateral grooves are obsolete; festoons are short and obscurely defined; coxa bear small spur or ridge-like outgrowth.................. [*H. aponommoides*]

Female

The basis capitulum is short, laterally convex, and lacks cornua; the palpi are narrowly elongate or clavate; segment III lacks a ventral spur; dental formula is 3/3; coxal spurs are small or obsolete; the body integument is leathery; the genital aperture is slit-like...................[*H. aponommoides*]

2.1.1 Haemaphysalis aponommoides

This species was described by Warburton in 1913 as a variety of *Haemaphysalis inermis*, a female specimen collected from Himalayan zebu cattle in West Bengal of India. Because of its structural similarity with *Aponomma*, Warburton named it *aponommoides* (Figure 2.1).

Male (Figures 2.2 and 2.3)[52]

Overall body length is 2.4 mm, while the width is 1.5 mm.

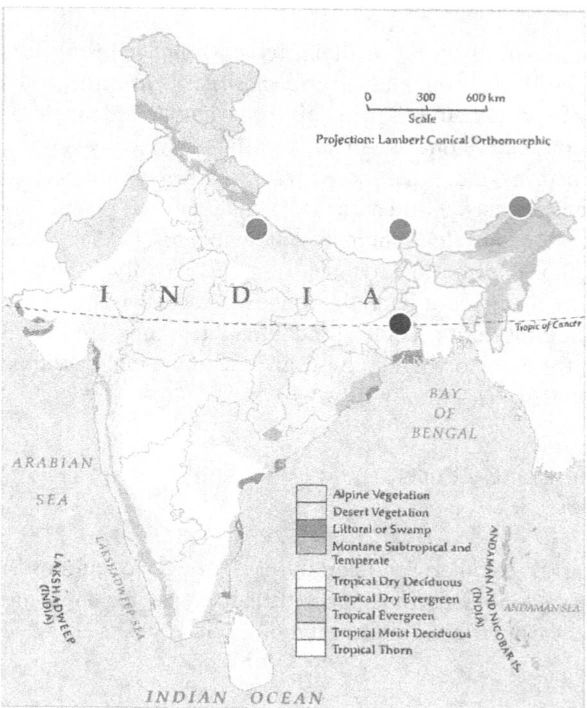

Figure 2.1 Distribution of *H. aponommoides* in India. (●) Place of first record. Map not to scale.

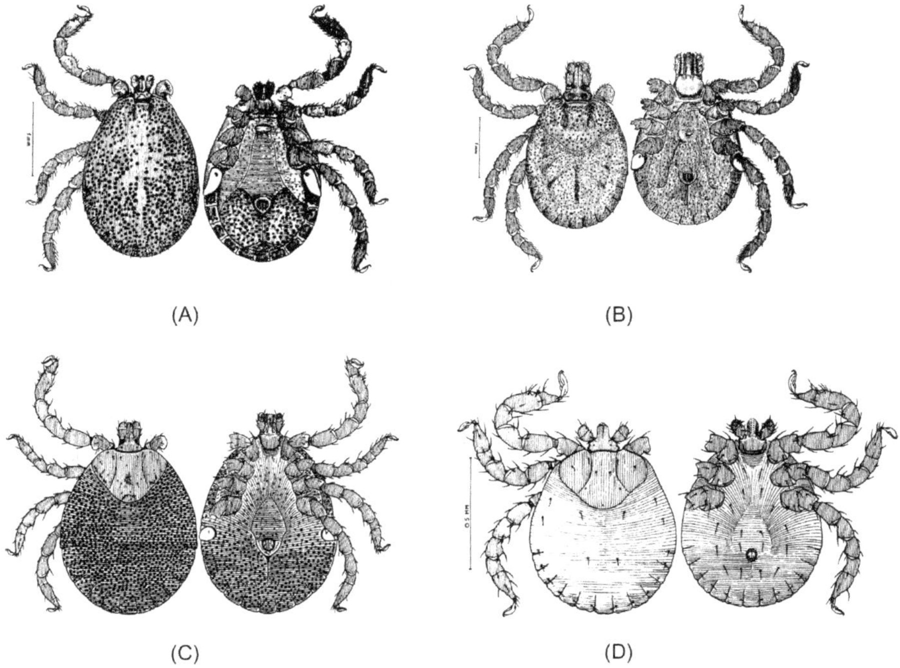

Figure 2.2 *Haemaphysalis aponommoides*. (A) Male, dorsal and ventral side; (B) female, dorsal and ventral side; (C) nymph, dorsal and ventral side; (D) larva, dorsal and ventral side. *Source*: Hoogstraal (1962), Hoogstraal and Mitchel (1971), Journal of Parasitology (vol 48 & vol 57), Allen press Publishing Services.

Capitulum: Capitulum is small, approximately 0.35 mm in length. Basis capitulum measures approximately 0.14 mm in length and 0.23 mm in width, approximately 1.7 times as wide as long. Hypostome is narrow with convex lateral margin, and dental formula is 2/2, with six to seven denticles in a file. Palpi are 0.2 mm in length with straight margins. Cornua are absent.

Scutum: Scutum is approximately 0.6 times as wide as long, widest at the level of spiracular plates. Lateral grooves are absent; cervical grooves are short and parallel. Punctations are numerous, shallow, irregularly sized, and closely spaced. Festoons number 11, are short and shallow. Genital apron is situated between coxa II.

Legs: Legs are moderately long and robust. Coxa are each with a very small triangular spur. Trochanters are lacking ventral spurs.

Females (Figures 2.2 and 2.3)[52]

Total body length measures approximately 2.6 mm, and width is 1.7 mm.

Capitulum: Basis capituli are dorsally approximately 2.2 times as wide as long and measure approximately 0.21 mm in length and 0.47 mm in width; porose areas are sub-circular, situated in saucer-like pits separated by area approximately

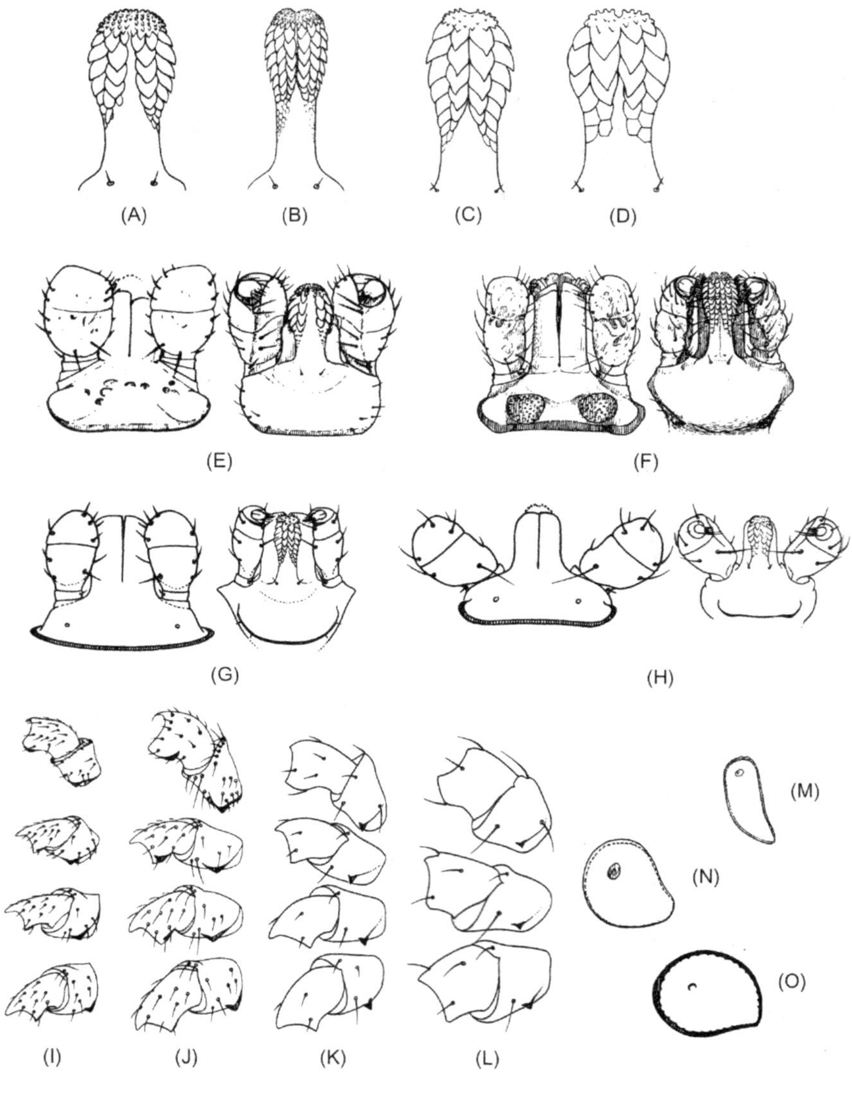

Figure 2.3 (A–D) Hypostome of male, female, nymph, and larva, respectively; (E–H) capitulum of male, female, nymph, and larva, respectively; (I–L) coxa of male, female, nymph, and larva, respectively; (M–O) spiracular plates of male, female, and nymph, respectively. *Source*: NIV. unpublished drawings.

equaling their own diameter. Hypostome is narrow with convex lateral margins. Dental formula is 3/3 with eight denticles in a file. Palpi are slightly less than 0.5 mm in length, with straight lateral margins. Cornua are small with blunt apex.

Scutum: Scutum of engorged specimen measures approximately 1.8 mm in length and 2.5 mm at greatest width. Cervical grooves are shallow and indistinct.

Punctations are numerous but shallow with irregular size and distribution. Genital aperture is situated between coxa III.

Legs: Legs are long and robust. Each coxa bears a moderate spur, while trochanters I to III show small spurs or ridges.

Nymph (Figures 2.2 and 2.3)[77]

Total body length is approximately 1.46 mm, breadth 1.07 mm.

Capitulum**:** Basis capitulum is dorsally approximately 4 times as broad as long; cornua are obsolete. Palpi are clavate and approximately 2.1 times as long as broad. Hypostome is almost as long as palpi.

Scutum**:** Scutum is approximately 1.4 times as broad as long. Cervical grooves are as deep arcs. Only few punctations are seen. Festoons number 11. Legs are robust and short.

Legs**:** Coxa are each with a small triangular spur reaching to (I, II) or beyond (III, IV) posterior margin.

Larva (Figures 2.2 and 2.3)[77]

Total body length is approximately 0.82 mm, breadth 0.64 mm.

Capitulum**:** Basis capitulum is dorsally approximately 3 times as broad as long; cornua are obsolete. Palpi are clavate, approximately 1.5 times as long as broad. Hypostome is approximately 2 times as long as broad; dental formula is 2/2, with five or six denticles in a file.

Scutum**:** Scutum is approximately 1.5 times as broad as long. Cervical grooves are as deep arcs. Punctations are few.

Legs**:** Legs are robust and short. Coxa bear short spur.

Hosts[26,52,120]

Immature stages: Lizards, shrews, rodents, birds, and from vegetation in Nepal.

Adults: Cattle, goat antelope, horse, man (biting), serow, flying squirrel, wild cat, black bear, barking deer, dog, zhum, yak, goat, sheep, wild boar.

Distribution[26,52,120]

India (Himalayan region, Arunachal Pradesh, West Bengal, Sikkim), Nepal, Japan, Taiwan, Iran.

Related Species[52]

H. aponommoides and phylogenetically primitive *H. inermis* are closely related species. These two species differ between themselves and widely from all other haemaphysalids by following characters. The female palpi of *H. aponommoides* are more narrowly elongate than those of *H. inermis*. In *H. inermis* females, palpi are approximately 1.5 times as long as the basis capituli; in *H. aponommoides* females they are slightly over twice as long as the basis capituli. Cornua, which are absent or very slightly suggested in *H. inermis* females, are short but distinct in

H. aponommoides females. Ventrally, the basis capituli are posteriorly narrower in *H. inermis* females than in *H. aponommoides* females.

2.2 Subgenus *Allophysalis*

General Characteristics

Only *H. warburtoni* and *H. garhwalensis* are found in India, representing this subgenus.

These species display combinations of SP structural conformity and of stepwise or abrupt advances in SP ranks. Female basis capitulum displays more primitiveness than male. Female *Allophysalis* palpi remain elongate, as in *Alloceraea*. Male palpi are shorter or compact but not broadened. The adult palpus for the first time bears a small ventral spur, which is, however, medially rather than posteriorly directed, displaying bird-parasitizing and primitive mammal parasitizing characteristic of structurally advanced (SA) subgenus *Ornithophysalis*. Certain nymphs and/or adults approach an SA pattern in part, but no larvae do so. Each immature *Allophysalis* has a short, broadly angular basis capitulum. Immature *Allophysalis* basis capitulum pattern ranges from short, remarkably laterally expanded type to posterolaterally convex and cornua bearing (*H. warburtoni* nymph). Immature palpi remain elongate, as in *Alloceraea*. The first ventral spur on immature palpi is seen in the *H. warburtoni* nymph. The *Allophysalis* adult basis capituli are variously shaped and armed with usually variable cornua. Adult coxal spurs are now moderately sized or fairly large and exhibit various triangular, spatulate, lanceolate, or hook-like forms, characteristic of haemaphysalines specialized for parasitizing Artiodactyla.

Keys to Identify Species of Subgenus Allophysalis

Males

Posteroventral spur on palpal segment III is small, broadly triangular, and slightly elevated, not reaching to the posterior margin of segment III; coxal spurs are blunt, stout, sub-equal in size, and gradually tapering; only trochanter I has a crest-like spur; basis capitulum is approximately 1.6 times as long as broad...................[*H. garhwalensis*]

Posteroventral spur on palpal segment III is small, broadly triangular, reaching only to anterior of palpal segment II and inwardly directed; coxal spurs are stout and blunt; all trochanters have crest-like spurs; spiracular plate is not circular and oblongate...................[*H. warburtoni*]

Females

Basis capitulum is dorsally 2.5 times as broad as long; palps are longer than in male; palpal segment II is dorsally 1.4 times as long as broad; coxal spurs are more or less equal in length in both sexes...................[*H. garhwalensis*]

Basis capitulum is dorsally 1.9 times as broad as long; coxal spurs are smaller than in male; spiracular plate is circular...................[*H. warburtoni*]

Comparative figures of *H. garhwalensis and H. warburtoni* are given below.

Comparison of capitulum of males of subgenus *Allophysalis*

garhwalensis *warburtoni*

Comparison of capitulum of females of subgenus *Allophysalis*

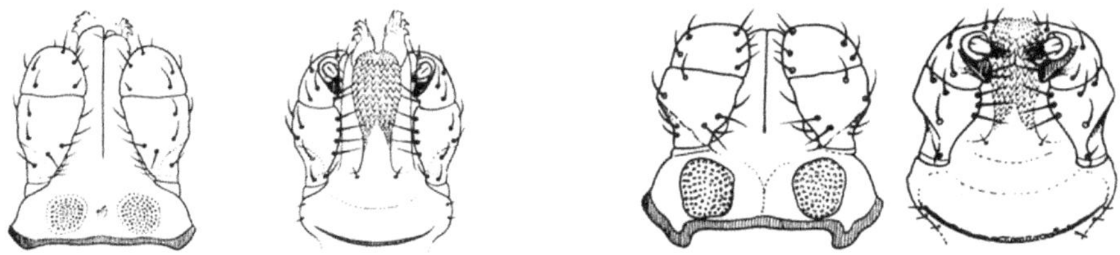

garhwalensis *warburtoni*

Comparison of capitulum of nymphs of subgenus *Allophysalis*

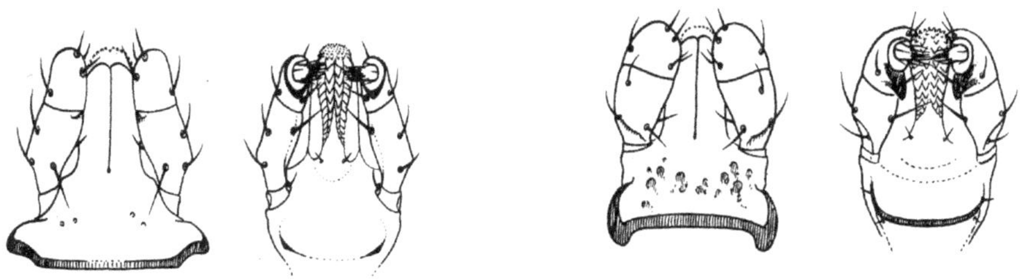

garhwalensis *warburtoni*

Comparison of capitulum of larvae of subgenus *Allophysalis*

garhwalensis *warburtoni*

Comparison of coxa

Male:

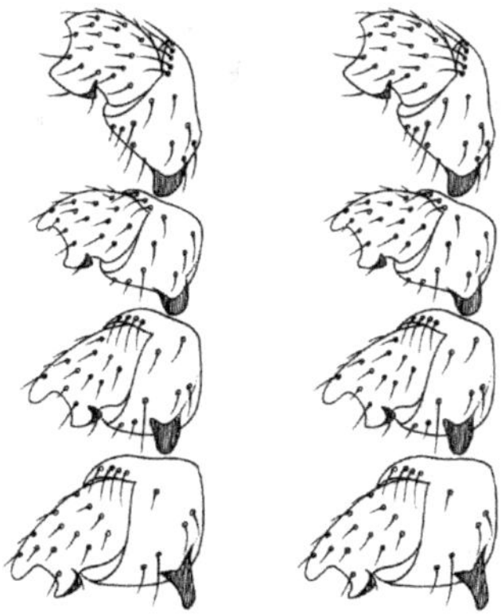

Female:

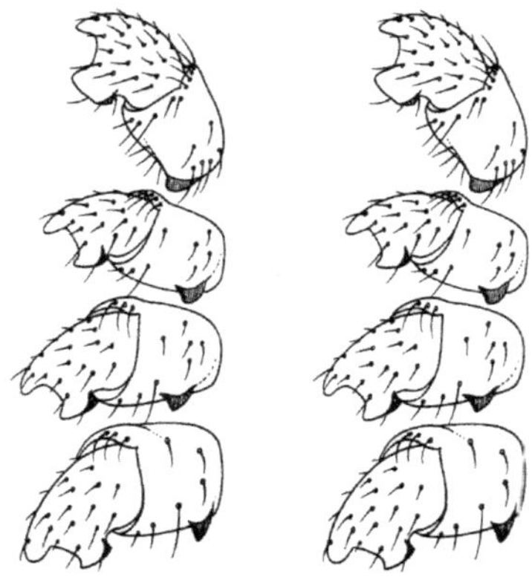

garhwalensis *warburtoni*

Nymph:

Larva:

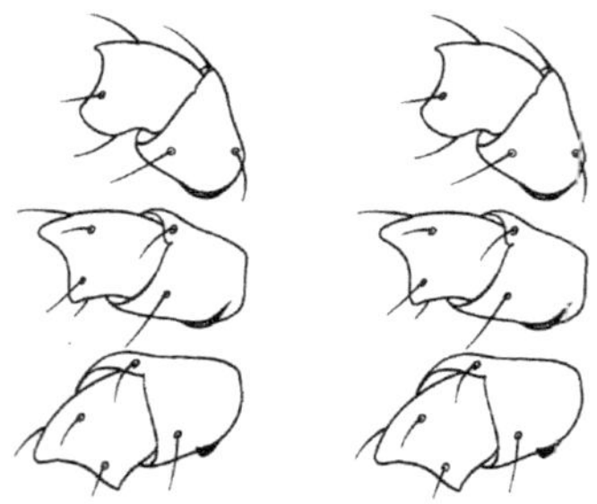

garhwalensis *warburtoni*

Comparison of hypostome

Male:

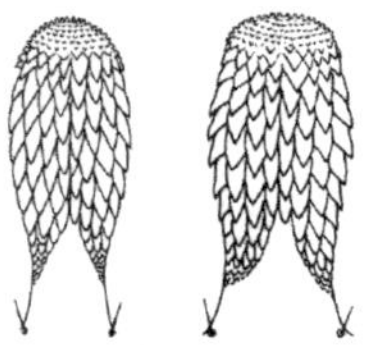

(A) (B)

Female:

(A) (B)

Nymph:

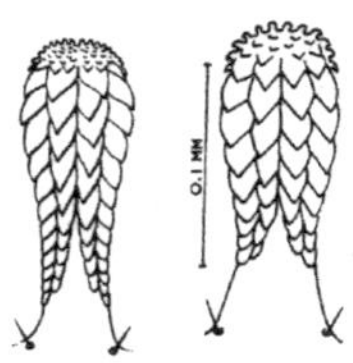

(A) (B)

Larva:

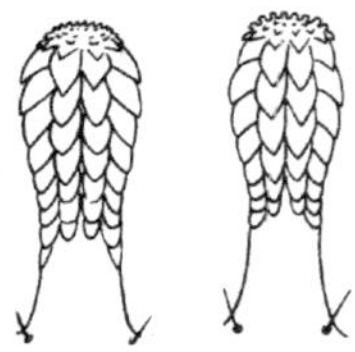

(A) (B)

(A) *garhwalensis*, (B) *warburtoni*

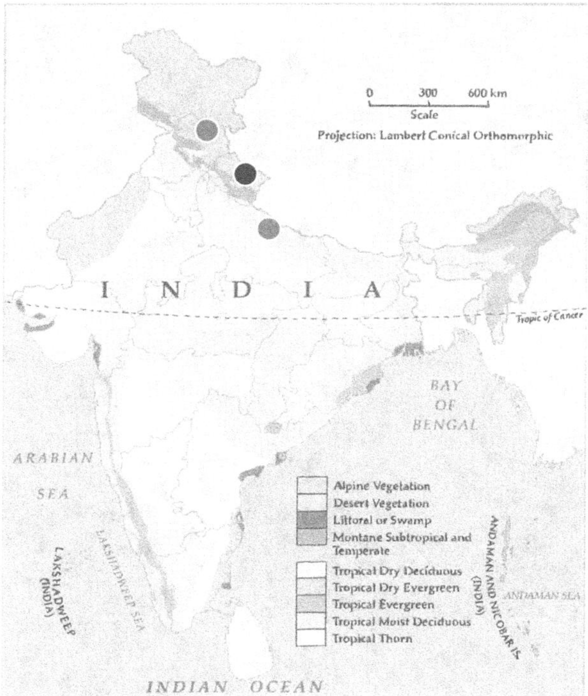

Figure 2.4 Distribution of *H. garhwalensis* in India. (●) Place of first record. Map not to scale.

2.2.1 Haemaphysalis garhwalensis

This is one of the high-elevation mountainous species from Asia. It was found in the Garhwal region of the Himalayan range in Uttar Pradesh in 1968, and hence the name *garhwalensis* (Figure 2.4).

Male (Figures 2.5 and 2.6)[28]

Capitulum: Basis capitulum is dorsally approximately 1.6 times as broad as long; cornua are short, with bluntly pointed apices. Palpi are compact, external plane is straight. Hypostome is shorter than palpi, clavate, and dental formula is 4/4, with maximum nine or ten denticles in a file.

Scutum: Cervical grooves are obsolete. Lateral grooves are distinct, extend to the level of anterior margin of coxa III, and enclose two festoons. Punctations are small, shallow, number moderate, and evenly distributed. Festoons number 11.

Legs: Coxa are each with a stout, sub-equal spur; trochanters do not have spurs.

Female (Figures 2.5 and 2.6)[28]

Capitulum: Basis capitulum is dorsally 2.5 times as broad as long; cornua are small with rounded apices; porose areas are large, oval, separated by an area equal to

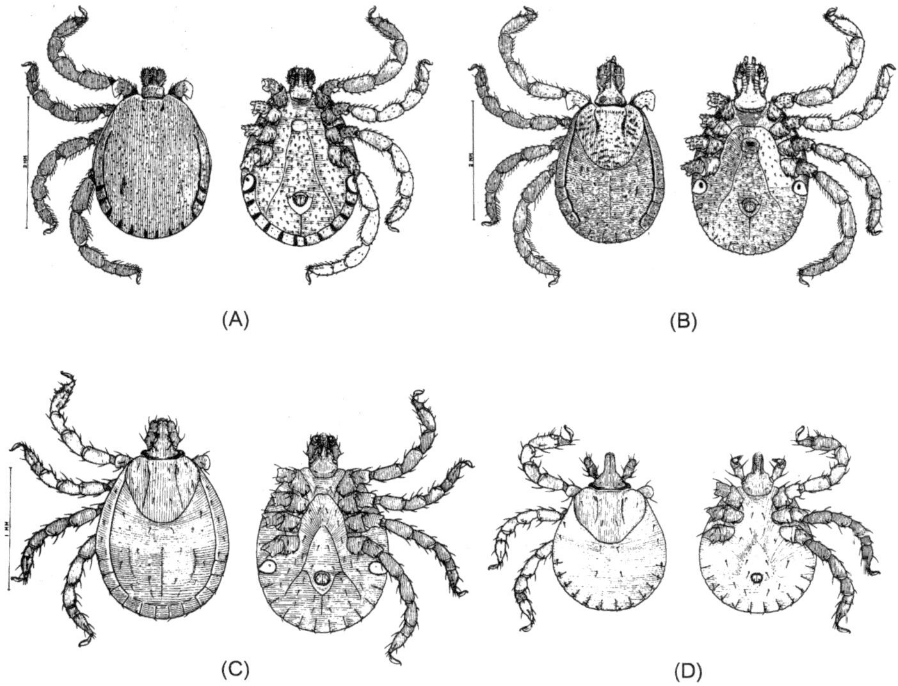

Figure 2.5 *Haemaphysalis garhwalensis*. (A) Male, dorsal and ventral side; (B) female, dorsal and ventral side; (C) nymph, dorsal and ventral side; (D) larva, dorsal and ventral side. *Source*: Dhanda and Bhat (1969), Journal of Parasitology (vol 54(4)), Allen press Publishing Services, Indian J Parasitol 1977; 1(1): 9–14.

length of one pit. Palpi are long, narrow, and external plane is straight. Hypostome is as long as palpi, external outline curving; dental formula is 4/4, with maximum 13–14 denticles in a file.

Scutum: Scutum is almost as broad as long. Cervical grooves are narrow and shallow. Punctations are numerous, slightly larger and deeper than in male.

Legs: Legs are similar to those of male.

Nymph (Figures 2.5 and 2.6)[33]

Length is approximately 1.6 mm, breadth 1.1 mm.

Capitulum: Basis capitulum is dorsally approximately 5 times as broad as long. Palpi are elongate and clavate. Hypostome is spatulate, longer than palpi; dental formula is 2/2, with 10–12 denticles in a file.

Scutum: Cervical grooves are linear, extending to almost posterior margin of scutum. Punctations are few, numbering eight to ten. Festoons are distinct and number 11.

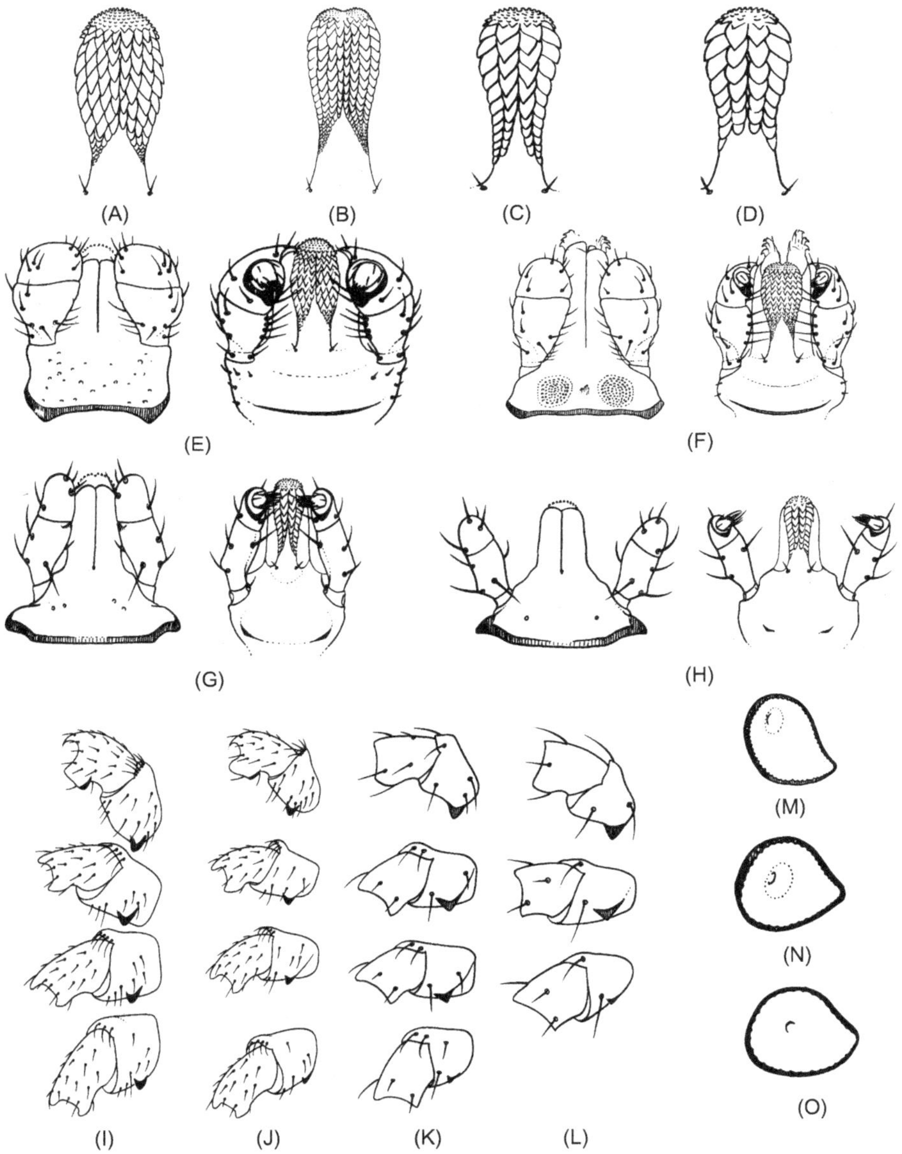

Figure 2.6 (A–D) Hypostome of male, female, nymph, and larva, respectively; (E–H) capitulum of male, female, nymph, and larva, respectively; (I–L) coxa of male, female, nymph, and larva, respectively; (M–O) spiracular plates of male, female, and nymph, respectively.
Source: NIV. unpublished drawings.

Legs: Legs are long and thin. Coxa are each with a broadly sub-triangular spur extending well (I) or slightly (II, III) beyond coxal margin; IV spur is small.

Larva (Figures 2.5 and 2.6)[33]

Length is approximately 0.9 mm, breadth 0.6 mm.

Capitulum: Basis capitulum is dorsally approximately 4 times as broad as long. Palpi are as in nymph, except that each is approximately 3 times as long as broad; segment III ventral spur is almost obsolete. Hypostome is spatulate, as long as palpi; dental formula is 2/2, with eight to ten denticles in a file.

Scutum: Scutum is 1.3 times as broad as long. Cervical grooves are narrow, extending a little beyond scutal mid-length. Punctations number approximately eight. Festoons number 11.

Legs: Coxa I is with spur as in nymph; II and III are with spur as in nymph II and IV, respectively.

Host[28,33,120]

Immature stages: Rats, field mice, mountain voles.

Adults: Sheep, cattle, yak, goat, rodents, horses, mountain voles, and from flag dragging through forest vegetation.

Distribution[28,33]

India (Himachal Pradesh, Uttar Pradesh, Uttarakhand), Nepal.

Ecology[33]

All larvae, nymphs, and adults were collected in the pre-monsoon months of May and June, mostly from 2,700 to 3,400 m alt. Adults from lower elevations (2,000 m) were from sheep that had descended from alpine pastures a day or two earlier. Domestic animals were heavily infested by adult *H.* (*A.*) *garhwalensis*, mostly in the ears, in May and early June but a few were found in late June. The Indian collecting localities were in narrow alpine valleys and in a narrow sub-alpine strip of the upper temperate zone of the Western Himalayas. This area has a severe winter and remains snowbound from November to April. After snow melts in late April, there is a rich growth of alpine and sub-alpine meadow grass. May and June are humid owing to frequent showers and hail; daytime temperatures are a few degrees above freezing and night time frost is frequent. July and August are mild with occasional showers and hail. The September–October climate is similar to that of May–June.

Related Species[28]

A closely related species is *H. warburtoni*. In *garhwalensis*, posteroventral spur on palpal segment III is small, broadly triangular, slightly elevated, not reaching to posterior margin of palpal segment III, whereas in *H. warburtoni* the spur is reaching to the anterior of palpal segment II and inwardly directed. Coxal spurs of *H. warburtoni* are stouter and stronger compared to *H. garhwalensis*. In *H. warburtoni*, all trochanters have crest-like spurs, while in *H. garhwalensis* only trochanter I has a crest-like spur.

H. garhwalensis also shows morphologic relationship with *H. pospelovashtromae* and *H. tibetensis*, which are not found in India.

2.2.2 Haemaphysalis warburtoni[64]

This species was first described by Nuttall in 1912 from a serow and goral in China (Figure 2.7).

Male (Figures 2.8 and 2.9)[79,104,110]

Overall body length is 2.20–2.90 mm, breadth 1.25–1.71 mm.

Capitulum: In basis capitulum, the external expansion is unapparent. The cornua are at least one-half as long as the base of the basis capitulum. The dental formula of Nepal specimens is chiefly 4/4, as in those from China, but the anterior 5/5 or 6/6 elements are fewer than in those from China.

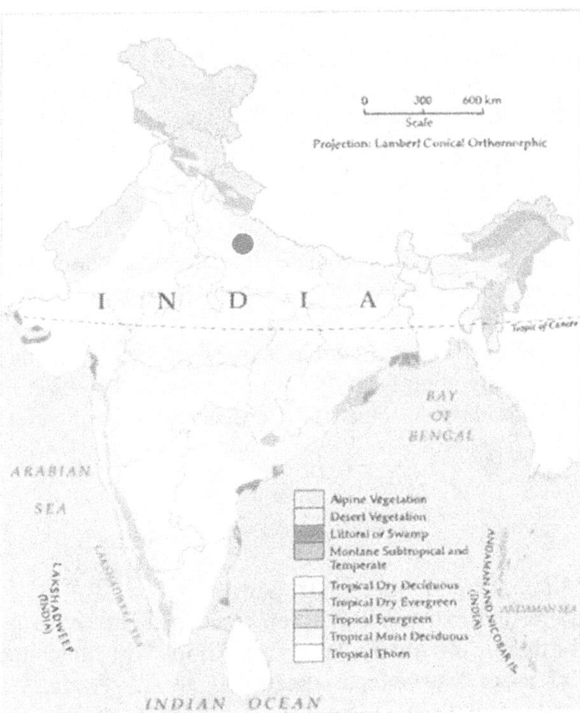

Figure 2.7 Distribution of *H. warburtoni* in India. (●) Place of first record. Map not to scale.

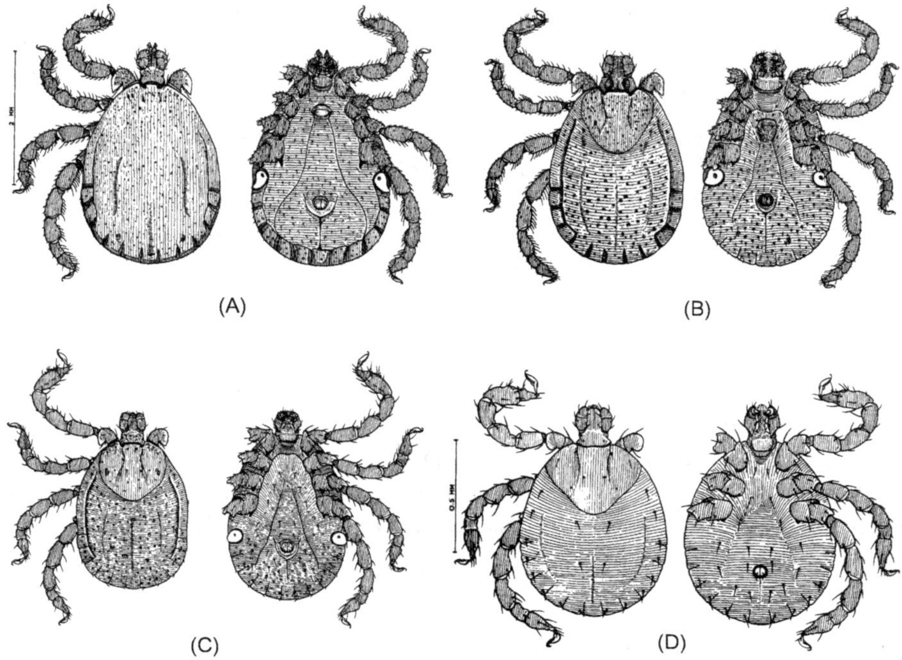

Figure 2.8 *Haemaphysalis warburtoni*. (A) Male, dorsal and ventral side; (B) female, dorsal and ventral side; (C) nymph, dorsal and ventral side; (D) larva, dorsal and ventral side. *Source*: Hoogstraal (1966 and 1971), Journal of Parasitology (vol 52(4) & vol 57), Allen press Publishing Services.

Scutum: The depth of lateral groove varies from shallow to deep, definitely encloses the first festoon, and faintly encloses the second festoon. Punctations number vary slightly.

Legs: Spur on coxae I is less elongate and more blunt in some specimens than in the others; spurs II and III vary only slightly in apical bluntness; spur IV varies rather remarkably, being more or less broadly lanceolate, curved apically, or broad and blunt apically.

Female (Figures 2.8 and 2.9)[79,104,110]

Total body length is 3.25–3.53 mm, breadth 1.82–2.05 mm.

Capitulum: In basis capitulum, the external expansion is unapparent or slight. The cornua vary less markedly than in the male; those from China and illustrated are much alike. The porose areas of most Nepal specimens are larger than in Chinese specimens. The dental formula of hypostome is consistently 4/4 in

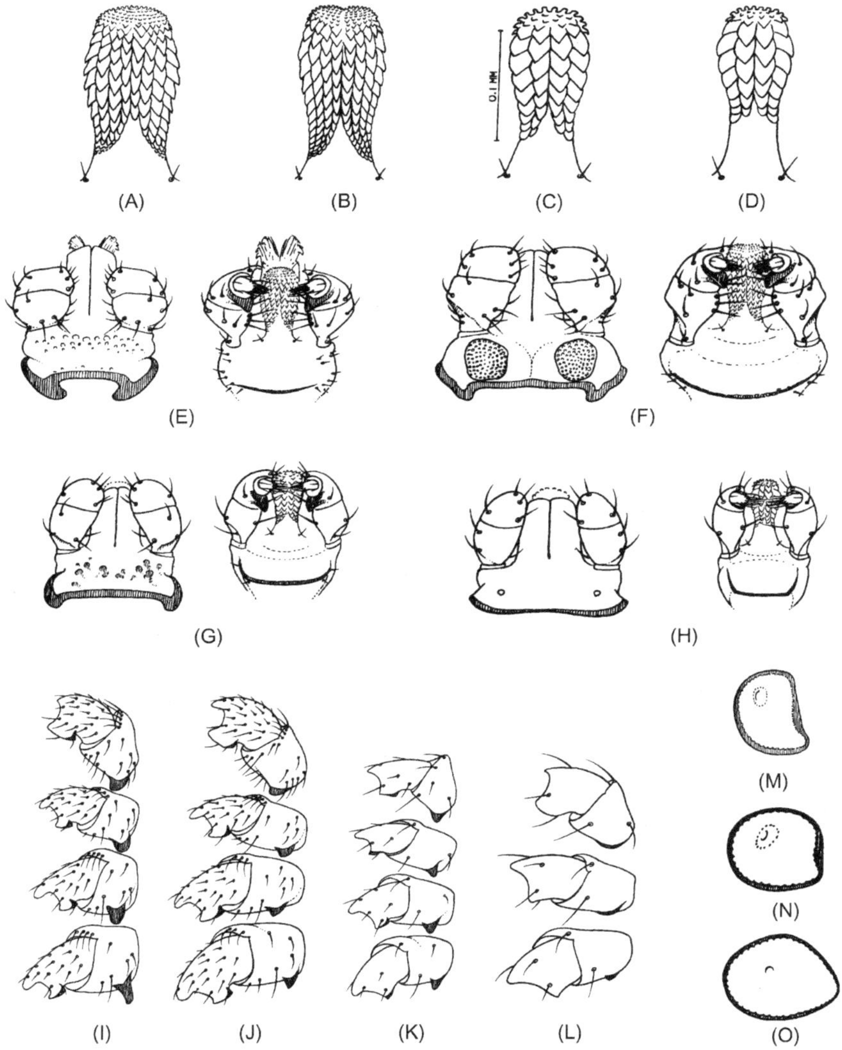

Figure 2.9 (A–D) Hypostome of male, female, nymph, and larva, respectively; (E–H) capitulum of male, female, nymph, and larva, respectively; (I–L) coxa of male, female, nymph, and larva, respectively; (M–O) spiracular plates of male, female, and nymph, respectively. *Source*: NIV. unpublished drawings.

specimens collected from Nepal; in Chinese specimens it is 6/6 anteriorly in females and 5/5 in most other rows.

Legs: Each coxal spur of Chinese specimens is somewhat larger than seen in Nepal material, among which these generally large spurs are quite reduced in IV specimens.

Nymph (Figures 2.8 and 2.9)[79]

Overall body length is approximately 1.6 mm, breadth 1.0 mm.

Capitulum: Basis capitulum is dorsally approximately twice as broad as long (including cornua); cornua are broad, approximately one-fifth as long as the base of basis capitulum, and with bluntly rounded apices. Palpi are elongate and clavate. Hypostome is as long as palpi, approximately 2.7 times as long as broad; apex is bluntly rounded; dental formula is 2/2, with six to eight denticles in a file, followed posteriorly by several (two to four) smaller denticles or hooklets.

Scutum: Scutum is approximately 1.1 times as broad as long. Cervical grooves are converging anteriorly, thence diverging to posterior one-third of scutal length. Punctations are few and obscure.

Legs: Legs are moderately long and narrow. Coxal spurs are each broadly triangular, apically blunted, sub-equal, extending well beyond posterior margin.

Larva (Figures 2.8 and 2.9)[79]

Overall body length is approximately 0.9 mm, breadth 0.6 mm.

Capitulum: Basis capitulum is dorsally approximately 2.5 times as broad as long; external margins are approximately straight anteriorly, and cornua are lacking. Palpi outlines and general features are approximately as in nymph, except as follows: segment III is lacking ventral spur. Hypostome is longer than palpi; hooklets and dental features are approximately as in nymph.

Scutum: Scutum is approximately 1.4 times as broad as long, broadest at anterior one-third of scutal length. Cervical grooves are as arcs extending somewhat beyond scutal mid-length. Punctations number three or four pairs. Legs are moderately long and narrow.

Legs: Coxa II and I are each with a broadly rounded marginal ridge replacing spur; III with spur very small, short, broadly rounded.

Host[79,110]

Immature stages: Small rodents, foxes, hares, wild goats, domestic cats, Himalayan thar, mice, domestic sheep, Himalayan Monal pheasant, mountain field mouse/vole, sweeping the ground vegetation.

Adults: Serow/goat antelope, Himalayan thar, artiodactyl mammals, brown goral, sheep, goats, snow goats, yak, cattle, horses, man, marmots (rodents), domestic zhum, goral, sweeping the ground vegetation.

Distribution[28,110]

India (Uttar Pradesh), Mongolia, Taiwan, China, Nepal.

Ecology[79]

The known distribution of this species in Nepal is from 8,100 to 12,500 ft altitude in the Himalayan range of the north central part of the country (north of Kathmandu) toward and near the Tibet border. Ecologic features of Langtang valley, Melumche, and Gosainkund, three of the five localities where *H.* (*A.*) *warburtoni* as collected in Nepal, have been described by Hoogstraal and Mitchell (1971).

Related Species

A closely related species is *H. garhwalensis*. Posteroventral spur on palpal segment III is small, broadly triangular, slightly elevated, not reaching to posterior margin of palpal segment III in *H. garhwalensis*, whereas in *H. warburtoni* the spur is reaching to the anterior of palpal segment II and inwardly directed. Coxal spurs of *H. warburtoni* are stouter and stronger compared to *H. garhwalensis*. In *H. warburtoni*, all trochanters have crest-like spurs, while in *H. garhwalensis* only trochanter I has a crest-like spur.

2.3 Subgenus *Aboimisalis*

General Characteristics

It includes four species, viz. *H. cornupunctata*, *H. punctata*, *H. chordeilis*, and *H. cinnabarina*. Among these only *H. cornupunctata* has been recorded in India. Immature palpi are compact, lacking salience in *H. cornupunctata* but are slightly salient in *H. punctata* and *H. chordeilis*, and the ventral spur is slight or obsolete. Adult palpi are quite compact, slightly broader than in the few males of earlier subgenera in which the palpi tended to be more compact. It can be observed that, under the structurally primitive (SP) *Haemaphysalis* category, adults of *Aboimisalis* are structurally more advanced and show a slight step in generic trend to broad palpi. Cornua exist in males of all species, but in females only in *H. cornupunctata*. Dental formula of adults is from 4/4 to 6/6 and in immatures 2/2. Coxal spurs are moderate sized in females, but male coxa I to III spurs are much reduced and coxa IV spur is very long and lanceolate.

Keys to Identify Species of Subgenus Aboimisalis

Male

Salience is medium as the convex basal margin of palpal segment II and compact palpi; basis capituli are rectangular; dental formula is 4/4 or 5/5; posterodorsal spur is absent; posteroventral spur is short, wide, blunt, reaching to intersegmental suture of palpal segment II and III; infrainternal setae are closely spaced, short, feathery, and number 11; punctations are few, superficial, and irregular in size and distribution; lateral grooves are long and superficial, reaching to coxa II, enclosing first two festoons; coxa IV bears a very long, lanceolate, and pointed spur, while coxa I to III spurs are reduced..................[*H. cornupunctata*]

Female

Salience is medium as the convex basal margin of palpal segment II and compact palpi; basis capitulum is rectangular, showing cornua, the only case in females of subgenus *Aboimisalis*; dental formula is 4/4; posterodorsal spur is absent; postero-ventral spur is short, wide, blunt, reaching to intersegmental suture of palpal segment II and III; infrainternal setae are closely spaced, short, feathery, and number 11; coxal spurs are moderate and increase in size from coxa I to IV; porose areas are circular...................[*H. cornupunctata*]

2.3.1 Haemaphysalis cornupunctata

This species was first collected from Kashmir by Varma in 1959, and was described jointly with Hoogstraal in 1962. Later on in 1965, Hoogstraal and McCarthy thought it as a primitive species and placed it in the subgenus *Aboimisalis*. The immature stages were described by Dhanda and Kulkarni in 1969 from the specimens collected in Himachal Pradesh, India (Figure 2.10).

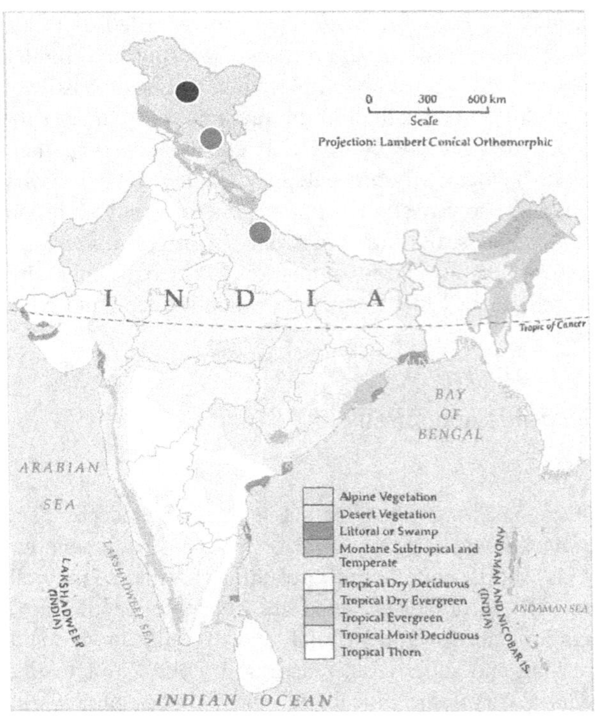

Figure 2.10 Distribution of *H. cornupunctata* in India. (●) Place of first record. Map not to scale.

Male (Figures 2.11 and 2.12)[51]

Total body length measures approximately 3.2 mm, width 1.6 mm.

Capitulum: Basis capitulum is rectangular; length is approximately two-thirds of width; cornua are robust, approximately half as long as the base, and widely triangular with less sharply pointed apices. Palpi have conical outline. Hypostome is moderately stout, and dental formula is 4/4, with six to eight denticles in a file.

Scutum: Scutum width is approximately two-thirds of length, widest at the level of coxa IV. Lateral grooves are long, superficial, extending from the level of coxa II to posterior margin of second festoon. Cervical grooves are deep. Punctations are superficial, comparatively few, and irregularly sized and spaced. Festoons number 11 and are clearly separated. Genital apron is situated between coxa II.

Legs: Coxa are with dissimilar spurs; spur of I is short, wide, and extending very slightly beyond coxal apex; spurs of II to IV arising adjacent to internoposterior corner, that of II slightly larger than I and that of III slightly larger than II; spur of IV is huge, sword-like, reaching to the level of mid-length of anus.

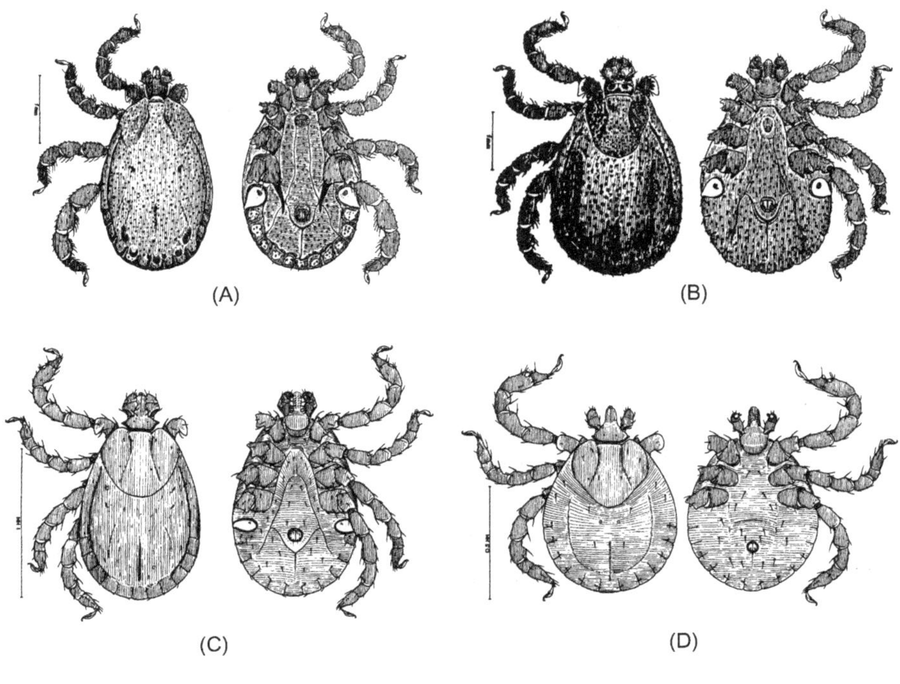

Figure 2.11 *Haemaphysalis cornupunctata*. (A) Male, dorsal and ventral side; (B) female, dorsal and ventral side; (C) nymph, dorsal and ventral side; (D) larva, dorsal and ventral side. *Source*: Hoogstraal and Varma (1962), Journal of Parasitology (vol 48(2)), Allen press Publishing Services, Dhanda and Kulkarni (1969), Oriental Insects, 3(1): 15–21.

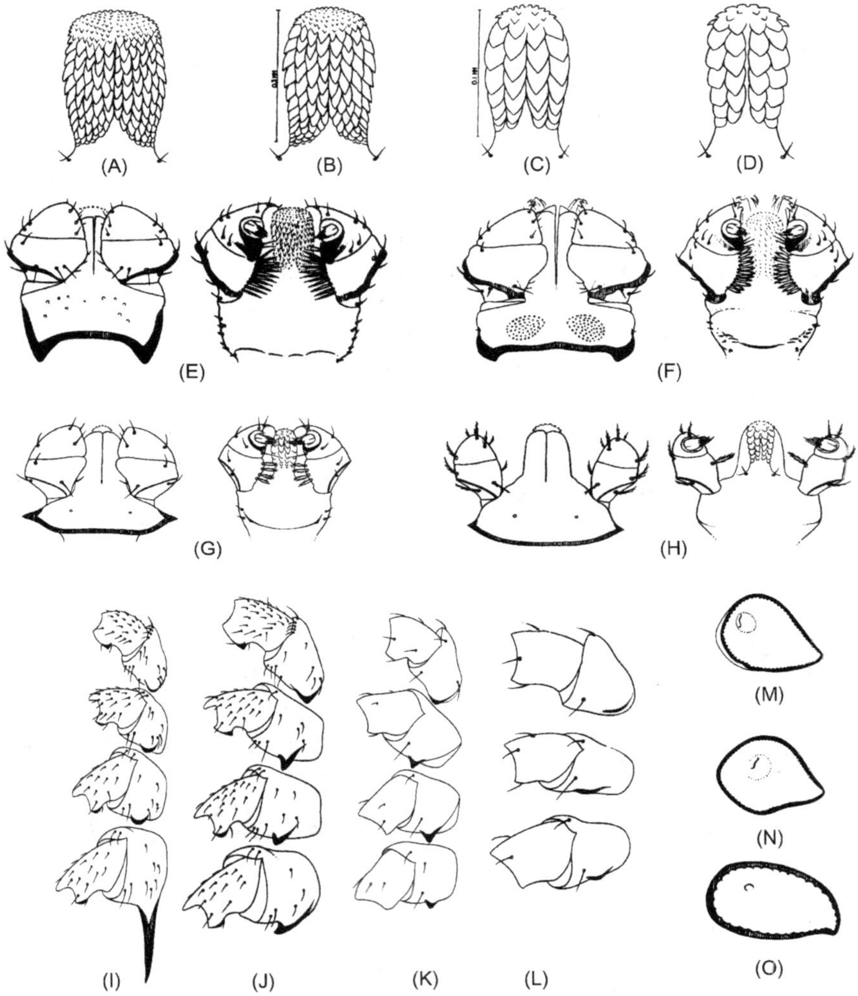

Figure 2.12 (A–D) Hypostome of male, female, nymph, and larva, respectively; (E–H) capitulum of male, female, nymph, and larva, respectively; (I–L) coxa of male, female, nymph, and larva, respectively; (M–O) spiracular plates of male, female, and nymph, respectively. *Source*: NIV. unpublished drawings.

Female (Figures 2.11 and 2.12)[51]

Overall body length measures approximately 3.7 mm, width approximately 2.2 mm.

Capitulum: Basis capitulum is with short, widely triangular cornua; porose areas are somewhat variable, moderate to very large, and sub-circular. Hypostome is with 4/4 dentition.

Scutum: Scutum is shield shaped, almost as wide as long. Cervical grooves are deep, concave, and extending slightly beyond mid-length of scutum. Punctations are larger and deeper than those of male and irregularly distributed.

Legs: Coxa are with spurs on I to III similar to but slightly larger than respective spurs of male; coxa IV is with a comparatively narrow spur approximating that of III in length.

Nymph (Figures 2.11 and 2.12)[29]

Total body length is approximately 1.7 mm, width 1.0 mm.

Capitulum: Capitulum is almost as wide as long (0.30 mm). Basis capitulum is dorsally approximately 2.4 times as wide as long; cornua are obsolete. Palpi compact, without lateral salience. Hypostome is about as long as palpi, 2–5 times as long as wide, and clavate; dental formula is 2/2, with seven to eight denticles in a file.

Scutum: Scutum is slightly longer than wide. Cervical grooves are concave, extend to three-fourths of scutal length. Punctations are rare.

Legs: Coxa are with a small, widely triangular spur on I, and slightly larger, pointed spurs on II to IV.

Larva (Figures 2.11 and 2.12)[29]

Overall body length is approximately 0.80 mm, width 0.60 mm.

Capitulum: Capitulum is slightly longer than wide. Basis capitulum is dorsally approximately 2.5 times as wide as long. Palpi are narrow and without lateral salience. Hypostome is slightly longer than palpi, 2.5 times as long as wide, lateral outline parallel; dental formula is 2/2, with six to seven denticles per file.

Legs: Legs are moderately long and stout.

Related Species[29]

All the three species, viz. *cornupunctata*, *chordeilis*, and *punctata*, included in the subgenus *Aboimisalis*, form a homogeneous group. The latter two species has not been recorded in India. *H. cornupunctata* is, however, closer to *H. punctata* than to *H. chordeilis*. The distinguishing characters of the adults have been amply discussed by Hoogstraal and Varma (1962). The nymphs and larvae of *H. punctata* can be readily separated from those of *H. cornupunctata* for having a distinct baso-lateral salience on palpal segment II. All the stages of *H. chordeilis* can also be distinguished from *H. cornupunctata* in having a distinct basolateral salience on palpal segment II. The adults can further be distinguished by their smaller cornua, and the male by the shape of scutum being ovoid, instead of elongate-oval.

Hosts[29,51,120]

Immature stages: Sheep, goat, domestic cattle, yak, horses, cow, cow calf, and from flag dragging in Kashmir (India).

Adults: Cattle, buffalo, yak, sheep, goat, rodents, insectivores, cow calf, horse, hare, and from flag dragging.

Distribution[29,51,120]

India (Himachal Pradesh, Jammu and Kashmir, Uttar Pradesh).

Ecology[29]

All *H. cornupunctata* collections from the above areas are during spring and fall season. The altitude of the localities in Himachal Pradesh, where *H. cornupunctata* was collected, ranges between 1,050 and 2,750 m. The total average yearly rainfall in the area varies from 100 to 150 cm. Most of the rainfall takes place between early July and mid-September. During the winter months, many places get moderate to heavy snowfall, but some localities at lower elevations such as Rakh (approximately 1,050 m) and Nogi (approximately 1,250 m) do not get any snowfall. Places at higher elevations, such as Kalpa (approximately 2,750 m) remain snowbound for more than six months in a year. The surrounding hills around Kalpa, distances of barely 3–5 km, remain snowbound all through the year. The vegetation of the different localities also varies greatly. Between 2,500 and 2,750 m, most of the localities have patches of coniferous forests, with very little undergrowth. Above 2,750 m, the terrain is mostly alpine.

2.4 Structurally Intermediate Haemaphysalines

2.4.1 *Subgenus* Herpetobia

General Characteristics

The subgenus *Herpetobia* is phylogenetically intermediate between the structurally primitive and advanced subgenera of *Haemaphysalis*. All *Herpetobia* immatures and adults have SA-pattern rectangular basis capituli. Their palpi, however, are compact, but are slightly salient in certain nymphs. They represent the forerunners of broad palpi characterizing SA haemaphysalines. Larval *Herpetobia* lacks cornua and discrete coxal spurs, retaining the features of their reptile-parasitizing progenitors. Adult cornua are moderately large in males and obsolete or small in females. The compact adult palpi are usually ridged posterodorsally in *H. sulcata* and *H. kashmirensis*, which was an evolutionary experiment that failed to survive (Hoogstraal and Kim).

Herpetobia is composed of four species and is a relict pivotal branch of *Haemaphysalis*. The type species is *H. sulcata* and also constitutes *H. kashmirensis*, *H. nepalensis*, and *H. sundrai*. Geographical distribution pattern of these species is much like those of SP subgenera *Alloceraea* and *Allophysalis*.

Keys to Identify Species of Subgenus Herpetobia

Males

1. Coxal and capitular spurs are absent or very small, only two infrainternal setae are present, and dental formula is 4/4...................[*H. nepalensis*]
Coxal and capitular spurs are moderate, six to nine infrainternal setae are present, and dental formula is 5/5 to 7/7...................[2]

2. Coxa IV spur is lanceolate; moderate spur on coxa I and conventionally triangular but variable spurs on coxa II and III; posteroventral spur on palpal segment III is short and inwardly directed; nine or more infrainternal setae are present; dental formula is 5/5 or 4/4...................[*H. sulcata*]
Coxal spurs are hook-like or conventionally triangular; capitular spur is short and broad; infrainternal setae number six to eight; dental formula is 5/5 or 6/6...................[3]

3. Coxal spurs are hook-like; dental formula is 6/6; infrainternal setae are seven or eight in number; body is elongated in shape; numerous; punctations are deep, lateral grooves are long and enclose festoon(s)...................[4]
Coxal spurs are triangular; dental formula is 5/5; infrainternal setae number six; body is oval in shape; number of punctations are moderate and are superficial; lateral grooves are small or absent...................[5]

4. Posteroventral spur on palpal segment III is broad and overlaps up to mid-length of palpal segment II; infrainternal setae number eight; cornua are long and pointed; all coxal spurs are hook-like and elongated...................[*H. kashmirensis*]
Posteroventral spur on palpal segment III is broad but just crossing the intersegmental suture between palpal segment II and III; infrainternal setae number seven; cornua are moderate but blunt or with rounded apical margin; all coxal spurs are hook-like but coxa I and IV spurs are elongated than coxa II and III...................[*H. kumaonensis*]

5. Coxal spurs are conventionally triangular on II to IV but appear to be spatulate or bluntly triangular on coxa I; posteroventral spur on palpal segment III is broadly triangular but reaching just short of suture between palpal segment II and III; punctations are numerous but irregularly sized and distributed; lateral grooves are very small and enclose just one festoon, reaching to the base of coxa IV...................
[*H. Himalaya*]
Coxal spurs are triangular but elongate on coxa I, while short but pointed on others; posteroventral spur on palpal segment III is broadly triangular but large, reaching to mid-length of segment II; punctations are numerous but regular in size and distribution; no lateral groove is present...................[*H. sundrai*]

Females

1. Coxal spurs are ridge-like or obsolete; posteroventral spur on palpal segment III is small or absent; infrainternal setae are two in number; dental formula is 4/4; porose areas are sub-circular in shape; external genitalia is sub-circular...................[*H. nepalensis*]
Coxal spurs are moderate to very small; posteroventral spurs on palpal segment III are moderate and inwardly directed; infrainternal setae numbering range from five to eleven; dental formula is 4/4 to 7/7; porose areas are circular or oval...................[3]

2. Coxal spurs are very small; posteroventral spur on palpal segment III is triangular but blunt and reaches to the suture between palpal segment II and III; infrainternal setae

number 11; dental formula is 4/4; porose areas are oval and moderate in size and quite spaced from each other; external genitalia are uniquely indented...................[*H. sulcata*]
Coxal spurs are moderate; posteroventral spur is moderate but blunt and inwardly directed; infrainternal setae number from five to eleven; dental formula is either 6/6 or 7/7; porose areas are elongately oval...................[3]

3. Coxal spurs are hook-like; posteroventral spur on palpal segment III is blunt or rounded and just crossing the suture between palpal segment II and III; porose areas are oval and inclined inwardly; dental formula is 6/6...................[4]
Coxal spurs are elongately triangular or conventionally triangular, posteroventral spur on palpal segment III is hook-like or inwardly directed and reaching to the suture between palpal segment II and III; porose areas are circular and small; dental formula is 7/7...................[5]

4. Porose areas are moderately sized, oval, and inclined inwardly; infrainternal setae number 11; cornua are small but blunt; all coxal spurs are almost equal in size and are hook-like; palps are loosely arranged...................[*H. kashmirensis*]
Porose areas are larger, oval, and inclined inwardly; infrainternal setae number seven; cornua are obsolete; coxal spurs decrease in size a little from coxa I to IV, except on coxa I where it is hook shaped, all others are triangular in shape; palps are compactly arranged...................[*H. kumaonensis*]

5. Coxal spurs are triangular, elongately spatulate on coxa I, while broad or ridge-like triangular on other coxa; posteroventral spur on palpal segment III is triangular, short, and inwardly directed; infrainternal setae are six in number; porose areas are small, circular, and well spaced; cornua are short and blunt...................[*H. himalaya*]
Spur on coxa I is elongately triangular and blunt, while ridge-like or plate-like on others; posteroventral spur on palpal segment III is hook-like and inwardly directed; infrainternal setae number five; porose areas are circular and moderate in size; cornua are broadly triangular, moderately sized, and pointed...................[*H. sundrai*]

2.4.2 Haemaphysalis nepalensis

This species was first described by Hoogstraal in 1962. The specimens were originally collected off white-bellied rats in the Leucopani area of Nepal in 1960 by R.L. Fleming; hence, the name of the species is after the place of collection (Figure 2.13).

Male (Figures 2.14 and 2.15)[52]

Overall body length measures approximately 2.6 mm, width 1.6 mm.

Capitulum: Capitulum is approximately 0.5 mm in length. Cornua are fairly widely triangular, approximately 0.35 as long as the base. Ventrally with short, broad, medially directed retrograde spur on segment III. Hypostome lateral margins are parallel; dental formula is 4/4, with maximum nine to ten denticles per file.

Scutum: Scutum is approximately 0.7 as wide as long, widest at the level of coxa IV. Lateral grooves are very short, extending only to the level of coxa IV and enclosing only first pair of festoons; cervical grooves are short, narrow, moderately deep, comma-shaped; punctations are superficial, irregularly sized and distributed,

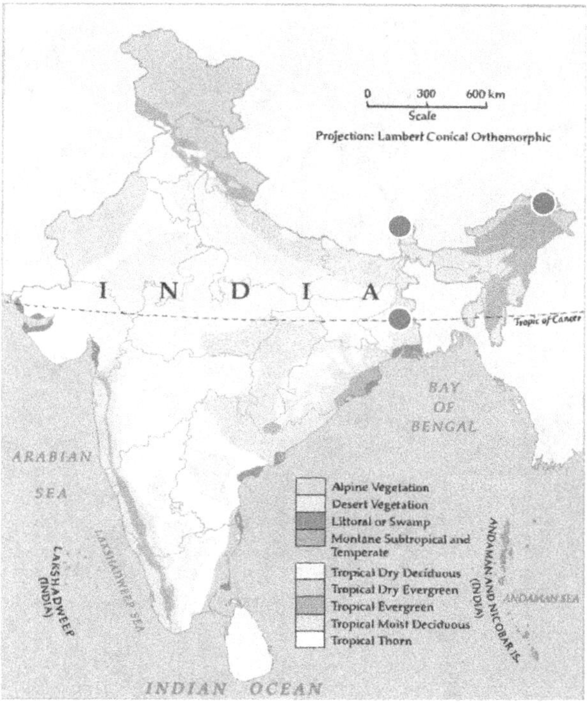

Figure 2.13 Distribution of *H. nepalensis* in India. (●) Place of first record. Map not to scale.

and fairly numerous; festoons number 11, are narrow, elongate, and clearly defined.

Legs: Coxa are each with a very small, triangular spur arising near posterointernal juncture. Trochanters are lacking ventral spurs.

Female (Figures 2.14 and 2.15)[52]

Total body length measures approximately 3.0 mm, width 1.6 mm.

Scutum: Porose areas are deep and triangular. Hypostome is with 4/4 dentition, but outline is basally somewhat narrow. Scutum is with length and width sub-equal, widest approximately at mid-length. Cervical grooves are narrow, shallow, and short. Punctations are moderately numerous, discrete, small, and moderately sized.

Legs: Coxal spurs are similar to those of male.

Nymph (Figures 2.14 and 2.15)[25]

Capitulum: Basis capitulum is dorsally approximately twice as wide as long. Cornua are triangular, approximately one-third as long as the base. Palpi are similar in outline to those of adult. One infrainternal seta is present. Hypostome is more

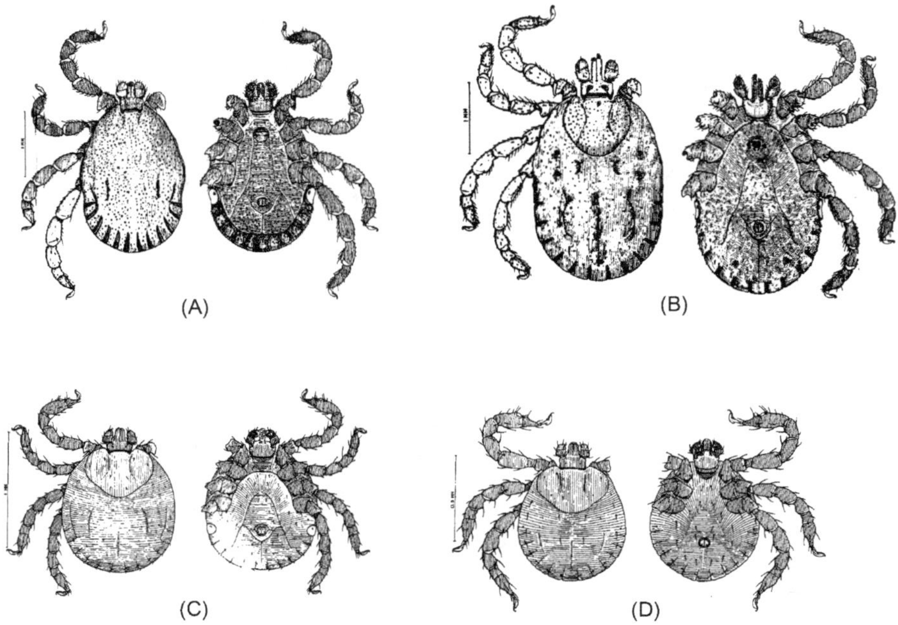

Figure 2.14 *Haemaphysalis nepalensis*. (A) Male, dorsal and ventral side; (B) female, dorsal and ventral side; (C) nymph, dorsal and ventral side; (D) larva, dorsal and ventral side.
Source: Hoogstraal (1962), Dhanda (1964), Journal of Parasitology (vol 48(2) & vol 50(6)), Allen press Publishing Services.

than twice as long as wide; lateral margins are parallel; dental formula is 2/2, with six or seven denticles in each row.

Scutum: Scutum is slightly wider than long. Cervical grooves are narrow and deep anteriorly, wide and shallow posteriorly, extending almost to posterior margin of scutum.

Legs: Coxa are each with a well-developed triangular spur; coxa I is with outer spur-like ridge. Trochanters are with a ventral ridge-like projection.

Larva (Figures 2.14 and 2.15)

The total body length is about 0.81 mm. The capitulum is 0.17 mm long.

Capitulum: The palpi are loosely arranged and are with very poor salience. No spur is present dorsally, but ventrally palpal segment III shows a little outgrowth or plate-like spur not even reaching to the basal margin of palpal segment III. Infrainternal seta is single. Hypostome is having 2/2 dental formula, with five to six denticles per file.

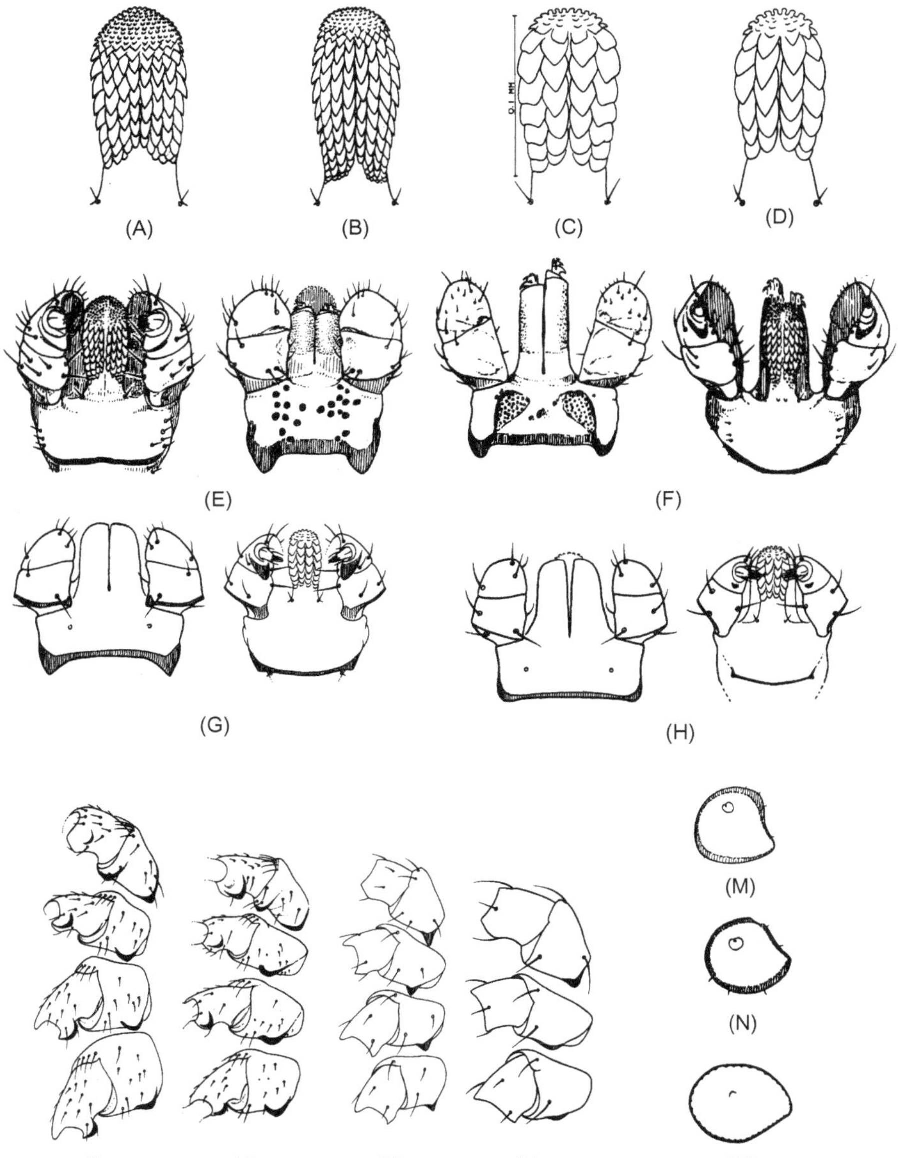

Figure 2.15 (A–D) Hypostome of male, female, nymph, and larva, respectively; (E–H) capitulum of male, female, nymph, and larva, respectively; (I–L) coxa of male, female, nymph, and larva, respectively; (M–O) spiracular plates of male, female, and nymph, respectively.
Source: NIV. unpublished drawings.

Scutum: Scutum is 1.6 times as wide as long. The punctations are equal to none. Cervical grooves are moderately deep and are reaching two-thirds of scutum.

Legs: Coxa are showing plate-like or broadly triangular spur.

Related Species[52]

Among Indian *Herpetobia* species, *H. nepalensis* is related to *H. kashmirensis* in appearances but differs in having only two narrow setae on the internal margin of palpal segment II ventrally and outer spur-like ridge on coxa I, short lateral grooves sub-circular female scutum, triangular porose areas, and ventral spur on female trochanter I. Each species except *H. nepalensis* has a prominent but variable ventral spur on palpal segment III. *H. nepalensis* is intermediate between and shares a number of peculiar structural characters with ticks in the subgenera *Alloceraea* (*H. aponommoides*) and *Allophysalis* (*H. warburtoni*).[61]

Hosts[25,26,120]

Immature stages: Sheep, and by flag dragging.

Adults: Sheep, cattle, yak, goat, buffalo, man, wild cat, bear, serow, black bear.

Distribution[25,26,120]

India (West Bengal, Sikkim, Arunachal Pradesh), Nepal.

Disease Relationship

Not known.

Ecology[25]

Haemaphysalis nepalensis was described by Hoogstraal (1962) from one male and two female specimens from a rodent (white-bellied rat) and man near Leucopani, about 5,000 ft altitude, in eastern Nepal. *H. nepalensis* has been collected subsequently from seven localities between 6,000 and 13,000 ft altitude in eastern Nepal.

2.4.3 Haemaphysalis sulcata

H. sulcata is a species which was described in 1877 by Canestrini and Fanzago from a nymph. Subsequently it was further placed into synonymy with *H. punctata* and was described several times anew under various names. However, the original name is still maintained (Figure 2.16).

Male (Figures 2.17 and 2.18)[110]

Capitulum: Dorsally basis capitulum is with very long and robust dorsal cornua; length of basis capitulum with cornua is nearly equal to width of basis capitulum. Palpal segment II is with arcuate lateral salience; dorsally posterior margin of

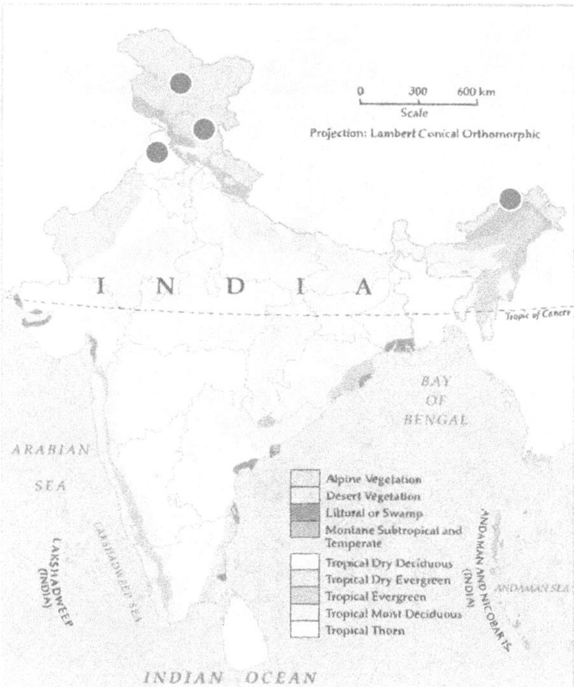

Figure 2.16 Distribution of *H. sulcata* in India. (●) Map not to scale.

palpal segment II is with a small arcuate denticle, which may not be developed in small specimens. Ventral spur of palpal segment III is moderate and directed posteriorly. Hypostome is with 6/6 longitudinal rows of denticles.

Scutum: Distinct marginal grooves are of normal length. Festoons number 11.

Legs: All coxa are with median spurs; spurs on coxa I are narrow and tapering; spurs on coxa II and III are very short and broad; spur on coxa IV is very long, but shorter than the length of coxa IV.

Female (Figures 2.17 and 2.18)[110,51]

Capitulum: Dorsally basis capitulum is with large, widely separated porose areas and with very short dorsal cornua or no dorsal cornua; width is 1.5 times as length. Lateral salience of palpal segment II is moderate, dorsally posterior margin of palpal segment II is with short arcuate projection; it may be absent in small specimens. Ventral spur on palpal segment III is moderate and directed posteriorly. Hypostome is with 4/4, rarely 5/5 or 3/3 longitudinal rows of denticles.

Scutum: Scutum is wide-oval or almost circular. Cervical grooves are narrow anteriorly, wider and shallower posteriorly, and two-thirds of scutum length. Punctations are large, deep, and dense. Genital operculum is U-shaped with thickened lateral margins.

Legs: Coxa I has a distinct median spur or slight tubercles.

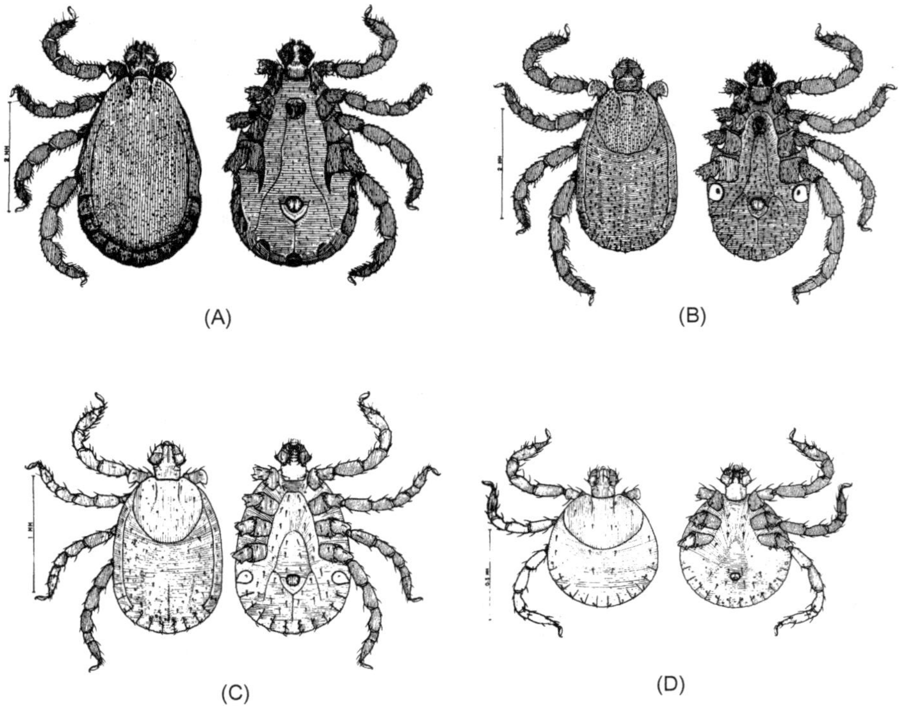

Figure 2.17 *Haemaphysalis sulcata*. (A) Male, dorsal and ventral side; (B) female, dorsal and ventral side; (C) nymph, dorsal and ventral side; (D) larva, dorsal and ventral side. *Source*: NIV. unpublished drawings.

Nymph (Figures 2.17 and 2.18)

Capitulum: Basis capitulum is nearly trapezium-like. Palpal segment II is with small but distinct lateral salience. Ventral spur of palpal segment III is lacking. Palpal segment II is with five to six ventromedian leaf-like setae. Hypostome is with 2/2 longitudinal rows of denticles.

Scutum: Scutum is sub-circular or slightly longer than width. Scutum is very short.

Legs: Coxa I is with very short and broad median spur; coxa II is with very short, arcuate, shifted to middle of posterior margin median spur; coxa III and IV are without median spur.

Larva (Figures 2.17 and 2.18)

Capitulum: Basis capitulum is similar to that of the nymph. Palpal segment II is with small blunt lateral salience. Ventral spur on palpal segment III is lacking.

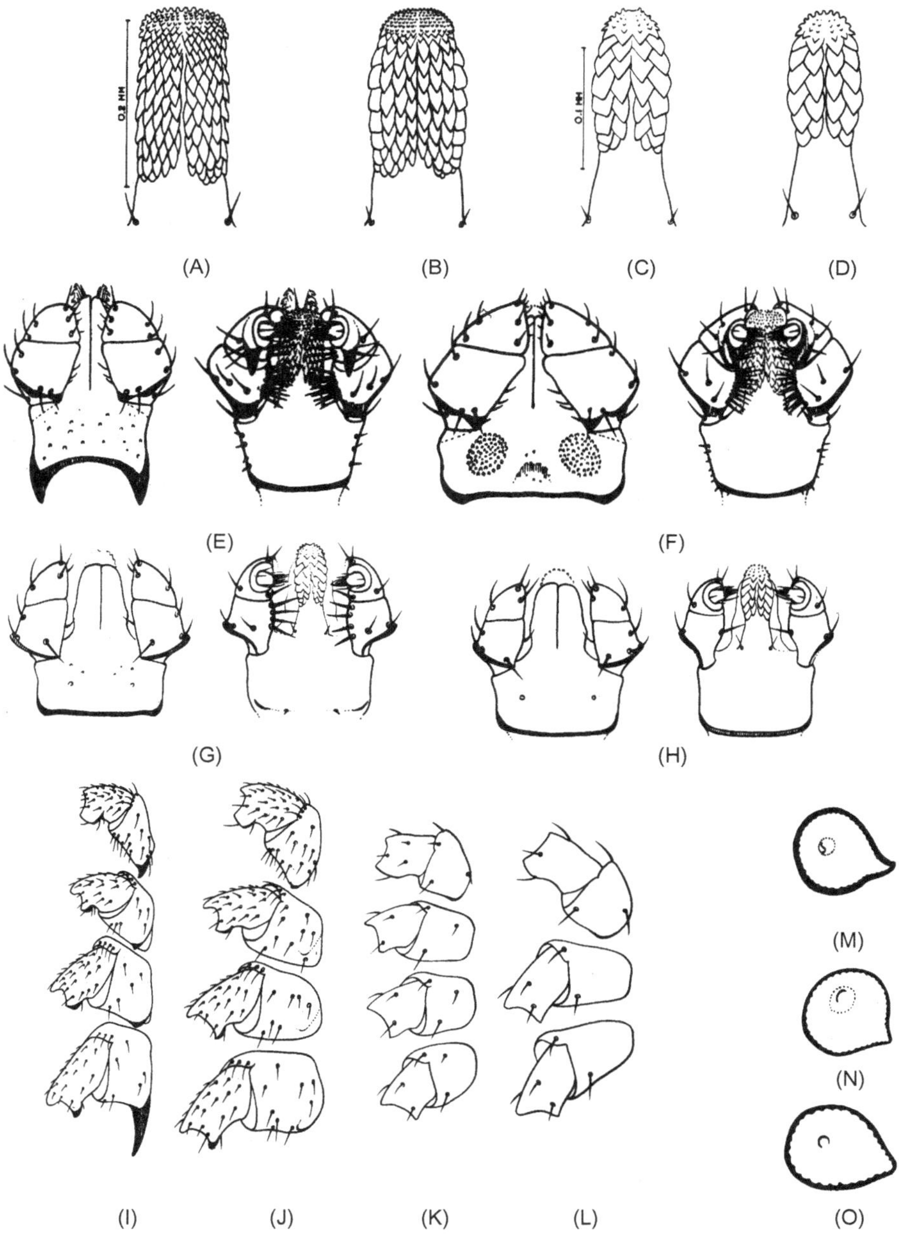

Figure 2.18 (A–D) Hypostome of male, female, nymph, and larva, respectively; (E–H) capitulum of male, female, nymph, and larva, respectively; (I–L) coxa of male, female, nymph, and larva, respectively; (M–O) spiracular plates of male, female, and nymph, respectively.

Source: NIV. unpublished drawings.

Ventromedian setae of palpal segment II are rather cone-like in contrast with those of nymph, denticulated. Hypostome is with 2/2 longitudinal rows of denticles.

Legs: All coxa are without spurs.

Host[120,51]

Immature stages: Goat.

Adults: Goat, sheep.

Distribution[51,110,120]

India (Punjab, Himachal Pradesh, Arunachal Pradesh, Jammu and Kashmir), South Russia, Moldova, Crimea, North Caucasus, Dagestan, Transcaucasia, Armenia, Azerbaijan, Kazakhstan, Kyrgyzstan, Uzbekistan, Tajikistan, Turkmenia, southern Europe, islands of Mediterranean sea, Turkey, Iran, Afghanistan, Pakistan, Western China, Syria, Lebanon, Israel, Iraq, Yemen, Morocco, Algeria, Tunisia, Italy, Palestine.

Ecology

This species mainly inhabits foothills and mountainous semi-deserts and dry steppe biotopes (extensive plain without trees), and occurs up to 2,200 m or even up to 2,500 m above sea level in valleys between mountain ridges. The species is characterized by pasture type of questing (a field covered with grass) and by three-host life cycle. Life cycle takes 1.5–2 years to complete.

Medical Importance[110]

The species is a vector of sheep anaplasmosis on the south of Kyrgyzstan and Kazakhstan.

2.4.4 Haemaphysalis kashmirensis

This species was collected by Varma in 1959 in Kashmir area of India during a collection trip for the Virus Research Center in Pune, India (now National Institute of Virology), along with *H. cornupunctata*. Both these species were observed to be new and were described by Hoogstraal and Varma in 1962 (Figure 2.19).

Male (Figures 2.20 and 2.21)[51]

Overall body length measures approximately 4.00 mm, width 2.45 mm.

Capitulum: Basis capitulum length is slightly greater than width; cornua are robust, elongate, triangular, almost as long as the base, and with fairly sharply pointed apices. Palpi have a conical outline. Hypostome is moderately stout, with slightly convex lateral margins; dental formula is 6/6, with approximately 9–12 denticles in a file.

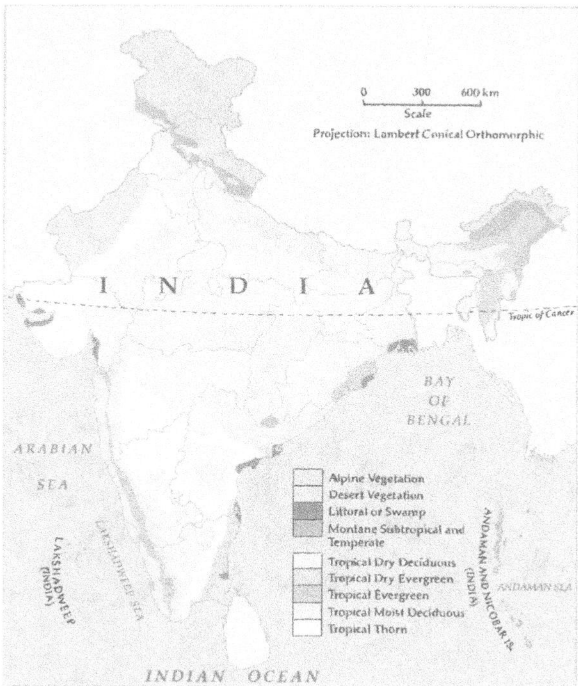

Figure 2.19 Distribution of *H. kashmirensis* in India. (●) Place of first record. Map not to scale.

Scutum: Scutum outline is elongate and oval, almost twice as long as wide, widest at the level of spiracular plates. Lateral grooves are deep, long, distinct, and extending from the level of coxa II to posterior margin of first festoon. Cervical grooves are fairly deep and very short. Punctations are few, inconspicuous, shallow, and irregularly distributed. Festoons number 11. Genital apron is rather large, subcircular, with minute serrations distally, and situated between coxa II.

Legs: Legs are moderately long and robust.

Female (Figures 2.20 and 2.21)[51]

Total body length measures approximately 4.10 mm, width 2.25 mm.

Capitulum: Basis capitulum is with short, widely triangular cornua; porose areas are shallow, of moderate size, tending to be elongate, triangular, and separated from each other by distance equaling width of one area. Hypostome is with 5/5 dentitions.

Scutum: Scutum is almost as wide as long; all margins are gradually rounded. Cervical grooves are deep and converging anteriorly, thence shallow, diverging, extending slightly beyond mid-length of scutum. Punctations are similar to those of anterior two-fifths of male scutum.

Legs: Coxal spurs, tarsi, and other characters are essentially as in male; tarsi, especially II and III, are more humped than in male.

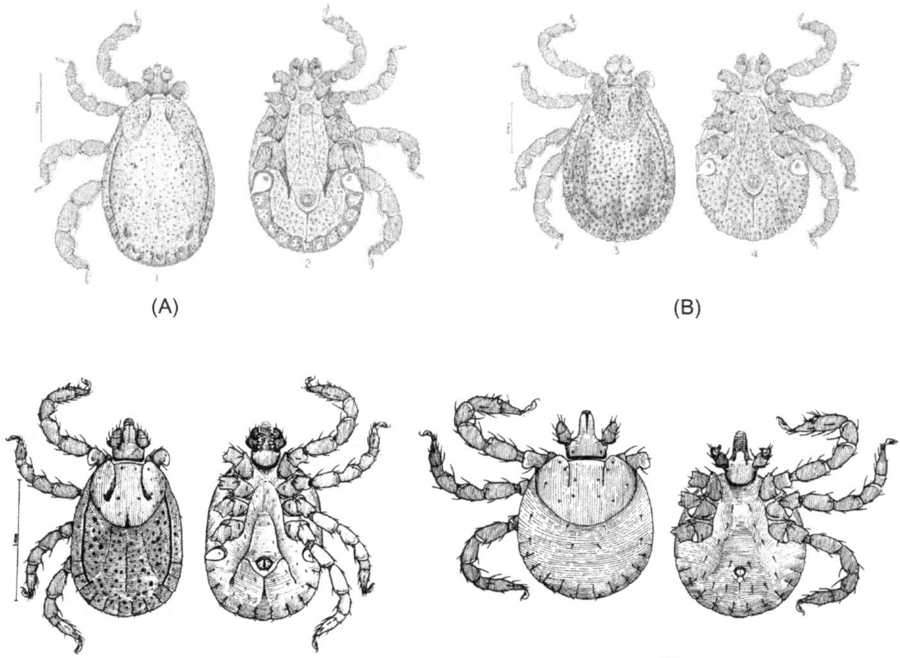

(A) (B) (C) (D)

Figure 2.20 *Haemaphysalis kashmirensis*. (A) Male, dorsal and ventral side; (B) female, dorsal and ventral side; (C) nymph, dorsal and ventral side; (D) larva, dorsal and ventral side. *Source*: Hoogstraal and McCarthy (1965), Hoogstraal and Varma (1962), Journal of Parasitology (vol 51 & vol 48(2)), Allen press Publishing Services.

Nymph (Figures 2.20 and 2.21)[60]

Overall body length is approximately 1.9 mm.

Capitulum: Basis capitulum is approximately 2.2 times as wide as long. Palpi compact with very slight lateral salience. Infrainternal setae number three. Hypostome is slightly longer than palpi, lateral outline is parallel; dental formula is 4/4, with seven (inner) to nine or ten (outer) denticles in a file.

Scutum: Scutum is slightly wider than long, widest at mid-length, all margins are gradually rounded; anterior emargination is wide and shallow. Cervical grooves are concave and extend to or slightly beyond scutal mid-length. Punctations are rare.

Legs: Coxa are with a very small, short, widely triangular spur on I and a slight sub-basal or basal ridge on II, III, and IV.

Larva (Figures 2.20 and 2.21)[60]

Overall length of body is approximately 0.6 mm.

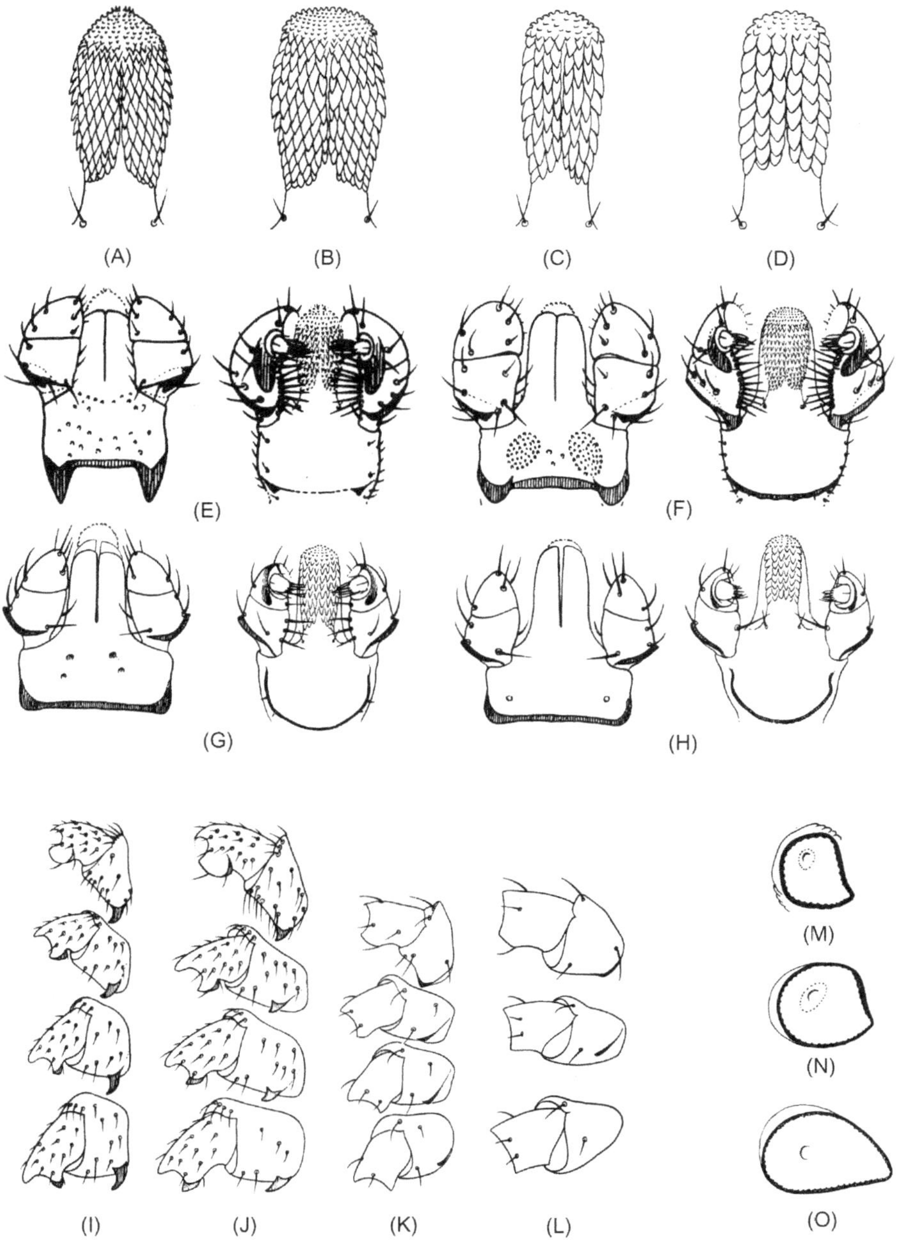

Figure 2.21 (A–D) Hypostome of male, female, nymph, and larva, respectively; (E–H) capitulum of male, female, nymph, and larva, respectively; (I–L) coxa of male, female, nymph, and larva, respectively; (M–O) spiracular plates of male, female, and nymph, respectively. *Source*: NIV. unpublished drawings.

Capitulum: Basis capitulum is twice as wide as long; outline as in nymph. Palpi are as in nymph. Infrainternal seta is single. Hypostome is essentially as in nymph, except that dental formula is 3/3, with seven or eight denticles in a file; margins are gradually rounded.

Scutum: Cervical grooves are parallel and extend to mid-length of scutum.

Legs: Coxa I is with faint basal ridge, II with faint subbed ridge, and III lacking spur or ridge.

Related Species

This species is morphologically related to *H. sundrai*, *H. nepalensis*, and *H. sulcata*. The curved coxal spurs of *H. kashmirensis* are unique to the species and different from the other two species in having very large cornua in male, distinctive basolateral palpal ridge, and large ventral spur of palpal segment III.

Ecology

Hoogstraal and McCarthy collected the specimens recorded above in the Shogran and Naran areas of Pakistan during the months of April and May (spring), August (summer), and September and October (autumn). Infestation patterns during other months of the year are not known. During spring, adults and large numbers of nymphs were taken from animals. In August, larvae and numerous nymphs were found. During autumn, males, females, nymphs, and larvae were collected on animals. These data suggest that immature stages may be most active during warmer months and those adults may feed chiefly during some of the cooler months of the year.

The known altitudinal range of *H. kashmirensis* in Pakistan is from 5,000 to 10,000 ft. In two collecting areas, Shogran and Naran, there is fairly heavy rain in late June and July. However, August and September are quite dry, and particularly at Naran the days are clear. At 10,000 ft and above, the terrain is arctic-alpine in nature, both at Shogran and at Naran. Snow remains on the ground at least 8 months of the year at Naran, and for probably 7 months at Shogran. The climate in both areas is much colder than in central Europe except at alpine elevations. For example, within a mile of Naran, at 7,800 ft elevation, glaciers are present throughout the year. The valley at Naran is an area of glacial till. At Naran, April is still winter, and so is most of May. Several of the collections of adult *H. kashmirensis* in Pakistan were made from domestic animals in narrow valleys through which nomadic tribesmen pass with large herds of goats, sheep, cattle, and camels. The lizard *Agama tuberculata* is common among large rocks near streams, on forested hillsides, and in cleared cultivated areas; it was infested by numerous larvae and nymphs of *H. kashmirensis* in each of these situations.

Host[51,60,120]

Immature stages: Agama lizards, rodents, flying squirrel, rat (*Rattus rattoides* Hodgson, 1845), mice (*Mus musculus homourus* Hodgson).

Adults: Goat, sheep, cattle.

Distribution[51,60,120]

India (Himachal Pradesh, Jammu and Kashmir, Uttar Pradesh), Pakistan.

Disease Relationship

Not known.

2.4.5 *Haemaphysalis himalaya*

This species was first described by Hoogstraal in 1966 from the tick collections made by Baron G.A. von during German India Expedition in Uttarakhand state, off Himalayan thar in 1956 (Figure 2.22).[65]

Male (Figures 2.23 and 2.24)[65]

Overall body length is approximately 2.4 mm, width 1.4 mm.

Capitulum: Basis capitulum is dorsally approximately 1.3 times as wide as long; cornua are triangular with pointed apices. Palpi are compact, each 1.5 times as long as wide. Hypostome is as long as palpi; external profile is convex; dental formula is 5/5, with approximately nine denticles in each file.

Scutum: Genital area is as illustrated.

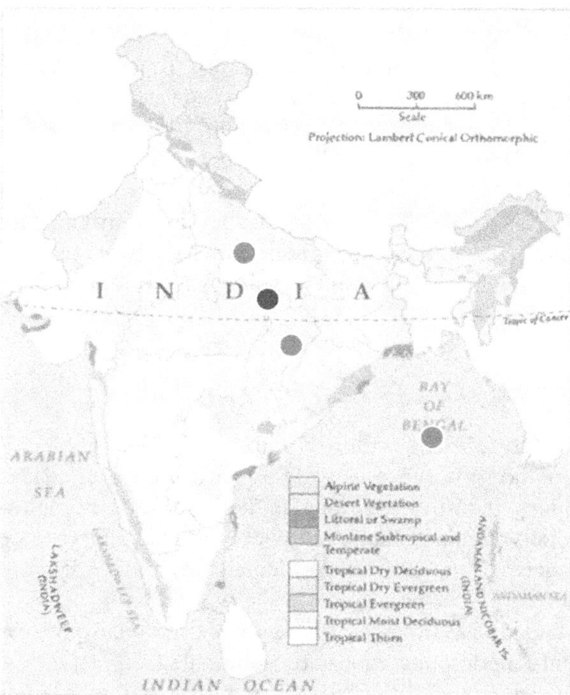

Figure 2.22 Distribution of *H. himalaya* in India. (●) Place of first record. Map not to scale.

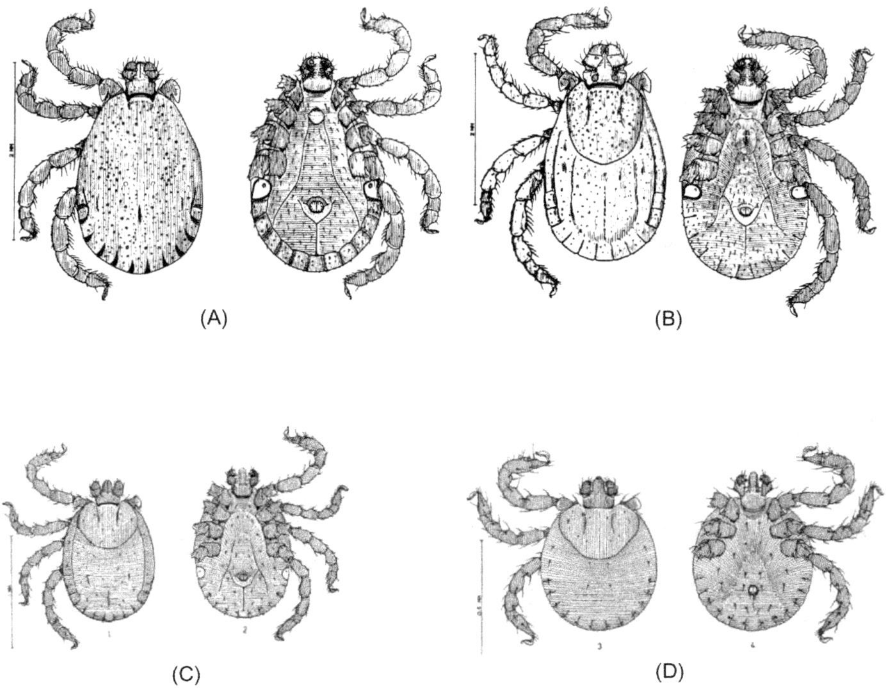

Figure 2.23 *Haemaphysalis himalaya.* (A) Male, dorsal and ventral side; (B) female, dorsal and ventral side; (C) nymph, dorsal and ventral side; (D) larva, dorsal and ventral side. *Source*: Hoogstraal (1966), Hoogstraal and Kammah (1970), Journal of Parasitology (vol 52 (4)) & 56(5), Allen press Publishing Services.

Legs: Legs are moderate. Coxa are each with a short spur of approximately equal length; spur of I is peg-like with rounded apex, approximately twice as long as wide; other spurs are widely triangular with pointed apices and extend a short distance beyond posterior margins of coxa.

Female (Figures 2.23 and 2.24)[65]

Overall body length is approximately 2.9 mm, width 1.7 mm.

Capitulum: Basis capitulum is dorsally twice as wide as long; cornua are triangular with narrowly rounded apices, 0.45 times as long as the base of basis capitulum; porose areas are elongate, sub-circular, small, and widely spaced. Hypostome is not quite as long as palpi, twice as long as in male; dental formula is 7/7, with 11 or 12 denticles in a close file.

Scutum: Scutum is shield shaped; 1.2 times as long as wide; cervical grooves are as short, narrow to scutal mid-length; punctations are as in male.

Legs: Coxa are increasing in size from I to IV, coxa I spur is moderately larger and other coxar with smaller or ridge-like spur.

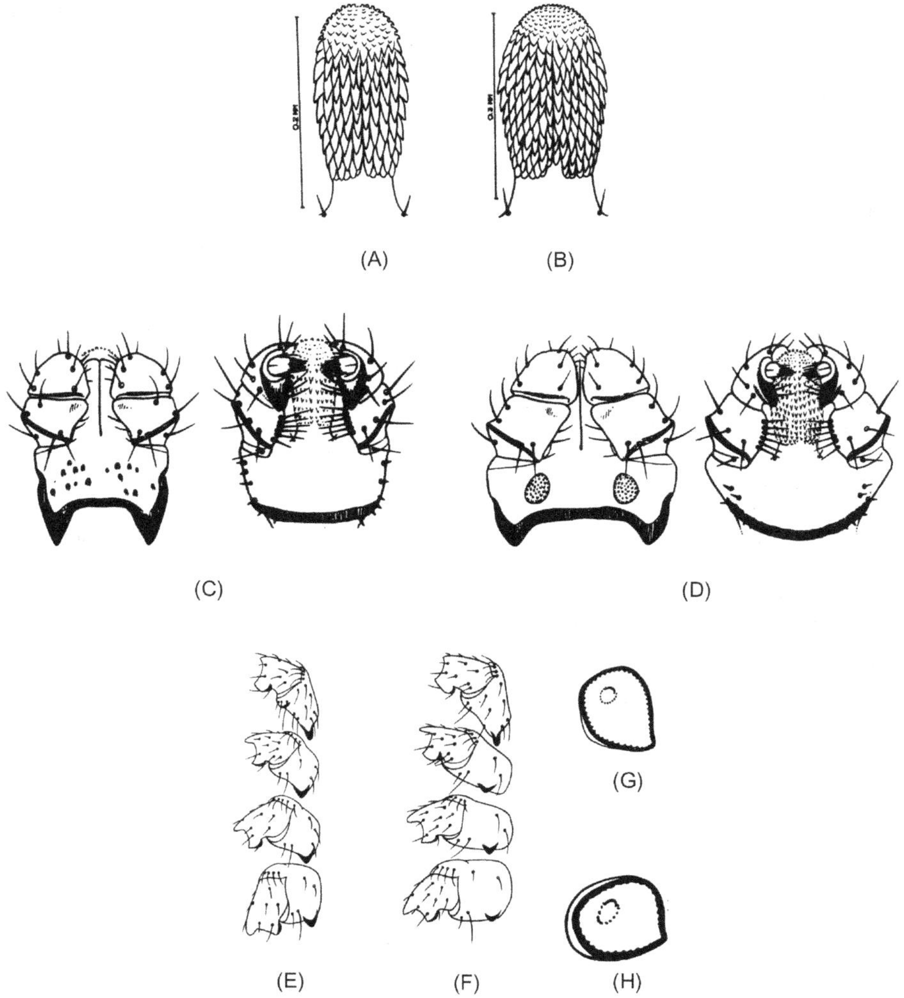

Figure 2.24 (A–B) Hypostome of male, female; (C–D) capitulum of male, female; (E–F) coxa of male, female; (G–H) spiracular plates of male, female.
Source: NIV. unpublished drawings.

Nymph (Figures 2.23 and 2.24)[72]

Overall body length is approximately 1.29 mm, breadth 0.75 mm.

Capitulum: Basis capitulum is dorsally approximately 2 times as broad as long; cornua are broadly triangular, approximately one-half as long as the base of basis capitulum. Palpi are compact. Hypostome is as long as palpi, and dental formula is 3/3, with six or seven denticles in a file.

Scutum: Scutum is approximately 1.2 times as broad as long. Cervical grooves are narrow. Punctations are obscure.

Legs: Legs are moderately long and thin. Coxa are each with a broadly triangular spur extending well beyond posterior margin; spurs are successively slightly decreasing in size from I to IV.

Larva (Figures 2.23 and 2.24)[72]

Length is approximately 0.75 mm, breadth 0.51 mm.

Capitulum: Basis capitulum is dorsally approximately 2 times as broad as long; cornua are obsolete. Palpi are campanulate. Hypostome is slightly longer than palpi, and hooklets in approximately five rows; dental formula is 2/2, with six or seven denticles in a file.

Scutum: Scutum is approximately 1.3 times as broad as long. Cervical grooves and punctations are approximately as in nymph.

Legs: Coxa I is with short, broadly rounded spur; II is with slight, broadly rounded ridge on posterior margin; and III is with minute ridge.

Related Species[65,72]

In the subgenus *Herpetobia*, *H. himalaya* is closely related to *H. nepalensis*, but differs chiefly as follows: male cornua 0.6 times as long as the base of basis capitulum (0.35 times as long in *nepalensis*), female porose areas small rather than large, posteroexternal juncture of female palpal segment II slightly flared (rounded in *nepalensis*), infra-internal setae number six rather than two (both sexes), female genital operculum longitudinally sub-quadrate (transversely oval in *nepalensis*). The dental formula is 5/5 in male and 7/7 in female of *himalaya*; in *nepalensis* it is 4/4 in both sexes.

The larva and nymph of *himalaya* are morphologically related to those of *kashmirensis* and *nepalensis*. The nymph of *kashmirensis* has a 4/4 dental formula as compared to 3/3 in *himalaya*; the ventral spur of palpal segment III in *kashmirensis* is smaller, while it is larger in *himalaya*; the basis capitulum lacks cornua in *kashmirensis* and *nepalensis* but *himalaya* possesses large cornua; the spiracular plates are broadly ovate in *himalaya* as compared to those in *kashmirensis* and *nepalensis*. The larva of *kashmirensis* has a 3/3 dental formula, while *himalaya* has 2/2. There is a definite spur on coxa I in *himalaya* but absent or obsolete in *kashmirensis* and *nepalensis*. The nymph of *nepalensis* has a 2/2 dental formula, a single ventrointernal seta (three in *himalaya*), and small pseudo-cornua (absent in *himalaya*). The larva of *nepalensis* can be differentiated chiefly by the absence of a ventral spur on palpal segment III (present in *himalaya*).

Host[65,72,120]

Immature stages: Indian muntjac, chital deer, black buck, domestic zhum, sambar deer, and from flag dragging.

Adults: Himalayan thar, barking deer, cattle, buffalo, sheep, goat, goral, chital deer, sambar deer, chital deer, black buck, and from flag dragging.

Distribution[65,72,120]

India (Uttar Pradesh, Uttarakhand, Himachal Pradesh, West Bengal), Nepal.

Disease Relationship

Not known.

Ecology[65]

H. himalaya is first known from locality, Molta, at 3,000 m altitude in Uttar Pradesh state, India. The 3,000 m elevation is at the upper end of the temperate zone and consists of mixed oak and conifer forest. The sub-alpine forest begins at 3,000 m and extends to 3,400 m; the alpine zone ranges from 3,400 to 4,200 m elevation. The type specimen is collected from Himalayan thar. The Himalayan thar, a wild goat variety, feeds in valleys in summer when snow melts from March to April, and later in the year these goats move upward.

It is known from localities between 3,200 and 9,750 ft altitude in northern India and Nepal.

2.4.6 Haemaphysalis sundrai

This species was first described by Sharif in 1928 from three female specimens collected on domestic sheep in Uttarakhand state (previously part of Uttar Pradesh state) of India. Later on Dhanda and Bhat described the male and redescribed the female in 1971, with ecologic observations (Figure 2.25).[51]

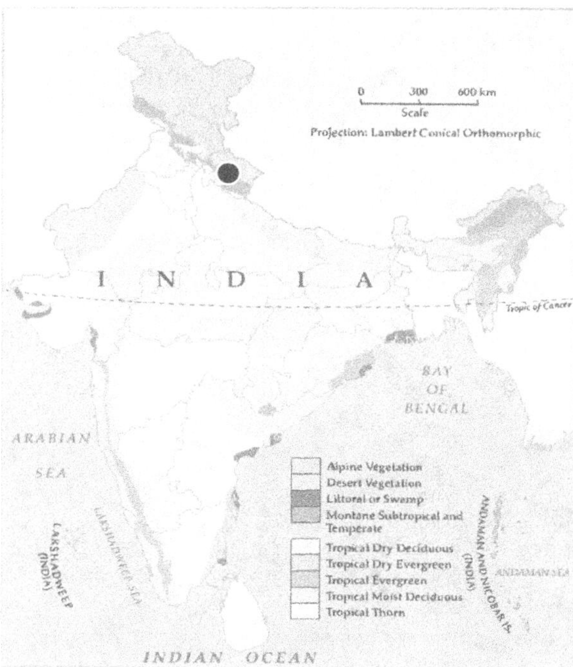

Figure 2.25 Distribution of *H. sundrai* in India. (●) Place of first record. Map not to scale.

Male (Figures 2.26 and 2.27)[32]

Overall body length is approximately 2.5 mm, breadth 1.3 mm.

Capitulum: Basis capitulum is dorsally approximately 1.3 times as broad as long; cornua are broad triangular, approximately one-half as long as the base of basis capitulum. Palpi are compact, most setae are relatively long. Hypostome is broad, shorter than palpi; dental formula is 5/5, with eight to ten denticles in a file.

Scutum: Scutum is approximately 1.45 times as long as broad; greatest breadth at the level of spiracular plates. Cervical grooves are narrow. Lateral grooves are obsolete. Punctations are few, faint, and small. Festoons number 11.

Legs: Legs are moderately long and narrow. Coxa are each with a triangular spur; I spur is elongate, almost reaching anterior margin of coxa II; II to IV spurs are broad and sub-equal, extending well beyond coxal margins.

Female (Figures 2.26 and 2.27)[32,51,127]

Capitulum: Basis capitulum is dorsally approximately 2 times as broad as long; cornua are broadly triangular, approximately two-thirds as long as the base of basis capitulum; porose areas are relatively small, elongately sub-ovate, and distantly

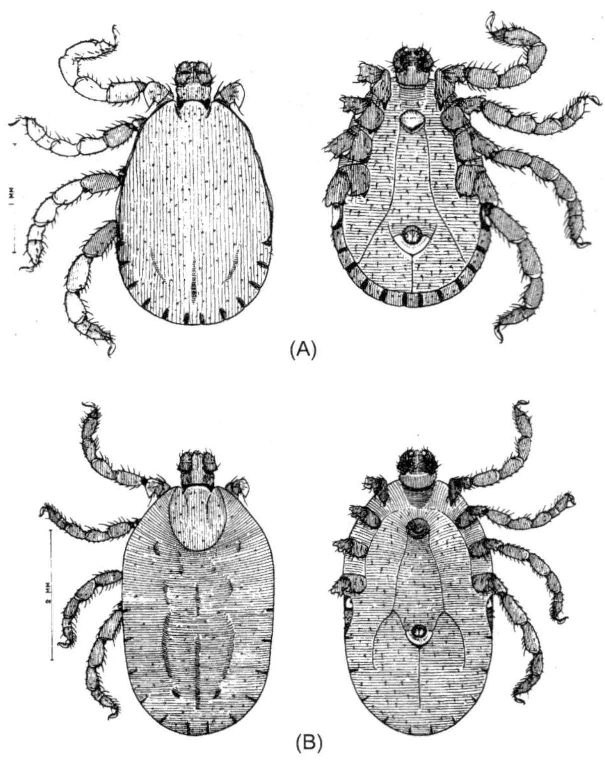

Figure 2.26 *Haemaphysalis sundrai*. (A) Male, dorsal and ventral view; (B) female, dorsal and ventral view. *Source*: Hoogstraal and Varma (1962), Dhanda and Bhat (1971), Journal of Parasitology (vol 48(2) & vol 57(3)), Allen press Publishing Services.

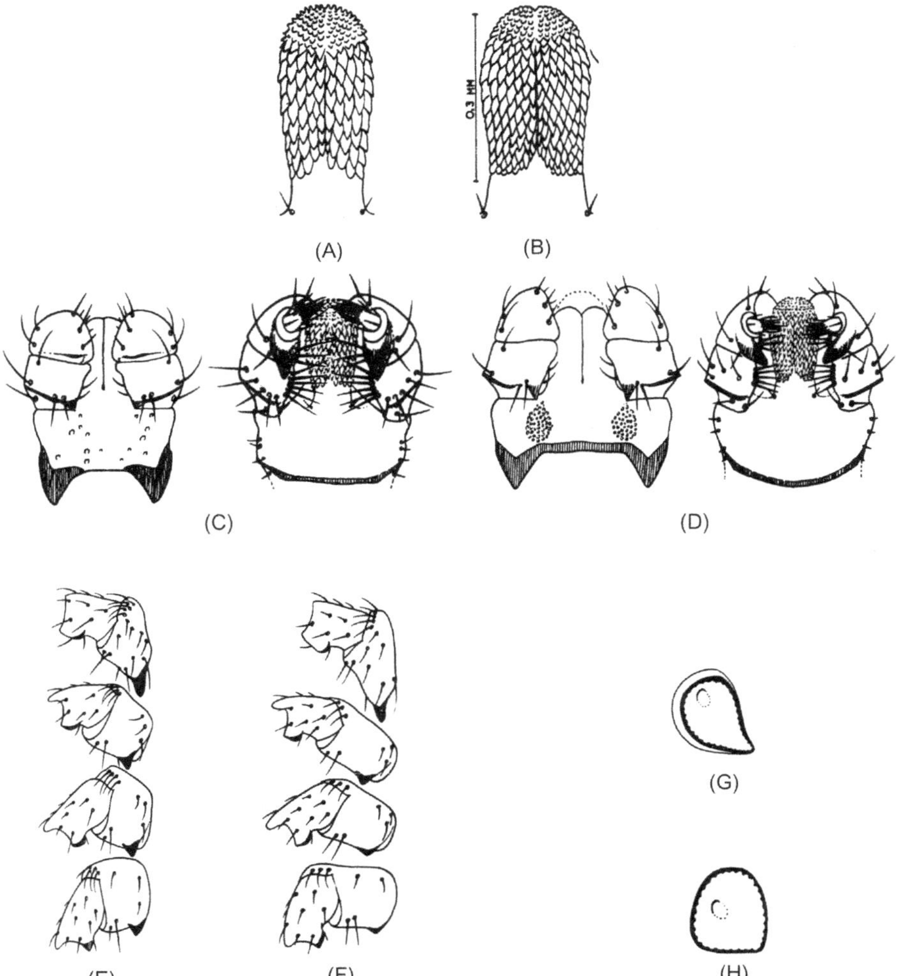

Figure 2.27 (A–B) Hypostome of male and female, respectively; (C–D) capitulum of male and female, respectively; (E–F) coxa of male and female, respectively; (G–H) spiracular plates of male and female, respectively.
Source: NIV. unpublished drawings.

spaced. Palpi are intermediate between compact and irregularly campanulate. Hypostome is broad and shorter than palpi; dental formula is 7/7, with eight to ten denticles in a file.

Scutum: Scutum is approximately 1.25 times as long as broad and shallow. Cervical grooves are like narrow arcs extending to scutal mid-length. Scutal punctations are of irregular size and distribution.

Legs: Legs are as in male except as follows: coxa I and II spurs are slightly smaller; III and IV spurs are much shorter (extend only to or slightly beyond coxal margins).

Related Species[32]

The subgenus *Herpetobia* Canestrini, redefined by Hoogstraal and McCarthy (1965), contains Indian species such as *H. sulcata*, *H. sundrai*, and *H. kashmirensis*, which are morphologically related to each other. According to these authors, *H. nepalensis* is a "borderline" species between the subgenera *Herpetobia* and *Haemaphysalis*. The *sundrai* male is the only one of this group in which lateral grooves are obsolete; its coxa lack a lanceolate spur of IV (present in *sulcata*) or curving spurs (*kashmirensis*) and have larger spurs than those of *nepalensis*. The female *sulcata* has much larger cornua than those of the other species, a 7/7 dental formula (4/4 in *sulcata* and *nepalensis*, 5/5 or 6/6 in *kashmirensis*), a different ventral spur on palpal segment III, different coxal spurs, etc.

Host[32,51,120]

Immature stages: Not available.
Adults: Sheep.

Distribution[32,51,120]

India (Uttarakhand).

Ecology[32]

Bhowali, the type locality of *H. sundrai,* lies between 5,000 and 6,000 ft altitude in an outer Himalayan valley. Some steep surrounding hills reach 7,000 ft. The climate is temperate with warm summers and an average annual rainfall of about 60 inches, mostly from late June and mid-September. Winter snow soon melts except in the higher hills. The forests are mostly pine with little undergrowth. Ravines have oak forests with scattered rhododendron trees. Uttarkashi, where *H. sundrai* was collected, is at approximately 3,400 ft altitude in the Bhagirathi River valley. The average annual rainfall is about 65 inches, and the climate is warmer than Bhowali. The plains beside the valley are intensely cultivated, but the surrounding hills are covered with scrub jungle and at higher altitudes, with pine forest. Until recently, these areas had a rich large-mammal fauna, consisting mostly of sambar deer, barking deer, tiger, leopard, Asiatic jackal, yellow-throated marten, rhesus macaque, Langur, Indian crested porcupine, and black-naped hare. With development of fruit orchards and increased human habitation, the deer, tiger, and leopard have become extremely rare. Domestic sheep and goats are common in the Uttarkashi area but rare around Bhowali. Cattle occasionally graze in forests in both areas.

2.5 Structurally Advanced Haemaphysalines

Subgenus *Ornithophysalis*
Subgenus *Haemaphysalis*
Subgenus *Kaiseriana*
Subgenus *Aborphysalis*
Subgenus *Segalia*
Subgenus *Rhipistoma*

General Characteristics

There are 11 subgenera under this category constituting approximately 132 species world over.

In this group, the basis capitulum is rectangular without any lateral expansion in both immatures and adults and usually bears posterior cornua.

The palpi of immatures and adults mostly show some basal broadening of segment II, and some retain compact form. The salience begins with the slight extension of the posterior breadth and reaches the broadly triangular outline characterizing most species of this genus. The dental formula of immatures is 2/2, rarely 3/3 or 4/4, and in adults 4/4, rarely 3/3 or 5/5 to 7/7.

No SA populations occur above 1,500 m altitude in the Himalayas or elsewhere.

Host-adapted structures are pronounced in this group, more in males than in females. These adaptations are less pronounced in nymphs and even less in larvae as these can slip between the host hairs and do not have to fight their way.

2.5.1 *Subgenus* Ornithophysalis

Ornithophysalis, with its broad palpi, appears to have evolved abruptly from SP subgenera with compact palpi when birds and mammals replaced reptiles as the world's dominant vertebrates. Significantly, of the twenty contemporary *Ornithophysalis* species, six parasitize only birds, five parasitize both birds and various mammals, three parasitize birds and marsupials or only marsupials, and two parasitize Oriental or Australian rodents. In India only five species have been recorded belonging to this subgenus.

Keys to Identify Species of Subgenus Ornithophysalis

Males

1. Basolateral margin of palpal segment II is convex or rounded...................[2]
Salience is sharp and ventrally bearing a prominent projection laterally...................[3]

2. Ventral spur of palpal segment III is elongately triangular, overlapping anterior half of palpal segment II; spurs of coxa I and trochanter I are large, having sharply pointed apex; infrainternal setae are long, lanceolate, and number 4...................[*H. ornithophila*]

Ventral spur of palpal segment III is broadly triangular, overlapping anterior one-third of palpal segment II; coxal spurs are small and blunt, decreasing in size from coxa I to IV; infrainternal setae are short, feathery, and eight in number...................[*H. howletti*]

3. Ventral spur of palpal segment III is very small and not reaching to intersegmental suture; infrainternal setae are very short and three to four in number; coxa I to IV are each with very short ridge-like projection...................[*H. megalaimae*]

   Ventral spur of palpal segment III extending slightly beyond intersegmental suture; infrainternal setae are six to eight in number, short, well spaced; coxa I to IV are each with short, broadly triangular spur...................[4]
4. Palpal segment II appearing somewhat broken at mid-length in lateral profile; ventral spur of palpal segment III is broadly triangular and extends slightly beyond intersegmental suture; infrainternal setae number six; spurs of coxa are short and blunt; trochanter bears no or obsolete spurs...................[*H. doenitzi*]

   Palpal segment II shows flared salience; ventral spur is short and extends only to intersegmental suture; infrainternal setae number seven; spurs of coxa and trochanters are moderate and apex is pointed...................[*H. minuta*]

Females

1. Basolateral margin of palpal segment II is rounded and the lateral profile of palpal segment II is little extended beyond the lateral margin of basis capituli...................[2]

   Basolateral margin is straight and the lateral margin of palpal segment II is extended more than half of its basal length beyond the lateral margin of the basis capitulum...................[3]
2. The ventrobasal spur of palpal segment III is very acute, sharp, and extending beyond half or two-thirds of palpal segment II; infrainternal setae are long, well spaced, lanceolate, and five to six in number; dental formula is 4/4; coxa I and trochanter I are larger, while the spurs on other coxa and trochanters are very short or obsolete...................
   [*H. ornithophila*]

   Palpi are very compact; the ventrobasal spur of palpal segment III is broadly triangular overlapping to half of palpal segment II; infrainternal setae are eight, short, feathery, and closely spaced; dental formula is 5/5; coxal and trochanter spurs are blunt and short...................[*H. howletti*]
3. Palpi are spread; salience is very broad; palpal segment II is extending laterally more than half of its basal breadth; ventrobasal spur on palpal segment III is very short or obsolete, infrainternal setae number three, short, and feathery; porose areas are circular and not inclined...................[*H. megalaimae*]

   Palpi are compact; salience is moderately broad; infrainternal setae number six to eight; ventrobasal spur of palpal segment III is moderate or long in length; porose areas are inclined to inner side...................[4]
4. Salience is flared and extends a little beyond lateral margin; the ventrobasal spur on palpal segment III is reaching to intersegmental suture; infrainternal setae are seven in number, short, and closely spaced with feathery outlook; porose areas are oval and more inclined to inner side...................[*H. minuta*]

   Palpi are compact and form a triangular-shaped capituli; salience is very much flared dorsally; ventrobasal spur of palpal segment III extends very short to intersegmental suture; infrainternal setae are seven or eight in number, closely spaced, feathery; porose areas are small and inclined a little to inner side margin...................[*H. doenitzi*]

2.5.2 Haemaphysalis ornithophila[81]

This species was originally described by Hoogstraal and Kohls in 1959 from the specimens collected in 1937 from Pitta birds in Thailand. The name refers to the avian ectoparasitic nature of the species (Figure 2.28).[50]

Male (Figures 2.29 and 2.30)[50]

Overall length is 2.1 mm, width 1.3 mm.

Capitulum: Basis capitulum is rectangular; cornua are gradually tapering to an acute apex, approximately one-fourth as long as basis capitulum. Palpi are broadly salient basally. Hypostome is narrowly elongate, lateral margins are slightly convex; dentition is 4/4, with eight or nine denticles in a file.

Capitulum: Lateral grooves are deep and broad, extending from the level of middle of coxa III and including one festoon on each side. Cervical grooves are deep, wide, and reaching approximately the same level as the anterior termination of lateral grooves. Festoons number 11, are clearly marked, and elongate. Punctations are quite numerous, fairly large, uniformly distributed, deep, and noncontiguous. Genital aperture is large, and situated at a level with mid-length of coxa II.

Legs: Coxa I has a strong, sharply pointed spur arising from middle of posterior margin and extending to anterior margin of coxa II; spur of II as an angled ridge

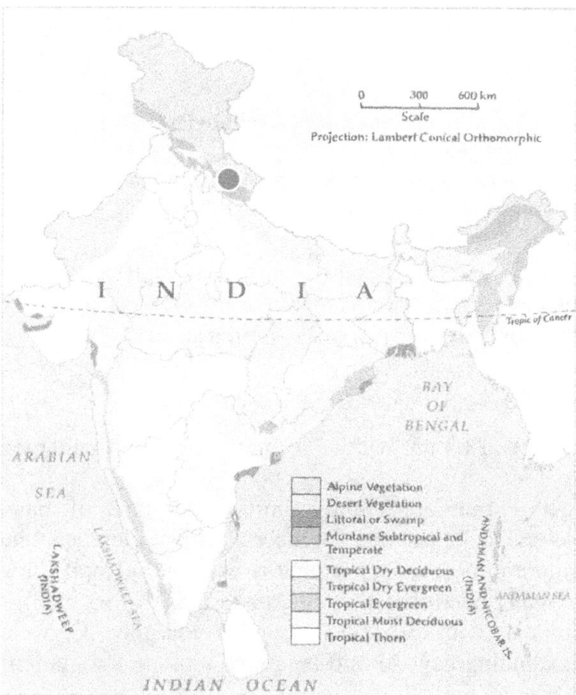

Figure 2.28 Distribution of *H. ornithophila* in India. (●) Place of first record. Map not to scale.

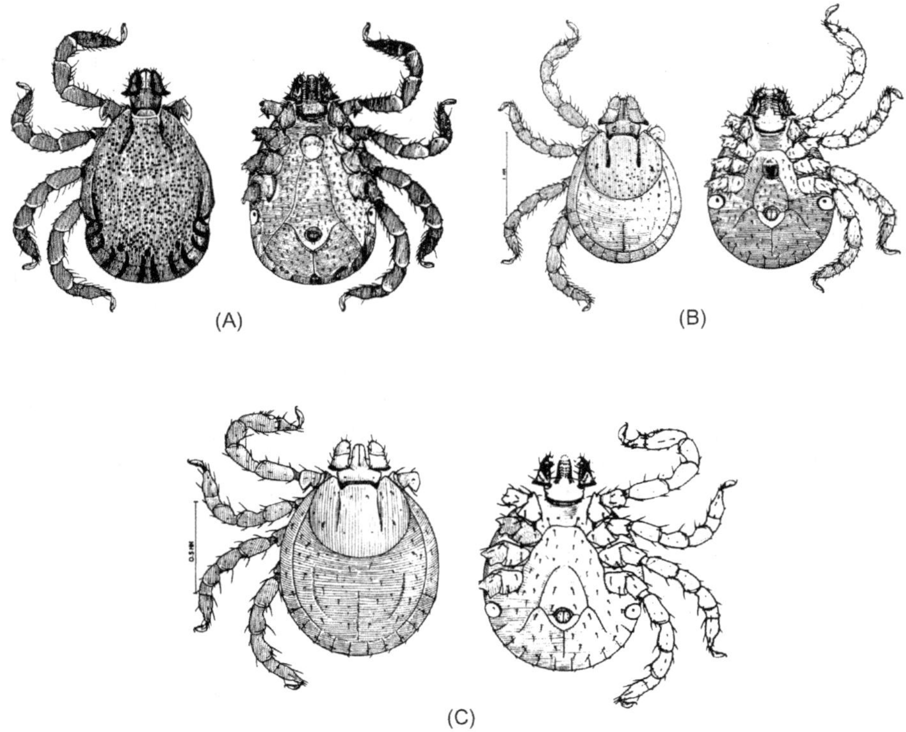

Figure 2.29 *Haemaphysalis ornithophila.* (A) Male, dorsal and ventral view; (B) female, dorsal and ventral view; (C) nymph, dorsal and ventral view.
Source: Hoogstraal and Kohls (1959), Journal of Parasitology (vol 45(4)), Allen press Publishing Services.

terminating at middle of posterior margin in a sharp point; spur of III similar to that of II but ridge only slightly diverging from inner margin of coxa and very slightly overlapping posterior margin of coxa; spur of IV similar to that of III.

Female (Figures 2.29 and 2.30)[50]

Overall body length is approximately 1.6 mm, width 1.1 mm; total length engorged is approximately 5.2 mm.

Capitulum: Basis capituli with cornua are slightly shorter than those of male; porose areas are large and sub-circular. Palpi are as in male but more elongate, and basolateral salience is slightly more elongate. Hypostome is with outline appearing as in male but slightly more elongate; dentition appearing to be 5/5.

Scutum: Scutum is sub-circular with gradually rounded margins. Cervical grooves are deep and straight, extending beyond mid-length of scutum. Punctations are as in male, irregularly distributed.

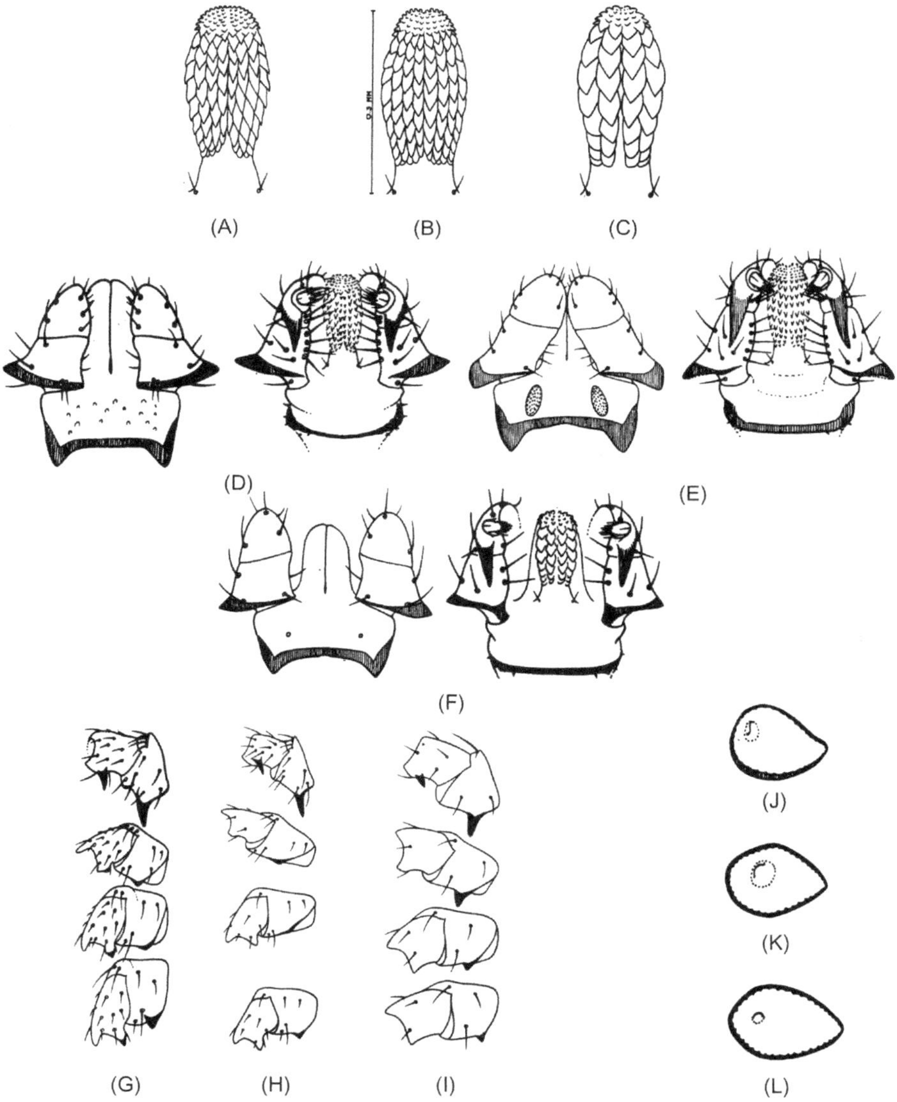

Figure 2.30 (A–C) Hypostome of male, female, and nymph, respectively; (D–F) capitulum of male, female, and nymph, respectively; (G–I) coxa of male, female, and nymph, respectively; (J–L) spiracular plates of male, female, and nymph, respectively.
Source: NIV. unpublished drawings.

Legs: Coxa are slightly more rectangular but otherwise similar to those of male; spurs are also similar, except that of II, which also extends slightly beyond basal margin of coxa, as on III and IV; spur I is equally large and arising from same position as in male.

Nymph (Figures 2.29 and 2.30)

Total body length is about 1.2 mm.

Capitulum: The capitulum is 0.2 mm in length with a moderate salience of palpal segment II. Palpal segment III shows a very large, pointed, triangular spur; overlapping almost two-thirds of palpal segment II. Infrainternal setae number two. Cornua are short and broadly triangular. Hypostome is with 2/2 dental formula and shows eight to nine denticles per file.

Scutum: The scutum is 1.3 times as wide as long. Cervical grooves are deep and extend to two-thirds of scutum. Festoons are marked very shallow.

Legs: Coxa are each with a spur; coxa I bears moderately long, blunt spur, while coxa II to IV bear broadly triangular, pointed spurs reducing in size consecutively.

Related Species

H. ornithophila is close to *H. megalaimae* and *H. doenitzi* in morphologic features. In *H. ornithophila*, the ventral spur of palpal segment III is elongately triangular, overlapping anterior half of palpal segment II, whereas in *H. megalaimae* the spur is very small and not reaching up to intersegmental suture. In *H. doenitzi*, infrainternal setae are long, lanceolate, well spaced, and five or six in number; in *H. megalaimae*, these are very short, closely spaced, feathery, and number only three or four. In *H. ornithophila*, these are long, lanceolate, well spaced, and four to five in number.

Host

Immature stages: Bird species.

Adults: Bird species (Burmese rufous-naped pitta, Gallus species like red jungle fowl, etc.).

Distribution[120]

India (Uttarakhand), Thailand, Burma.

Disease Relationship

Not known.

2.5.3 Haemaphysalis howletti

Haemaphysalis howletti was first described by Warburton in 1913, collected from a hill pony in Rawalpindi area of Pakistan. Subsequently, it was redescribed by Nuttall and Warburton in 1915. In India it was collected from Pune area of Maharashtra state in 1962 by Dhanda from rodents and birds. The taxonomic status of these specimens was confirmed by Hoogstraal. Dhanda in 1964 described the immatures along with the redescription of adults (Figure 2.31).[143]

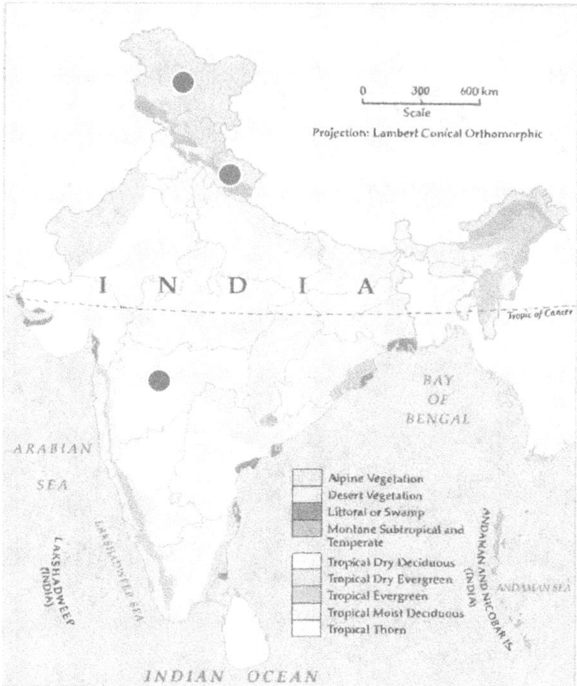

Figure 2.31 Distribution of *H. howletti* in India. (●) Place of first record.
Map not to scale.

Male (Figures 2.32 and 2.33)[24]

Capitulum: Basis capitulum is dorsally approximately 1.7 times as wide as long; cornua are moderate, triangular, length equals basal width, and with somewhat pointed apex. Hypostome is stout, club-shaped, 1.5 times as long as wide, and dental formula is 5/5 to 6/6 with minute denticles, 9–12 in each file.

Scutum: Scutum is ovoid; widest slightly posterior to the level of coxa IV. Punctations are numerous, irregularly sized, and distributed. Lateral grooves are prominent, long, beginning at the level of coxa III anteriorly, enclosing first festoon posteriorly. Cervical grooves are moderate, deep, and narrow anteriorly. Festoons are long, clearly marked, number 11. Genital operculum is large, oval, with minute serrations at the posterior margin, and situated between coxa II.

Legs: Coxa I is with moderate, elongately triangular spur; coxa II and III are each with bladelike spurs; coxa IV is with slightly longer and narrower spur than those of coxa II and III.

Female (Figures 2.32 and 2.33)[24]

Capitulum: Basis capitulum is dorsally approximately 2.6 times as wide as long; cornua are smaller than in male and rounded posteriorly, half as long as the base; porose areas are ovoid, with narrow ends pointing anteromedially, separated from each other by a space approximately twice the least diameter of one porose area.

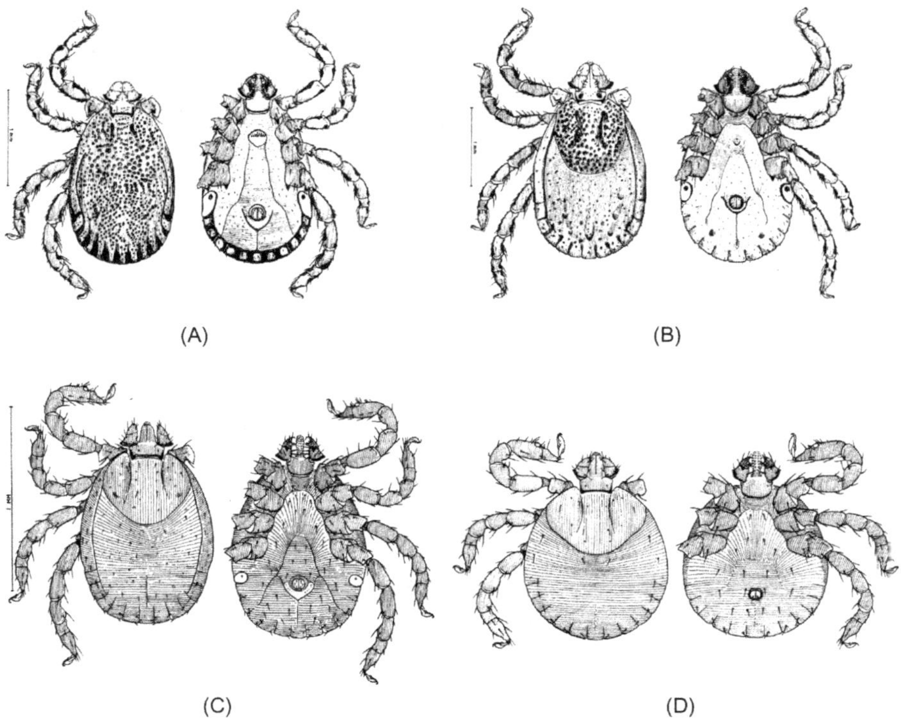

Figure 2.32 *Haemaphysalis howletti.* (A) Male, dorsal and ventral side; (B) female, dorsal and ventral side; (C) nymph, dorsal and ventral side; (D) larva, dorsal and ventral side. *Source*: Dhanda (1964), Journal of Parasitology (vol 50(2)), Allen press Publishing Services.

Palpi are longer than in male, length (including scapulae) slightly greater than width.

Scutum: Punctations are numerous, large, sub-equal, distributed all over scutum. Cervical grooves are distinct, deep and narrow anteriorly, shallow and wide posteriorly, extending more than half the length of scutum. Dorsal integumental area is with distinct marginal grooves, extending to scutum anteriorly, enclosing one festoon on each side posteriorly. Festoons number 11. Genital operculum is longer than wide, rounded posteriorly, and situated at the level of interval between coxa II and III.

Legs: Legs are similar to male, except that coxal spurs II and III are more rounded and tarsi narrower.

Nymph (Figures 2.32 and 2.33)[24]

Length is 1.09 mm, width 0.65 mm.

Capitulum: Basis capitulum is dorsally more than twice as wide as long; cornua are well developed, pointed, and each approximately as long as its base.

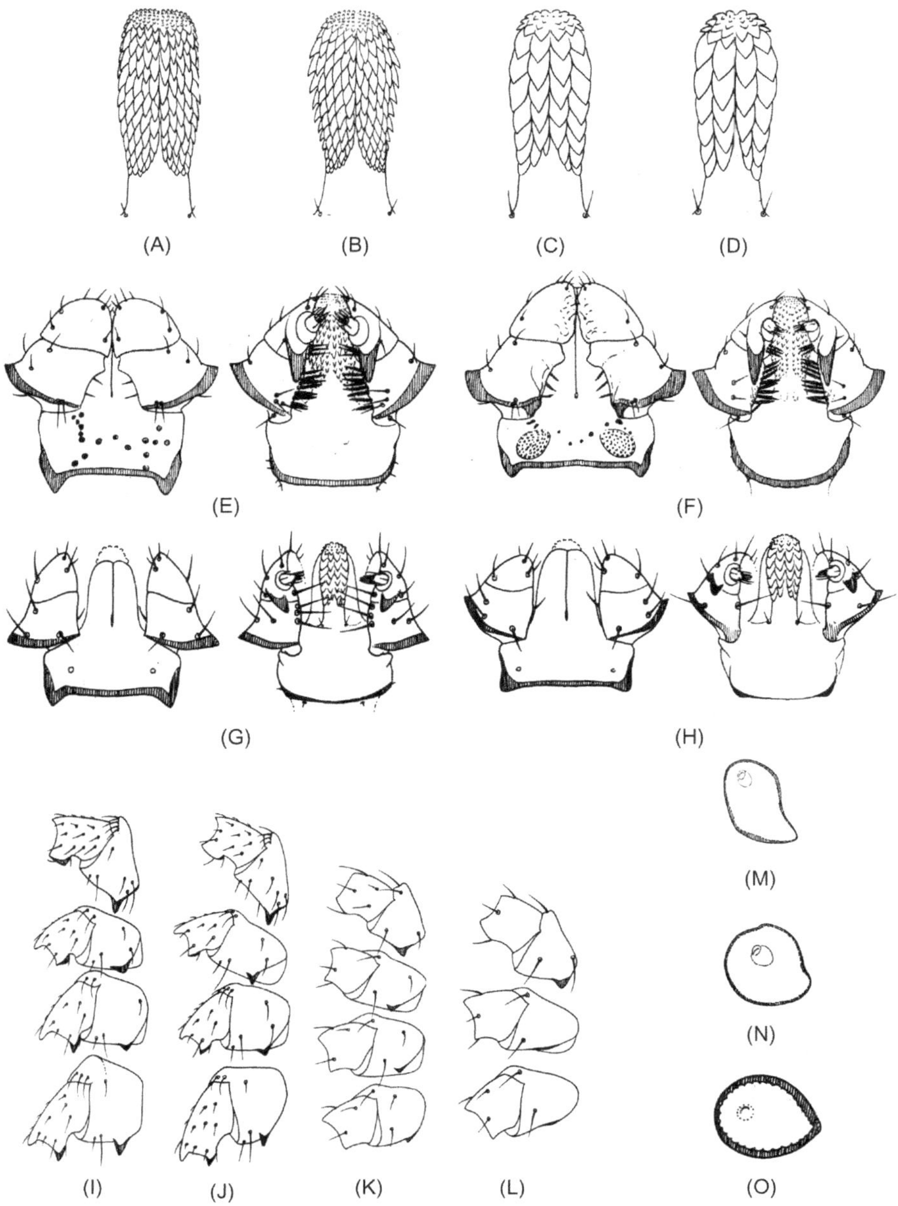

Figure 2.33 (A–D) Hypostome of male, female, nymph, and larva, respectively; (E–H) capitulum of male, female, nymph, and larva, respectively; (I–L) coxa of male, female, nymph, and larva, respectively; (M–O) spiracular plates of male, female, and nymph, respectively.
Source: NIV. unpublished drawings.

Hypostome is long and narrow; corona is moderate; dental formula is 2/2 or 3/3, with seven or eight denticles in each file.

Scutum: Scutum is rounded, slightly wider than long. Cervical grooves are subparallel, extending more than half-length of scutum. Only nine distinct festoons are present. Genital and anal grooves are much as in female; posterior arm of anal groove does not reach festoons.

Larva (Figures 2.32 and 2.33)[24]

Length is 0.55 mm, width 0.41 mm.

Capitulum: Basis capitulum is dorsally twice as wide as long; cornua are not apparent; sub-rectangular ventrally with posterior margin convex. Hypostome is narrow; dental formula is 2/2, with six or seven denticles in each file; corona lacking.

Scutum: Scutum is 1.6 times as wide as long. Outline, number, and disposition of setae are as illustrated. Dorsal integument is with 11 distinct festoons.

Related Species[24]

Among known Indian species, *H. howletti* appears to be closest to *H. campanulata*. Hoogstraal refers to the long, narrow ventral spur of palpal segment III of *H. howletti* as an excellent characteristic to distinguish it from *H. campanulata*. Other distinguishing characters are the expanded inner margin of palpal segment II dorsally, and the outline of the female scutum. *H. howletti* differs from *H. silvafelis*, which has a dorsobasal spur on palpal segment III. From *H. doenitzi* and *H. megalaimae*, *H. howletti* can be easily distinguished by lesser salience of palpal segment II and the long, narrow ventral spur on palpal segment III. In the nymphal stage, *H. howletti* is closest to *H. doenitzi* on one side and to *H. kutchensis* on the other. From *H. doenitzi* it differs in having a lesser salience, and more curved posterior and lateral margins of palpal segment II and by the larger cornua. From *H. kutchensis* it is easily separated by greater salience of palpal segment II and smaller ventral spur on palpal segment III. In addition, the coxal spurs of *H. howletti* are blunt and broad, while those of *H. kutchensis* are somewhat pointed and narrower. The larva also closely resembles *H. kutchensis*. The chief difference is the slightly wider salience of palpal segment II.

Host[24,120]

Immature stages: Rodents, birds, rats, field mice, crow pheasant, palm squirrel.

Adults: Mongoose, rodents, bird species, hill pony, rat, field mice.

Distribution[24,120]

India (Jammu and Kashmir, Maharashtra, Uttarakhand), Pakistan.

Disease Relationship

Not known.

2.5.4 Haemaphysalis megalaimae

This species was first described by Rajagopalan in 1963 from the tick samples collected in Shimoga district of Karnataka state of India, on the bird "small green barbet (*Megalaimae virdis*)" in 1960. The name of the species is derived from the scientific name of the host species (Figure 2.34).[112]

Male (Figures 2.35 and 2.36)[112]

Overall body length is 1.62 mm, width 1.11 mm.

Capitulum: Length of basis is 0.17 mm, width 0.31 mm, rectangular, and width about twice the length. Length of cornua is 0.37 mm, moderate, and sharp. Hypostomal teeth is 5/5 apically, diminishing to 3/3 proximally, with about ten denticles per file. Length of hypostome is 0.17 mm.

Scutum: Punctations are fine, numerous, and evenly distributed. Lateral grooves are prominent, beginning at a point between legs II and III and continuing posteriorly across two festoons on either side. Cervical grooves are moderate, deep

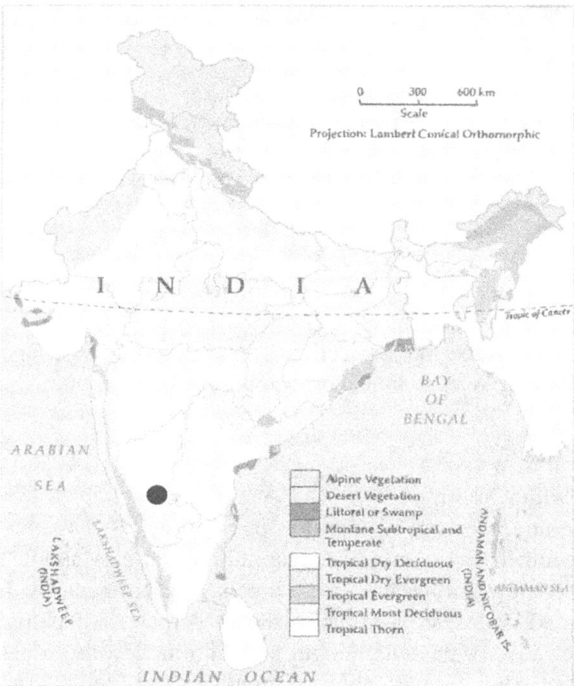

Figure 2.34 Distribution of *H. megalaimae* in India. (●) Place of first record. Map not to scale.

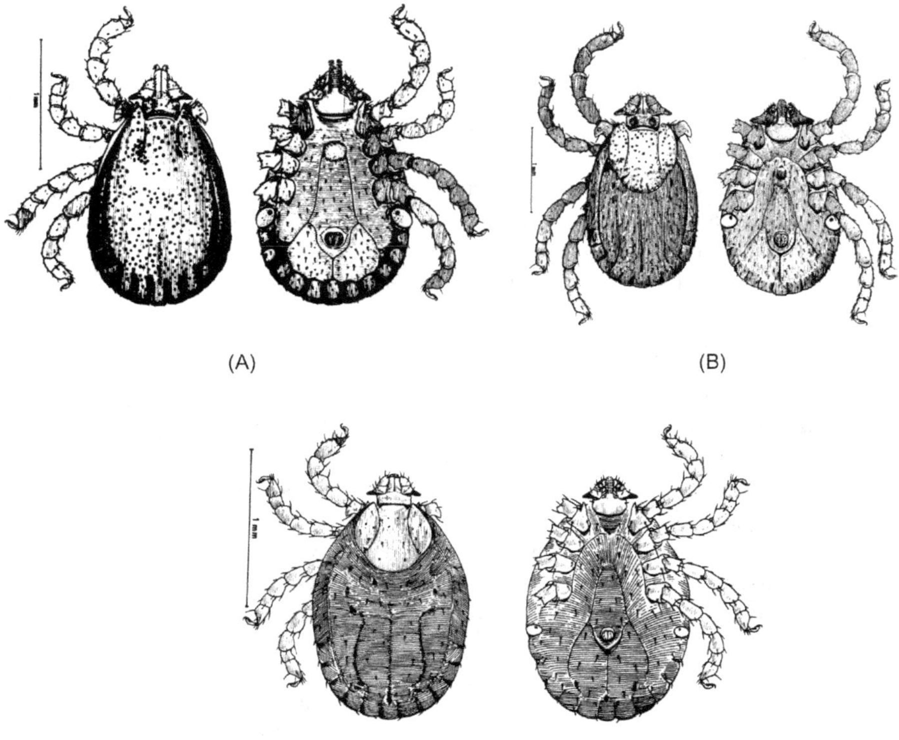

(A) (B)

(C)

Figure 2.35 *Haemaphysalis megalaimae*. (A) Male, dorsal and ventral view; (B) female, dorsal and ventral view; (C) nymph, dorsal and ventral view.
Source: Rajagopalan (1963), Journal of Parasitology (vol 49(2)), Allen press Publishing Services.

anteriorly, and slightly diverging posteriorly. Festoons are long, well separated, and 11 in number. Genital aperture is at the level of coxa II.

Legs: Coxa I to III are with short ridge-like projections, that of coxa I larger and more prominent than II and III. Coxa IV is with a short triangular spur.

Female (Figures 2.35 and 2.36)[112]

Total body length is 2.54 mm, width 1.43 mm.

Capitulum: Length of capitulum from tips of palpi to tips of cornua is 0.40 mm. Length of basis is 0.16 mm; width is 0.43 mm. Basis capitulum width is about 3 times the length. Cornua are moderate and blunt. Porose areas are prominent, oval, slightly depressed, well separated. Palps are similar to those of male, but palpal article III is shorter than II ventrally. Hypostome length is 0.24 mm. Teeth 5/5 per file distally, diminishing to 4/4 proximally, with about 10 denticles per file.

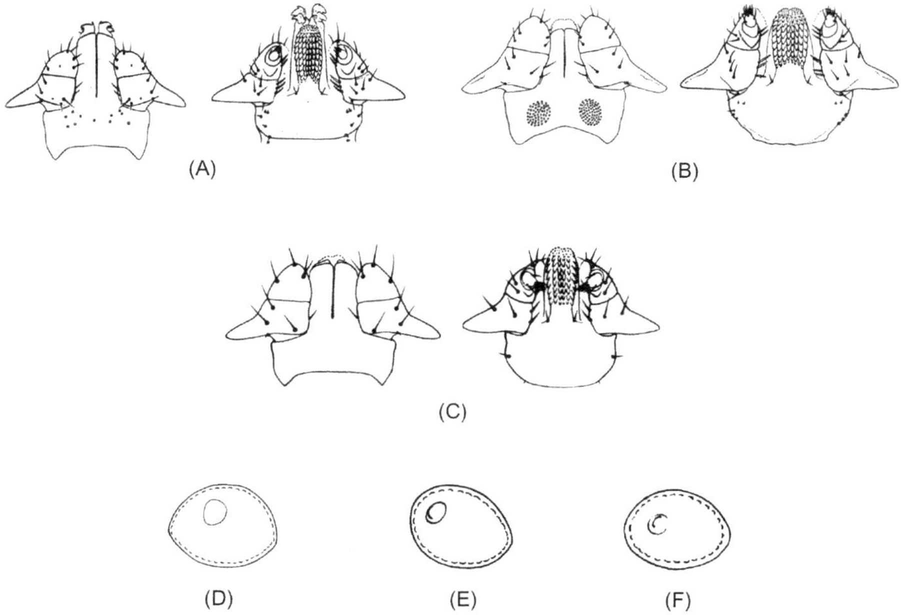

Figure 2.36 (A–C) Capitulum of male, female, and nymph, respectively; (D–F) spiracular plates of male, female, and nymph, respectively.
Source: NIV. unpublished drawings.

Scutum: Scutum length is 0.96 mm; width is 0.87 mm. Cervical grooves are deep, extending to base to one-third of the length of scutum. Festoons are distinct, about as wide as long, and intervals between them dark. Genital aperture is at the level of coxa III.

Legs: Coxa I and IV are with broad ridge-like spurs, that of coxa I is more prominent. Coxa II and III are with broad ridge-like projections.

Nymph (Figures 2.35 and 2.36)[112]

Length of partially fed specimen is approximately 1.6 mm.

Capitulum: Dorsal basis capitulum is rectangular. Cornua are present but weak. Posterior margin of ventral basis is semicircular in outline. Palps are about as broad as long. Teeth are 3/3, about eight per file.

Scutum: Cervical grooves are concave and deep anteriorly.

Legs: Coxa are with short, broad, and ridge-like projections; that of coxa I is most prominent.

Larva:[112]

Capitulum: Dorsal basis is without cornua, about twice as wide as long, posterior margin is straight. Palps are short with slightly rounded lateral contour.

Scutum: Scutum is as in nymph.

Legs: Coxa I is with a very slight ridge-like projection, coxa II and III are unarmed.

Related Species[112]

This species is closest to *H. doenitzi* from which it differs as follows: In males, the lateral salience of palpal article II is semicircular in outline in *H. megalaimae* whereas it is broadly angular in *H. doenitzi*; there are three or four infrainternal bristles on palpal article II in the males of *H. megalaimae,* as compared to six or seven in *H. doenitzi*; the width of the dorsal basis is twice its length in *H. megalaimae*, whereas the width is only about 1½ times the length in *H. doenitzi*. In females of *H. megalaimae*, there are three infrainternal bristles on palpal article II as compared to eight to ten bristles in *H. doenitzi*.

Ecology[112,118]

This species has been originally recorded from only a single-host bird species in Kyasanur Forest Disease (KFD) area, *Megalaima virdis*, which is exclusively arboreal and nests in tree holes. It seems that the life cycle of this tick must be completing on this host bird in nest hole. The ticks may be dropping and molting in the nest only and attaching on to the bird host while it is in the nest. This tick has never been collected on vegetation or on ground drags, unlike most of the other KFD ticks.

Host[112]

Immature stages: Small green barbets.

Adults: Small green barbet and many other bird species from KFD area.

Distribution[112]

India (Karnataka).

2.5.5 Haemaphysalis doenitzi

The Indian specimens of this species were first described as a separate species, *Haemaphysalis centropi*, by Kohls in 1949 from birds, and later on, it was synonymized with *H. doenitzi* Warburton and Nuttall, 1909 on the basis of morphologic similarities (Figure 2.37).

Male (Figures 2.38 and 2.39)[35,97,104]

Overall body length is 2.62–2.82 mm, width 1.35–1.68 mm; average 2.57 mm × 1.53 mm, ovoid, widest at about the fourth coxa.

Capitulum: Capitulum length from tips of palpi to tips of cornua is about 0.54 mm; length of basis is 0.21 mm; width of basis is 0.36 mm. Cornua are moderate and sharp. Hypostome is widest at about the middle, blunt, and slightly notched.

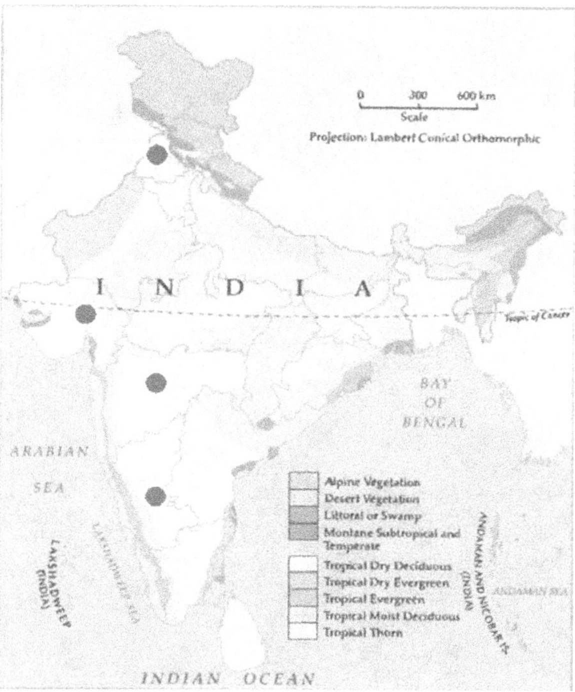

Figure 2.37 Distribution of *H. doenitzi* in India. (●) Map not to scale.

Dentition is 4/4 with trace of 5/5 near the base, with six to eight large teeth in the principal files.

Scutum: Punctations are numerous, moderately large, and rather evenly distributed. Lateral grooves are prominent, beginning at the level of interval between coxa II and III and continuing across two festoons on each side. Cervical grooves are moderate, deep anteriorly, and concave externally. Festoons are long, well marked, and the intervals usually dark. Genital aperture is situated between coxa II.

Female (Figures 2.38 and 2.39)[35,97,104]

Overall length is approximately between 2.88 and 3.0 mm, width 1.68 and 1.74 mm, widest at about the fourth coxa.

Capitulum: Length of capitulum from tips of palpi to tips of cornua is 0.54–0.63 mm; length of basis is 0.18–0.24 mm; width of basis is 0.45–0.54 mm. Cornua bluntly pointed and about as long as the width at the base. Porose areas are oval, well separated, mildly depressed. Palps greatly resemble those of the male. Hypostome is broad. Dentition is usually 4/4, with principal denticles all of about equal size and eight to ten in each file; in some specimens the arrangement is 4/5 and in others 5/5.

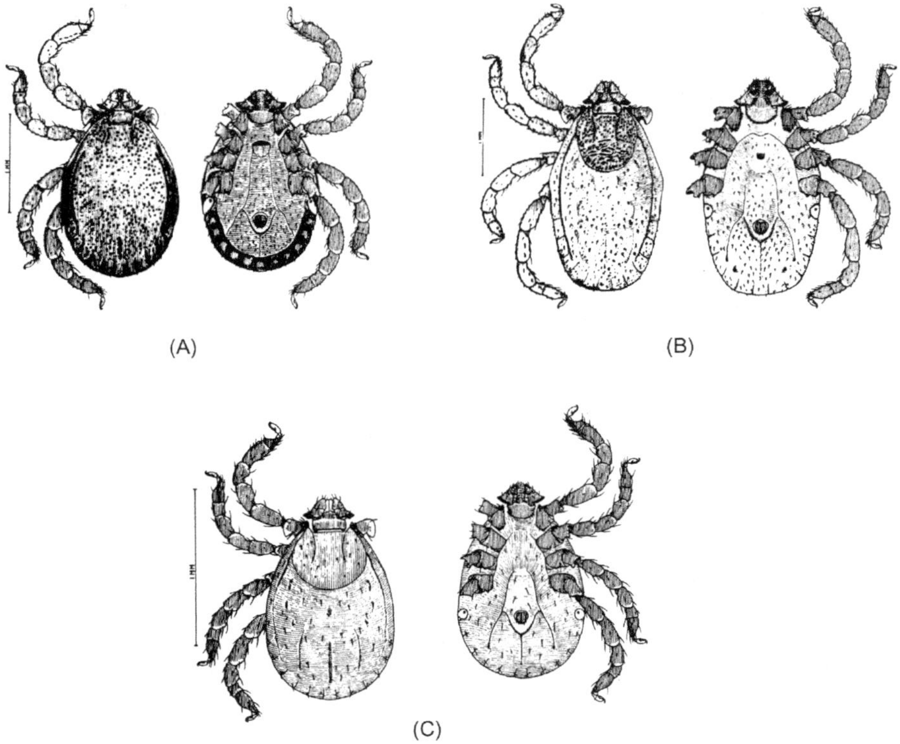

Figure 2.38 *Haemaphysalis doenitzi*. (A) Male, dorsal and ventral side; (B) female, dorsal and ventral side; (C) nymph, dorsal and ventral side.
Source: Kohls (1949), Journal of Parasitology (vol 35(3)), Allen press Publishing Services; NIV unpublished drawings.

Scutum: Scutum is nearly circular; 0.90–1.20 mm in length and width is from 0.90 mm to 1.20 mm. Cervical grooves are sub-parallel and deeper anteriorly. Punctations are moderately large and evenly distributed. Marginal grooves are distinct, beginning at about the level of coxa II and continuing across two festoons on each side. Festoons are about as wide as long, intervals between them dark. Genital aperture is at opposite the interval between coxa II and III.

Nymph (Figures 2.38 and 2.39)

Capitulum: Salience is broad and flared. Ventrobasal external margin of palpal segment II is without retroverted spur. Basal margin of palpal segment II is rounded both dorsally and ventrally; palpal segment III is with ventral distinct spur.

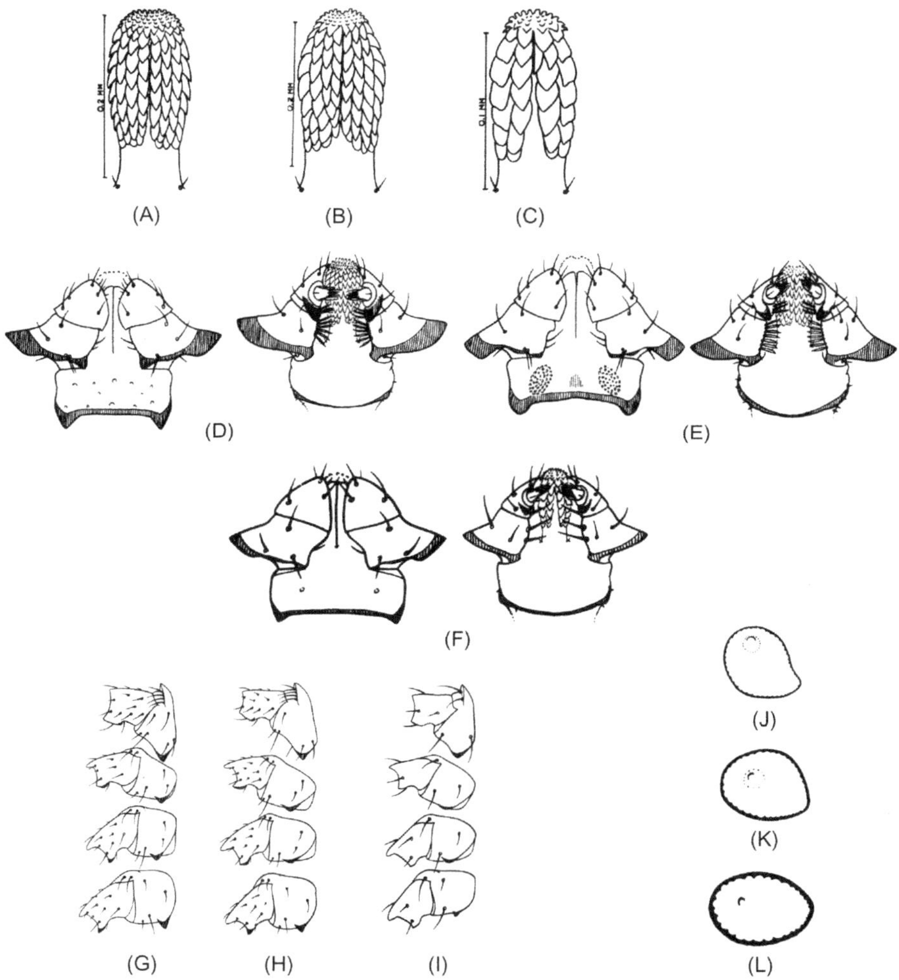

Figure 2.39 (A–C) Hypostome of male, female, and nymph, respectively; (D–F) capitulum of male, female, and nymph, respectively; (G–I) coxa of male, female, and nymph, respectively; (J–L) spiracular plates of male, female, and nymph, respectively.
Source: NIV. unpublished drawings.

Legs: Coxa III and IV are with small sharp spurs.

Host[97,143]

Immature stages: Common myna, crow pheasant.

Adults: Lesser coucal, crow pheasant birds, jungle fowl, bulbul bird, common myna, small white throat bird, pied harrier, Australian pheasant.

Distribution[35,97,143]

India (Karnataka, Punjab, Gujarat, Maharashtra), Burma, Malaya, Philippines, Papua New Guinea, Australia.

Related Species

Species relationship is discussed under *H. megalaimae.*

2.5.6 Haemaphysalis minuta

Kohls first described this species in 1950 from males off the jungle fowl, *Gallus lafayettei* in Wilpattu of Sri Lanka. This tick was commonly found on the jungle fowl in the Wilpattu sanctuary in Ceylon in association with *H. intermedia.* The name perhaps refers to the comparative smaller size of this tick (Figure 2.40).[143]

Male (Figures 2.41 and 2.42)[99]

Being very small, the length from the tip of palpi to the posterior margin of the body ranges from 1.27 to 1.57 mm and width from 0.88 to 1.02 mm, widest ahead of the spiracles.

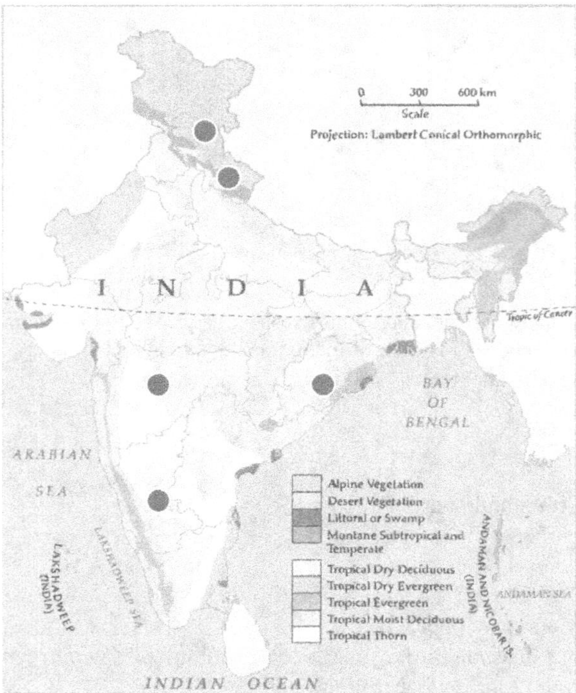

Figure 2.40 Distribution of *H. minuta* in India. (●) Place of first record.
Map not to scale.

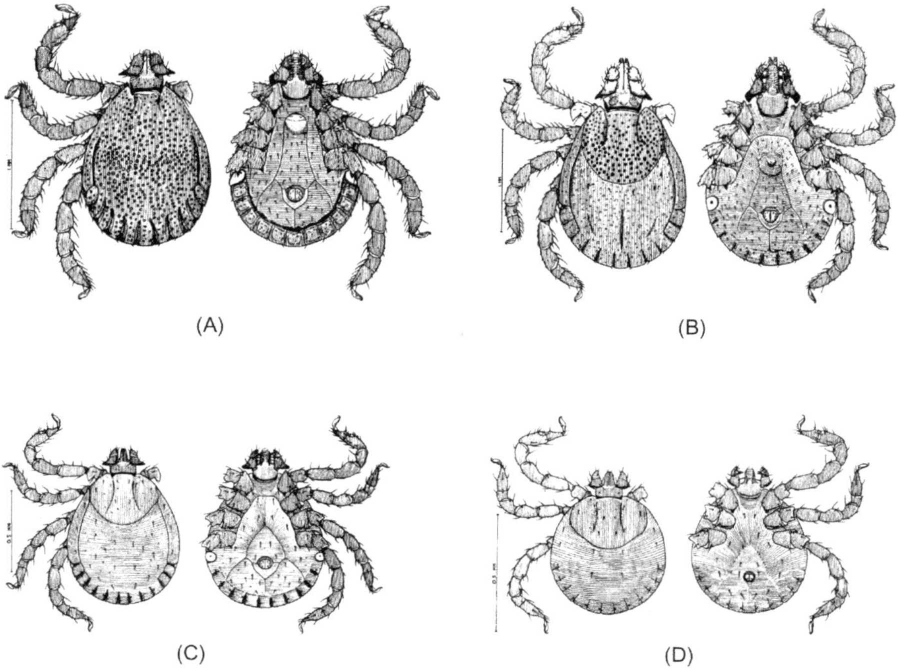

Figure 2.41 *Haemaphysalis minuta*. (A) Male, dorsal and ventral side; (B) female, dorsal and ventral side; (C) nymph, dorsal and ventral side; (D) larva, dorsal and ventral side. *Source*: Kohls (1950), Journal of Parasitology (vol 36(4)), Allen press Publishing Services; NIV unpublished drawings.

Capitulum: Length of capitulum from tips of palpi to tips of cornua is about 0.31 mm; width of basis is 0.22–0.25 mm. Cornua are moderate and pointed. Palpal article II is strongly salient laterally and longer than article III. Article III is without dorsal spur. Palpal article II is with a blunt retrograde spur at the ventrolateral angle. Article III has a very short, pointed retrograde spur. Dentition in 4/4 pattern, with seven or eight denticles in each file.

Scutum: Length of scutum from tips of scapulae to posterior margin ranges from 1.09 to 1.33 mm. Lateral grooves well defined and enclosing one festoon on each side. Cervical grooves are short and inconspicuous. Punctations are coarse and deep, and evenly distributed. Festoons are longer than broad.

Legs: A moderate spur on coxa I; spurs are progressively shorter on II, III, and IV. All trochanters have short, pointed ventral spurs, which diminish in size from I to IV.

Female (Figures 2.41 and 2.42)[124]

Very small, and length from tips of palpi to posterior margin of body is about 1.30 mm.

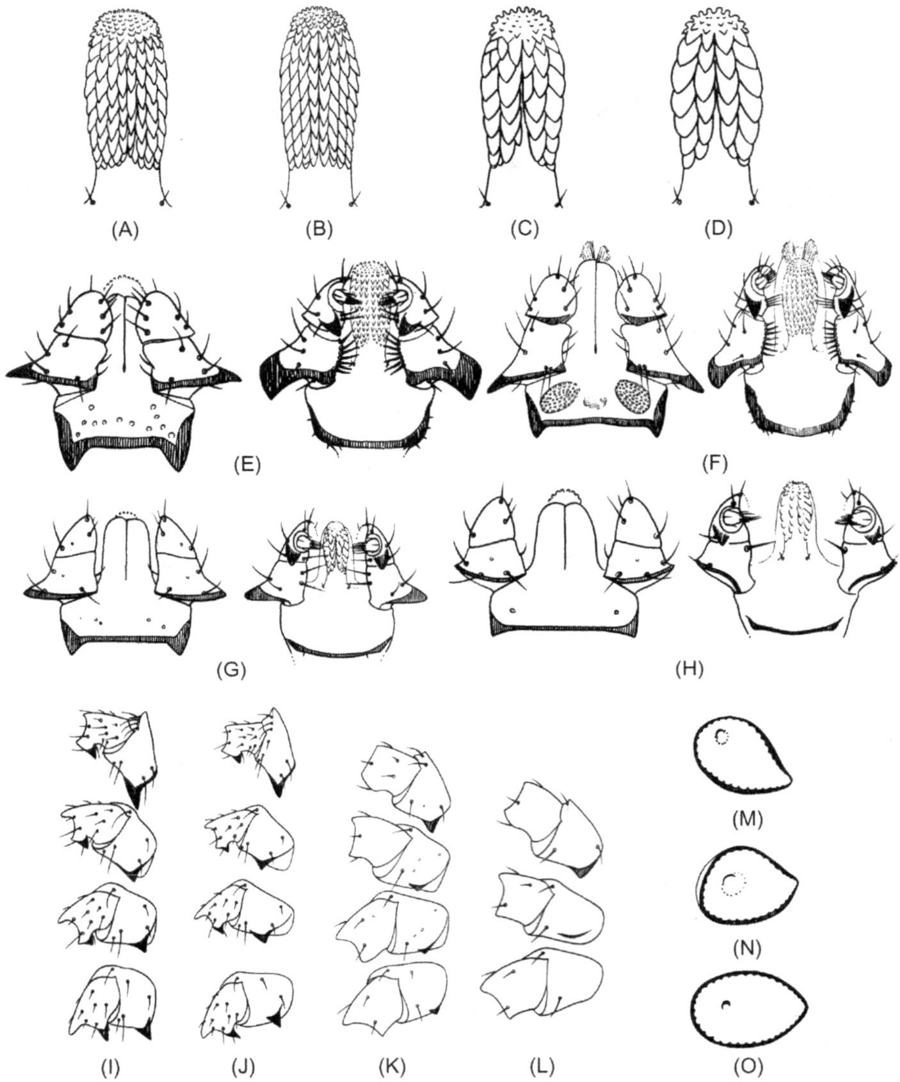

Figure 2.42 (A–D) Hypostome of male, female, nymph, and larva, respectively; (E–H) capitulum of male, female, nymph, and larva, respectively; (I–L) coxa of male, female, nymph, and larva, respectively; (M–O) spiracular plates of male, female, and nymph, respectively. *Source*: NIV. unpublished drawings.

Capitulum: Length of capitulum from tips of palpi to tip of cornua is about 0.4 mm. Porose areas are oval and are inclined toward hypostome. Cornua are moderately long and pointed. Dorsobasal margin of palpal segment III is with median retroverted spur. Palpal segment II is with a blunt retrograde spur at the ventrolateral angle. Infrainternal setae number six or seven. Palpal segment III on ventral side bears a retroverted and reduced spur. Hypostome is slightly longer than the male, measuring about 0.2 mm in length. The dental formula is 4/4 and each file contains nine or ten denticles.

Scutum: Length of the scutum from tip of the scapulae to the posterior margin is about 1–1.5 mm. Cervical grooves are short and inconspicuous. Punctations are rather coarse and deep, and evenly distributed. Festoons are 11 in number.

Legs: On coxa I a moderate spur with blunt apices is there, which diminishes on coxa II and III, and again a moderate spur on coxa IV. All trochanters have small blunt spurs. Spiracular plates are little oval with conical apices. Genital aperture is located between the level of coxa II and III.

Nymph (Figures 2.41 and 2.42)

Average length is about 0.8 mm.

Capitulum: The length of capitulum is about 0.3 mm, lateral margins are little rounded and converging posteriorly. Cornua are short and triangular. Palpal segment II is broadly salient and has no spurs, infrainternal setae are two in number. Palpal segment III bears a retroverted, median, blunt spur on ventrobasal margin. Hypostome is about 0.1 mm in length. Dental formula is 2/2 toward the base and 3/3 toward the apex. Each file contains six or seven denticles.

Scutum: Little longer cervical grooves are present. Punctations are negligible in number. Festoons are deep and as long as broad.

Legs: Coxa I has a moderate bluntly triangular spur, decreasing in size from coxa II to IV. Trochanters do not have any spur.

Larva (Figures 2.41 and 2.42)

Larva is about 0.7 mm in length from apex of palpi to the posterior margin of the scutum.

Capitulum: Length of the capitulum is about 0.2 mm. Cornua are equal to none or very short. Palpal segment II is with broad salience, having one infrainternal seta. It shows no spurs on its posterior margin. Palpal segment III is with small median retroverted spur. Hypostome length is about 0.07 mm; dental formula is 2/2, with five or six denticles per file.

Scutum: Scutum is circular in shape and is about 0.5 mm long. Cervical grooves are moderate in length. Punctations are equal to none. Festoons are as long as broad and are not deep.

Legs: Coxa I bears a moderate blunt spur, no spur on coxa II and III, and trochanters do not show any spur.

Hosts[93,114,120,139,143]

Immature stages: Langur monkey, buffalo, rat (9), crested porcupine, and from flag dragging.

Adults: Sheep, goat, mongoose, bird species (jungle fowl), and from flag dragging.

Distribution[93,113,114,120,143]

India (Himachal Pradesh, Karnataka, Orissa, Uttarakhand, Maharashtra), Sri Lanka.

Related Species

This is a very small species and resembles *Haemaphysalis doenitzi* Warburton and Nuttall, 1909. Comparison of *H. minuta* with *H. doenitzi* reveals that *H. minuta* differs in several respects, including the presence of the retrograde spur on palpal segment II ventrally; shorter lateral grooves, which cross one instead of two festoons on each side; longer spurs on coxa I; and spurs on the trochanters ventrally.

Disease Relationship[131]

In KFD area collection, only six KFD virus isolations have been reported from nymphal stages of this species during the period from 1961 to 1972. This species has proved its efficiency to transmit the KFD virus (in laboratory transmission studies).

2.6 Subgenus *Haemaphysalis*

General Characteristics

According to Hoogstraal and Kim, this subgenus may have apparently evolved from *Herpetobia* in Indian subregion of Oriental region. It contains 16 contemporary species in six species groups, world over. Wild and domestic Artiodactyla (various deer, wild pigs, the serow, goral, blackbuck, etc.) are the chief hosts of most adults (11 of 16 species). Carnivores are secondary hosts of some of these 11 species. The basis capitulum of each stage, except some larvae, bears moderate to fairly large cornua. Small but distinctive ventral cornua are present in immatures of *H. indoflava* and *H. birmaniae* and also appear in a less developed form in the immatures of a few other species in this subgenus. Ventral cornua are rare in the genus *Haemaphysalis*. Immatures of *H. birmaniae* are also unusual in that they like adults parasitize Artiodactyla (rather than Rodentia). Palpi of *Haemaphysalis* immatures and adults are only slightly advanced from the compact from characterizing *Herpetobia* and the SP group *H. punctata*. These palpi are slightly more elongate, either broader posteriorly (as in some *Herpetobia* nymph) or elongate with moderate flange, but are not broadly expanded (as in *Ornithophysalis* and *Rhipistoma*). Except for the ventral spur of palpal segment III and some posterior broadening, the palpi of the subgenus *Haemaphysalis* are not generally specialized for mammal-hair hooking.

Keys to Identify the Subgenus Haemaphysalis

Male

1. Coxal spurs are crest or ridge-like; salience is very low, which is not bulging posteriorly...................[*H. indoflava*]
Coxal spurs are elongated or sub-triangular...................[2]
2. Coxa I spur is acute, lanceolate, and approximately 2 times as long as other three coxal spurs, which are broadly triangular and sub-equal in length...................[*H. darjeeling*]
Coxa I spur is not 2 times as big as other sub-equal spurs; each coxa is with a prominent spur, which is widely triangular...................[*H. birmaniae*]

Female

1. Coxal spurs are crest or ridge-like..................[*H. indoflava*]
Coxal spurs are elongated or sub-triangular...................[2]
2. Coxa I spur is elongated, while coxa II to IV possess short and broadly triangular but pointed spurs...................[*H. darjeeling*]
Coxa I spur is short, while spurs on coxa II to IV are ridge-like...................[*H. birmaniae*]

2.6.1 Haemaphysalis campanulata

This species was first described by Warburton in 1908 from numerous specimens taken off Chinese dogs in Mongolia collected by Major M.L. Hearn in 1906. This common dog tick is distinct from other haemaphysalid in that palpal article II is strongly salient laterally, the posterior margin is flared upward to become "bell-shaped;" hence the name "*campanulata*" (Figure 2.43).[104,143,147]

Male (Figures 2.44 and 2.45)[104,127,146]

Capitulum: The capitulum is 0.43 mm in length, base rectangular, dorsally barely twice as broad as long, punctate all over, and with short blunt cornua. The palps are 1½ times as long as broad and without any dorsal spines. Article II is as broad as long and has a strong lateral salience about the middle of its length with two or three infrainternal long setae. Article III is half as long as the second and is 1½ times as broad as long. It possesses only a single short blunt retroverted spur. Hypostome is 4/4, with files of broad teeth, about eight teeth per file.

Scutum: Scutum is with numerous fine punctations; cervical grooves are deep, sub-parallel, and slightly convex externally; lateral grooves are beginning about the anterior third of the body and ending behind the spiracles; festoons are long and narrow, and well marked. Sexual orifice is broad, between coxa II.

Legs: Coxa are each with a short blunt spur; tarsus IV is very short and thick, ending in a somewhat abrupt slope, the false articulation about the middle of the article.

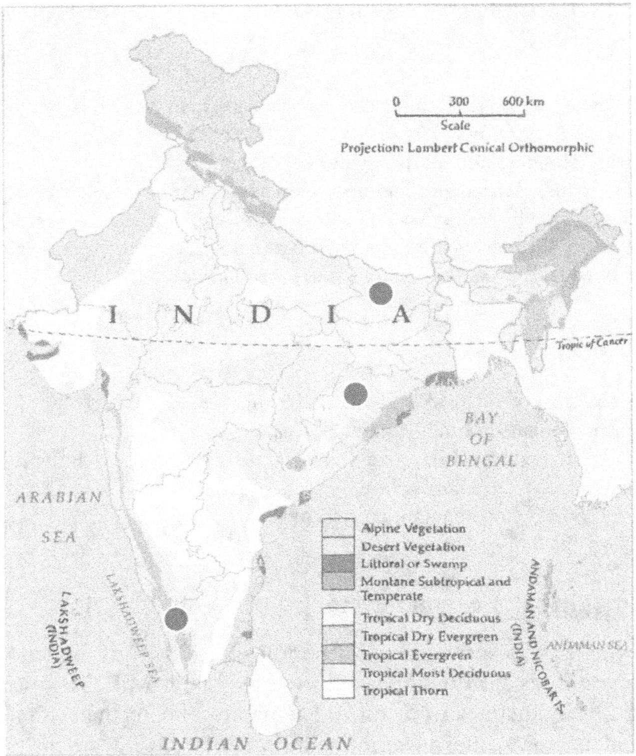

Figure 2.43 Distribution of *H. campanulata* in India. (●) Place of first record. Map not to scale.

Female (Figures 2.44 and 2.45)[104,146]

Capitulum: Capitulum base is broader and shorter with less distinct cornua; porose areas are oval, their axes converging forward. Hypostome dental formula is 4/4, with teeth narrower.

Scutum: Scutum is glossy-brown or yellow, cordiform, and finely punctate; cervical grooves are well marked, far apart, at first convex outwardly, then turning outward.

Nymph (Figures 2.44 and 2.45)

Body is comparatively short.

Capitulum: Similar to the female, but longer; hypostome dentition is 3/3, with eight teeth per file. Venter: Spiracles are sub-triangular.

Scutum: Scutum is cordiform, about 0.48 mm × 0.45 mm, with a small number of indistinguishable punctations in the posterior and the lateral fields; cervical grooves clearly visible, parallel anteriorly and diverging posteriorly to extend barely to the posterior margin.

Legs: Coxa are moderate; tarsi are considerably long with tapered tips.

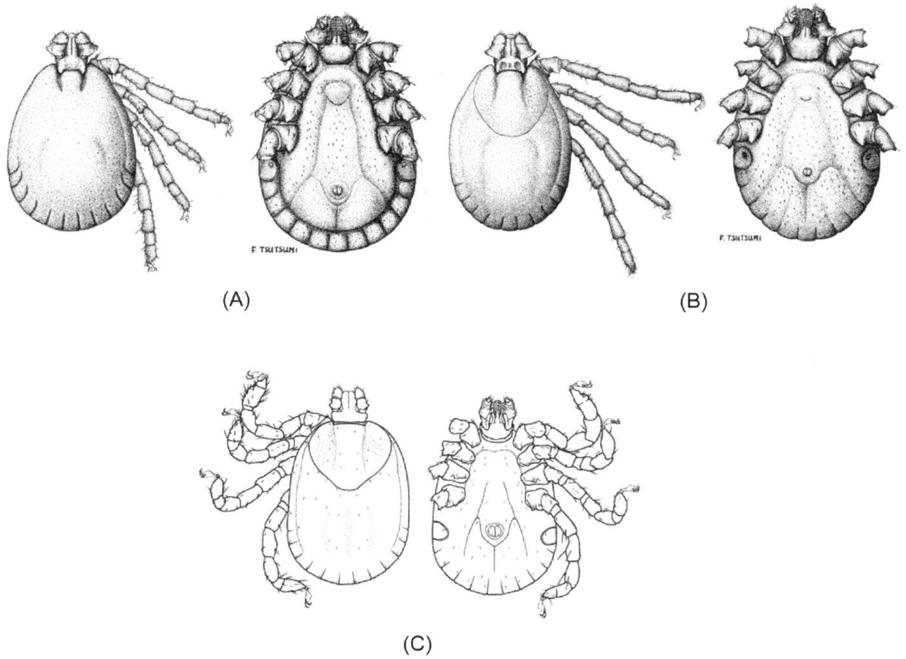

Figure 2.44 *Haemaphysalis campanulata*. (A) Male, dorsal and ventral side; (B) female, dorsal and ventral side; (C) nymph, dorsal and ventral side.
Source: From Yamaguti, Tipton, Keegan, Toshioka, 1971.

Larva

Unfed specimens are about 0.65 mm long.

Capitulum: Basis capitulum is rectangular; posterior edge is linear; no cornua are present; palps are similar to those of the nymph, but article III with a more indistinct ventral spine; hypostome dentition is 2/2 with six teeth per file.

Scutum: Broader than long, about 0.3 mm × 0.25 mm, posterior margin is uneven; cervical groove is defined, diverging posteriorly.

Legs: Each coxa is with a spur; tarsi are fairly long, attenuate in talus.

Host[104,143,147]

Immature stages: Not available.

Adults: Sambar deer, wolf, cows, horses, dogs, rats, cattle, and man.

Distribution[143,147]

India (Kerala, Bihar, Orissa), China, Japan, Mongolia, Korea (North and South).

Biology[147]

Both immature and adult ticks are very common on Japanese dogs, even in urban areas. All stages in the life cycle may be found on the ground around dog kennels.

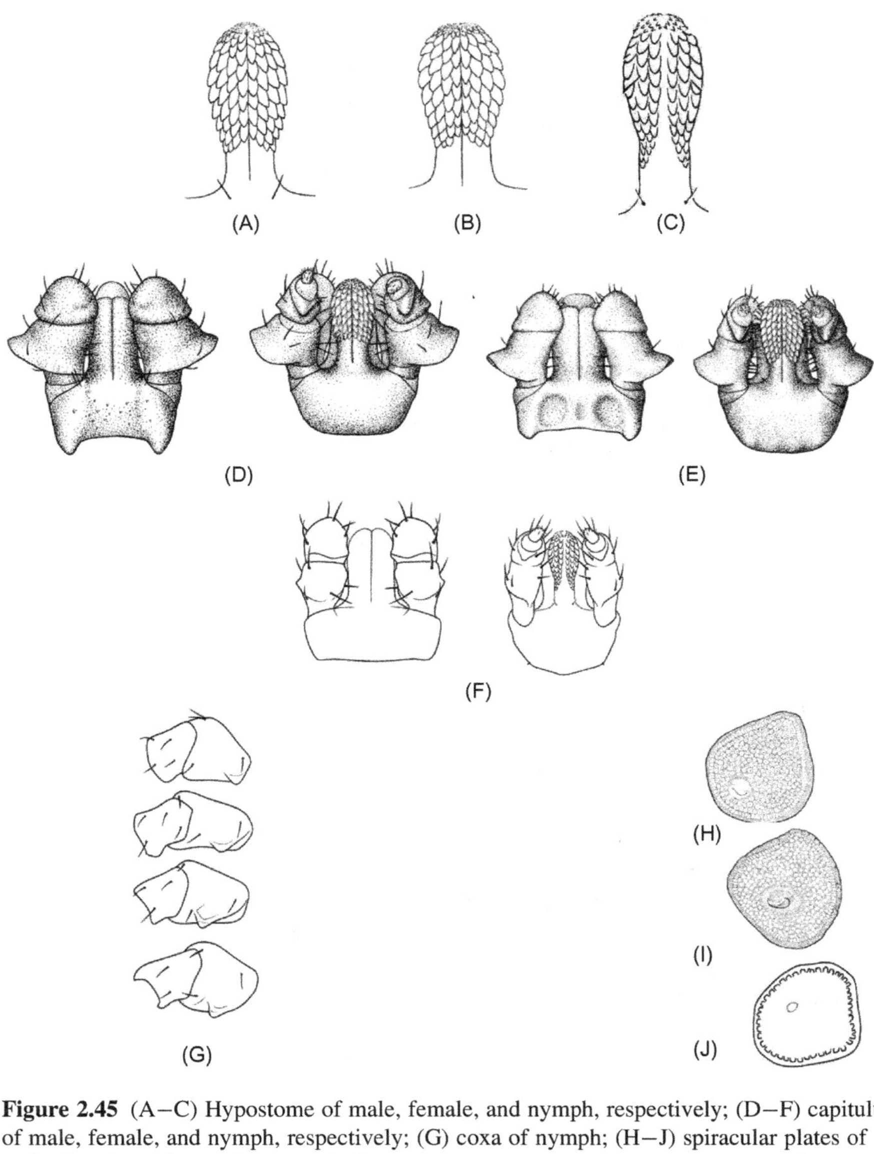

Figure 2.45 (A–C) Hypostome of male, female, and nymph, respectively; (D–F) capitulum of male, female, and nymph, respectively; (G) coxa of nymph; (H–J) spiracular plates of male, female, and nymph, respectively.
Source: From Yamaguti, Tipton, Keegan, Toshioka, 1971.

Life Cycle[147]

Rearing was attempted at the 406th Medical Laboratory. A single female was obtained from a dog in Chiba Prefecture, June 12, 1967, and 12 females of the F_1 progeny were fed on rabbit ears. Ticks were reared in facilities at the 406th Medical Laboratory in which temperature was maintained at approximately 25 °C and relative humidity fluctuated from 10% to 70% (Table 2.1).

Table 2.1 Laboratory Life Cycle of *Haemaphysalis campanulata* (Reared on Rabbits)

Generation and Stage	Phase	Period in Days
F1 adult	Feeding	9–14
F1 adult	Pre-oviposition	4–7
F1 adult	Oviposition	12–24
F2 egg	Incubation	25–29
F2 larva	Feeding	2–10
F2 larva	Post-parasitic (pre-molting)	10–13
F2 nymph	Feeding	3–7
F2 nymph	Post-parasitic (pre-molting)	11–27

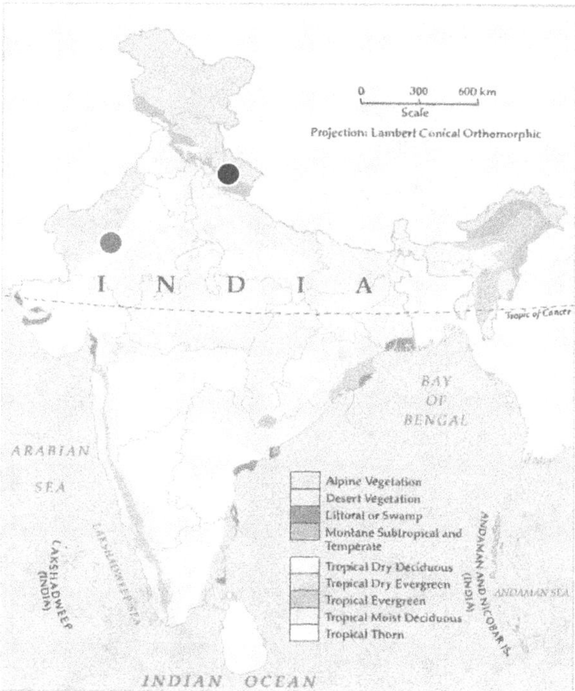

Figure 2.46 Distribution of *H. indoflava* in India. (●) Place of first record.
Map not to scale.

2.6.2 Haemaphysalis indoflava

This species was first collected and described by Bhat and Dhanda in 1968. The collection was made in Uttarakhand state (previously part of Uttar Pradesh state), and all the samples were collected by flag dragging. Previously this species was misidentified as *Haemaphysalis flava* of Japan, which was originally described by Nuttall and Warburton (Figure 2.46).[27]

Male (Figures 2.47 and 2.48)[27]

Capitulum is slightly longer than broad.

Capitulum: Basis capitulum is dorsally approximately 1.3 times as broad as long; cornua are triangular, approximately one-third as long as basis capitulum, and with pointed apices; dorsal surface is with a few fine punctations. Palpi are short, compact, with a small posteroexternal salience; outline is trapezoidal. Hypostome is shorter than palpi, and dental formula is 4/4, with maximum of nine or ten denticles in a file.

Scutum: Scutum is approximately 1.4 times as long as broad, broadest at the level of spiracular apices; about two-thirds as long as basis capitulum. Cervical grooves are obsolete. Lateral grooves are short; anteriorly reaching mid-length of scutum. Punctations are small, shallow, some indistinct, numbers moderate, and evenly distributed. Festoons number 11.

Legs: Coxa I is with spur small, sub-triangular, with rounded apex; II and III are each with spur somewhat larger, triangular, with bluntly pointed apex; spur of III is longer and sharper than that of II; IV is with spur elongate, tapering, narrowly pointed, and about one-half as long as coxa.

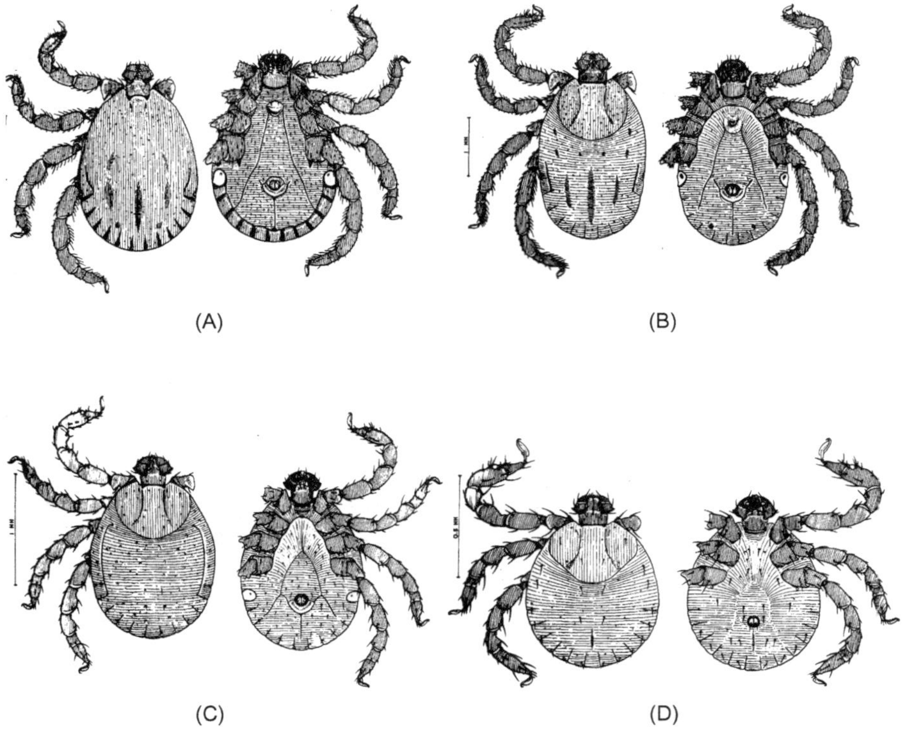

Figure 2.47 *Haemaphysalis indoflava*. (A) Male, dorsal and ventral side; (B) female, dorsal and ventral side; (C) nymph, dorsal and ventral side; (D) larva, dorsal and ventral side. *Source*: Indian J Parasitol 1968; 54(5): 1063-67, Oriental Insects 1970; 4: 253–58.

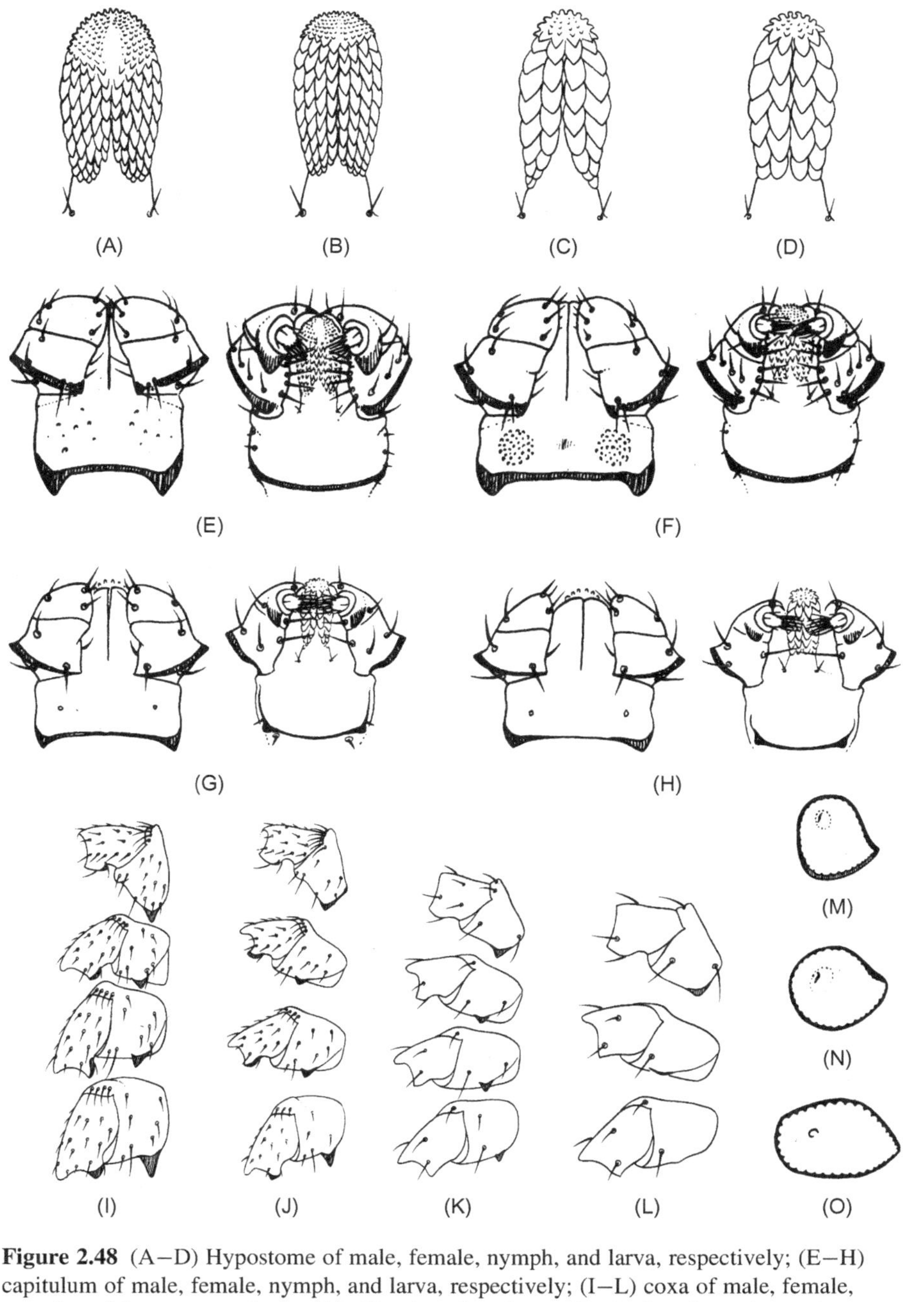

Figure 2.48 (A–D) Hypostome of male, female, nymph, and larva, respectively; (E–H) capitulum of male, female, nymph, and larva, respectively; (I–L) coxa of male, female, nymph, and larva, respectively; (M–O) spiracular plates of male, female, and nymph, respectively.
Source: NIV. unpublished drawings.

Female (Figures 2.47 and 2.48)[27]

Capitulum: Basis capitulum is dorsally twice as broad as long; cornua are small, one-fifth as long as basis capitulum, and with rounded apices; porose areas are large, broadly oval, and tilted. Palpi are short, compact, and sub-conical. Hypostome is shorter than palpi, 2.2 times as long as broad; dental formula is 4/4, with maximum 10–11 denticles in a file.

Scutum: Scutum is almost as broad as long; outline is rounded, widest at about mid-length. Cervical grooves are curving narrow anteriorly, and wide and shallow posteriorly, extending beyond mid-length of scutum. Punctations are numerous, slightly larger, and deeper than in male. Genital area is with aperture broadly "U" shaped.

Legs: Coxa are with spurs slightly smaller than in male, particularly that of IV.

Nymph (Figures 2.47 and 2.48)[30]

It measures 1.67 mm long, 1.02 mm wide.

Capitulum: Capitulum is almost as long as broad (0.31 mm). Basis capitulum is dorsally approximately twice as broad as long; cornua are triangular, approximately one-fifth as long as basis capitulum. Palpi are short, with small posteroexternal salience; outline is trapezoid; dorsally each palpus is approximately 0.20 mm long, 0.11 mm wide. Hypostome is almost as long as palpi, dental formula is 2/2 with about seven teeth per file.

Scutum: Scutum is 0.57 mm long and 0.62 mm wide; outline is rounded. Cervical grooves are shallow, sometimes obsolete. Marginal grooves are short, shallow, anterior end reaching the level of posterior margin of coxa II, and posterior end enclosing first festoon.

Legs: Coxa are with small triangular spur.

Larva (Figures 2.47 and 2.48)[30]

It measures about 0.85 mm long, 0.60 mm wide.

Capitulum: Capitulum is slightly broader than long (0.17 mm × 0.19 mm). Basis capitulum is dorsally approximately 2.3 times as broad as long (0.06 mm × 0.14 mm); cornua are small with rounded apex. Palpi are with small posterolateral salience; dorsally each palpus is approximately 0.12 mm long, 0.07 mm broad. Hypostome extends slightly anterior to apex of palpi; dental formula is 2/2 with about seven teeth per file.

Scutum: Scutum is 0.33 mm long, 0.38 mm wide, and widest near mid-length. Cervical grooves are shallow concave.

Legs: Coxa I has a small sub-triangular spur rounded posteriorly; II and III are with small ridge-like spurs.

Ecology[30]

The altitude of localities in Uttar Pradesh where *H. indoflava* has been recorded is between 1,800 and 2,300 m and has a temperate climate. In winter the temperature often reaches below freezing point, while in summer it rarely exceeds 30 °C.

The area receives southeasterly monsoon from the Bay of Bengal. The average yearly rainfall ranges from 150 to 200 cm. The rainy season starts in the middle of June and lasts till the end of September, with July and August as peak monsoon months. There is considerable snowfall in winter, although the snow does not last long. Most of the type specimens were collected from the forest vegetation, either by handpicking or by flag dragging.

Species Relationship[27]

H. indoflava is close to *H. flava*, from which their larger size, smaller coxal and trochantal spurs, and shorter ventral spur on palpal segment III can easily distinguish the sexes. The male can further be distinguished by the reduced posteroexternal salience of palpal segment II, and the female in having rounded instead of pointed cornua. *H. indoflava* can easily be separated from all other species in this subgenus except *H. flava* by the fact that coxal spurs III and IV are larger than I.

As in the case of adults, the immature stages of *H. indoflava* show a marked resemblance with those of *H. flava* of Japan. The nymphs as well as the larvae of *H. flava* are slightly smaller than those of *H. indoflava.* Besides, the cornua and the first coxal spurs of the former are slightly longer and more pointed than those of *H. indoflava.*

Host[27,30,120]

Immature stages: Flag dragging in north India.

Adults: Dog, sheep, jackal, bullock, wild boar, man, fox, and from flag dragging in forest vegetation in north India.

Distribution[27,120]

India (Uttarakhand, Tamil Nadu, Rajasthan).

Disease Relationship

Not known.

2.6.3 Haemaphysalis birmaniae

Supino first described the male and female of this species in 1897 from an Indian muntjac and brush-tailed porcupine from Burma (Figure 2.49).

Male (Figures 2.50 and 2.51)[58,73,104,127]

Overall body length is approximately 2.0 mm, width 1.2 mm.

Capitulum: Basis capitulum is approximately twice as wide as long. Palpi are short and compact, lacking basolateral salience. Infrainternal setae are undifferentiated, number two or three. Hypostome not reaching anterior margin of palpi; lateral

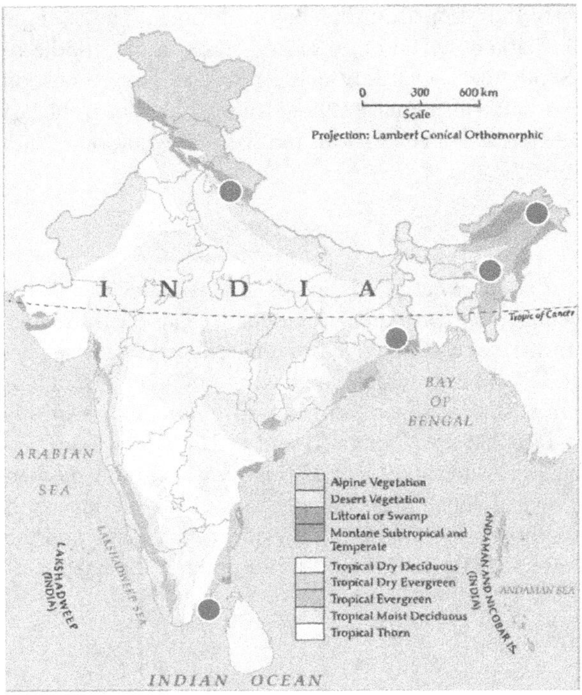

Figure 2.49 Distribution of *H. birmaniae* in India. (●) Place of first record. Map not to scale.

outline is slightly convex, and broadly rounded apically; dental formula is 4/4, with seven to nine denticles in a file.

Scutum: Scutum is glossy, outline moderately wide, approximately 1.35 times as long as wide, widest just anterior of the level of spiracular plates. Lateral grooves are obsolete or faintly suggested by depression no longer than the width of first festoon. Cervical grooves are obsolete; pits deep and short. Punctations are small, few may be medium size, shallow, moderately numerous, rather evenly distributed, and noncontiguous. Festoons number 11 and are distant. Spiracular plates are small and oval; dorsal projection short, widely triangular.

Legs: Legs are moderately long and robust. Coxa are each with a prominent spur.

Female (Figures 2.50 and 2.51)[58,73,127]

Total length of body is approximately 2.2 mm, width 1.4 mm.

Capitulum: Basis capitulum is approximately 2.3 times as wide as long; cornua are triangular, with bluntly rounded apex, and approximately one-third as long as the base; porose areas are moderately large, oval, and widely spaced. Palpi similar to those of male except for slight lateral extension of basal margin forming a small, and uptilted basolateral point. Hypostome is slightly larger and more elongate than male with dental formula 4/4 and with 9–12 denticles in each file.

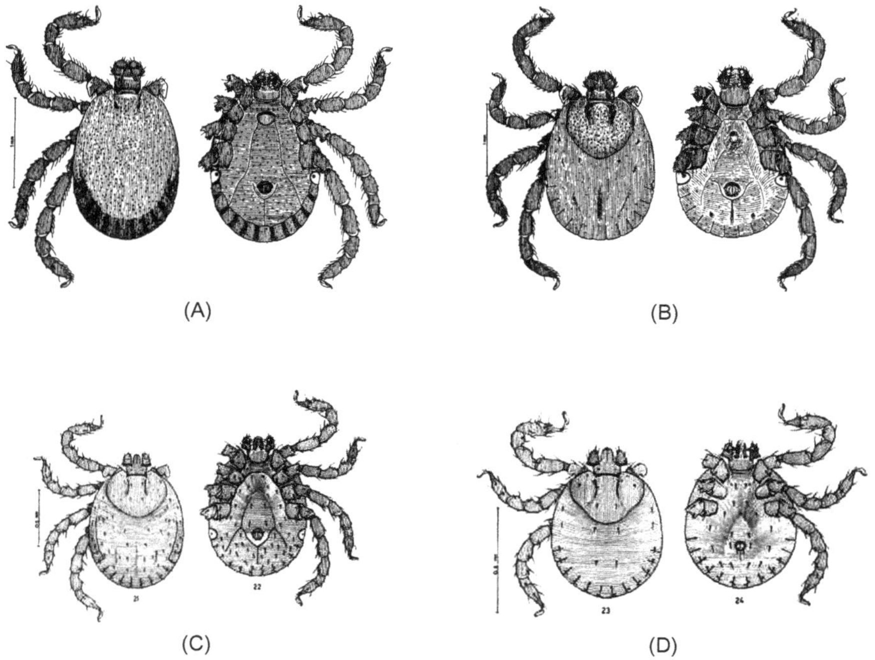

Figure 2.50 *Haemaphysalis birmaniae*. (A) Male, dorsal and ventral side; (B) female, dorsal and ventral side; (C) nymph, dorsal and ventral side; (D) larva, dorsal and ventral side. *Source*: Hoogstraal et al. (1965), Hoogstraal (1970), Journal of Parasitology (vol 51(1) & vol 56(6)), Allen press Publishing Services.

Scutum: Scutum is very slightly wider than long. Cervical grooves consist of narrow elongate, deep pits tapering to a point at mid-length of scutum and separated from anterior margin by a short shallow groove. Punctations are essentially as in male. Genital operculum is approximately twice as wide as long. Spiracular plates are ovoid.

Legs: Coxa and trochanters are with spurs similar to those of male.

Nymph (Figure 2.50)[73]

Length is approximately 1.5 mm, breadth 1.1 mm.

Capitulum: Basis capitulum is dorsally approximately 2.2 times as broad as long; external margins are straight; cornua are broadly triangular, approximately one-fourth as long as the base of basis capitulum. Palpi are campanulate. Hypostome is as long as palpi, and dental formula is 2/2, with seven to nine denticles in a file.

Scutum: Scutum is approximately 1.3 times as broad as long; margin is broadly convex, broadest approximately at mid-length. Cervical grooves are narrow. Punctations are obscure, in five to eight pairs. Spiracular plates are narrow oval.

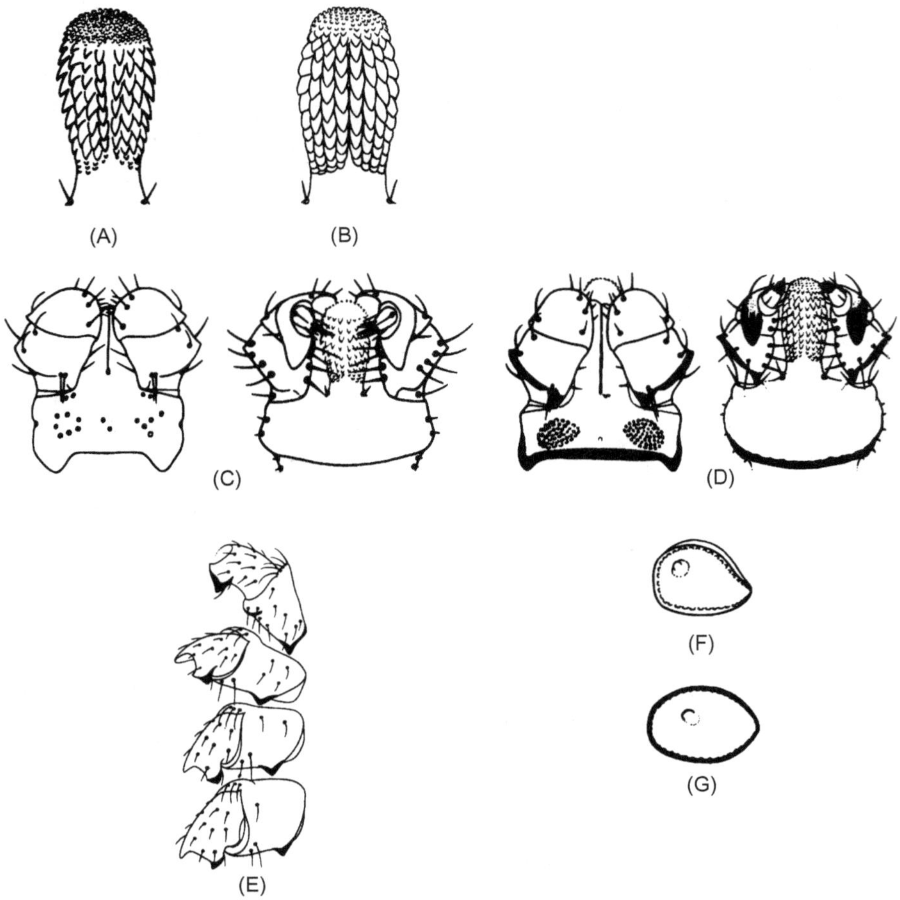

Figure 2.51 (A–B) Hypostome of male, female; (C–D) capitulum of male, female; (E) coxa of male; (F–G) spiracular plates of male and female.
Source: NIV. unpublished drawings.

Legs: Legs are moderately long and thin. Coxa are each with a broadly triangular spur extending well beyond posterior margin.

Larva (Figure 2.50)[73]

Length is approximately 1.0 mm, breadth 0.7 mm.

Capitulum: Basis capitulum is dorsally approximately 2.3 times as broad as long; external margins are straight; cornua are broadly triangular, very short, and may appear as slight extension of posteroexternal junctures. Palpi are campanulate. Hypostome is as long as palpi; dental formula is 2/2, with seven or eight denticles in a file.

Scutum: Scutum is approximately 1.3 times as broad as long. Cervical grooves and punctations are approximately as in nymph.

Hosts[26,58,73]

Immature stages: Chital deer, antelope, domestic cow, human (feeding), cattle, goral, barking deer, nymphs collected by flag dragging in Arunachal Pradesh state of India.

Adults: Chital deer, serow, domestic zhum, domestic cow, cattle, barking deer, Asiatic brush-tailed porcupine, black buck, yak.

Distribution[26,58,73,120]

India (Arunachal Pradesh, Assam, Uttar Pradesh (Dehradun), West Bengal, Tamil Nadu), Nepal, Burma, Malaysia.

Related Species

H. birmaniae belongs to *birmaniae* group, which consists of *H. atherurus, darjeeling*, and *capricornis*. *H. atherurus* and *H. capricornis* are not found in India. *H. birmaniae* differs from the closely related *H. darjeeling* by absence of projections on posterodorsal and posteroventral margins of palpal segment II. Also, presence of punctations in the central area of scutum and 4/4 dental formula distinguish *H. birmaniae* from *H. darjeeling*. *H. birmaniae* is smaller, paler, more yellowish than *H. darjeeling*.

The female of *H. darjeeling*, notably larger (2.8 mm long) than that of *H. birmaniae* (2.3 mm long), is readily recognized by its broadly U-shaped genital operculum (broadly rectangular in *H. birmaniae*), much larger coxal spurs, absence of ventral spurs on trochanters, broad scutum, etc.

Ecology

H. birmaniae is known from a mountain area between 3,150 and 4,500 ft altitude in Burma, 4,200–7,900 ft altitude in Nepal, and 500–3,200 ft altitude in northeastern India.

2.6.4 Haemaphysalis darjeeling

This species belongs to *birmaniae* group, described by Hoogstraal and Dhanda for first time in 1970 from the specimens collected on Himalayan artiodactyls and porcupines in the Himalayan forests of Darjeeling, West Bengal, India. The species has been named after the place of its first collection (Figure 2.52).

Male (Figures 2.53 and 2.54)[69]

Overall body length is approximately 2.35 mm, breadth 1.5 mm.

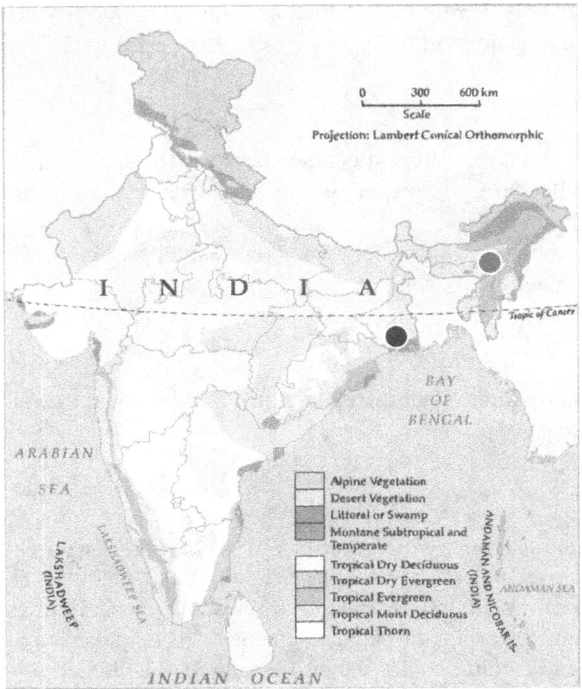

Figure 2.52 Distribution of *H. darjeeling* in India. (●) Place of first record. Map not to scale.

Capitulum: Basis capitulum is dorsally approximately 1.6 times as broad as long. Palpi are moderately salient. Hypostome is short, broad, reaching to or almost anterior margin of pit of palpal segment IV, approximately 2 times as long as broad posteriorly; dental formula is 5/5, with five (internal) to nine (external) denticles in a file.

Scutum: Scutum is approximately 1.3 times as long as broad, margins broadly rounded; surface is convex and shiny. Cervical grooves are obsolete or faint. Cervical pits are small, deep, and converging. Lateral grooves are obsolete. Punctations are rare, small, superficial, and widely scattered. Festoons number 11. Spiracular plates are transversely oval, and dorsal projection is continuous with plate.

Legs: Coxa are each with a prominent spur; I is with spur lanceolate, approximately 2 times as long as other spurs; II to IV are each with a broadly triangular spur extending well beyond coxal margin, length sub-equal.

Female (Figures 2.53 and 2.54)[69]

Overall body length is approximately 2.8 mm, breadth 1.9 mm.

Capitulum: Basis capitulum is approximately 2 times as broad as long, margins as in male; cornua are broadly triangular, one-third as long as the base of basis capitulum; porose areas are comparatively small, tilted, widely spaced, and ovate.

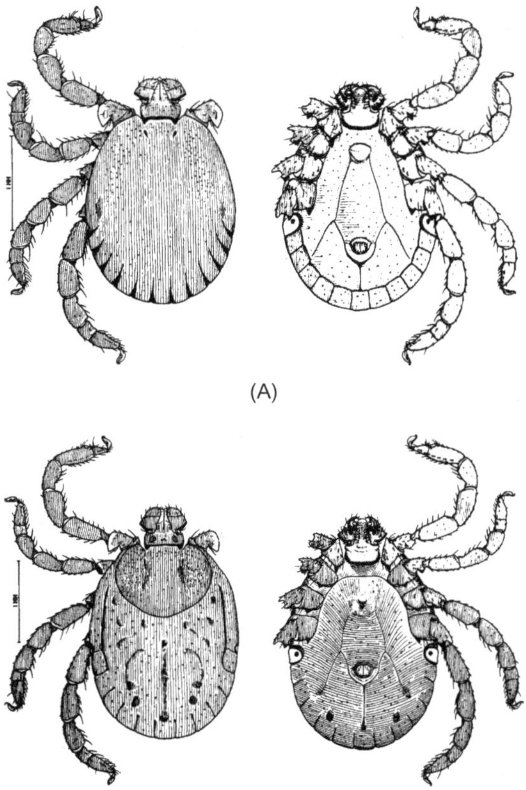

(A)

(B)

Figure 2.53 *Haemaphysalis darjeeling*. (A) Male, dorsal and ventral side; (B) female, dorsal and ventral side.
Source: Hoogstraal and Dhanda (1970), Journal of Parasitology (vol (56(1)), Allen press Publishing Services.

Palpi are as in male, longer than in male, reaching almost to the level of palpal apices; dental formula is 4/4, with eight (internal) to ten (external) denticles in a file.

Scutum: Scutum is approximately 1.35 times as broad as long. Cervical grooves are shallow and faint. Cervical pits are narrow, linear, and sub-parallel. Punctations are small, shallow, and obscure; widely distributed in external fields, rare or obsolete in median field. Genital operculum is broadly U-shaped with external margins diverging anteriorly.

Legs: Legs are same as in males except coxa with spurs of II and I slightly shorter.

Related Species[69]

H. darjeeling is most closely related to *H. birmaniae*. Both sexes of *birmaniae* lack the projections of the posterodorsal and posteroventral margins of palpal segment II, which characterize *darjeeling*. In the *birmaniae* male, the absence of a pincer-like form of palpal segment III, 4/4 dental formula, presence of punctations in the

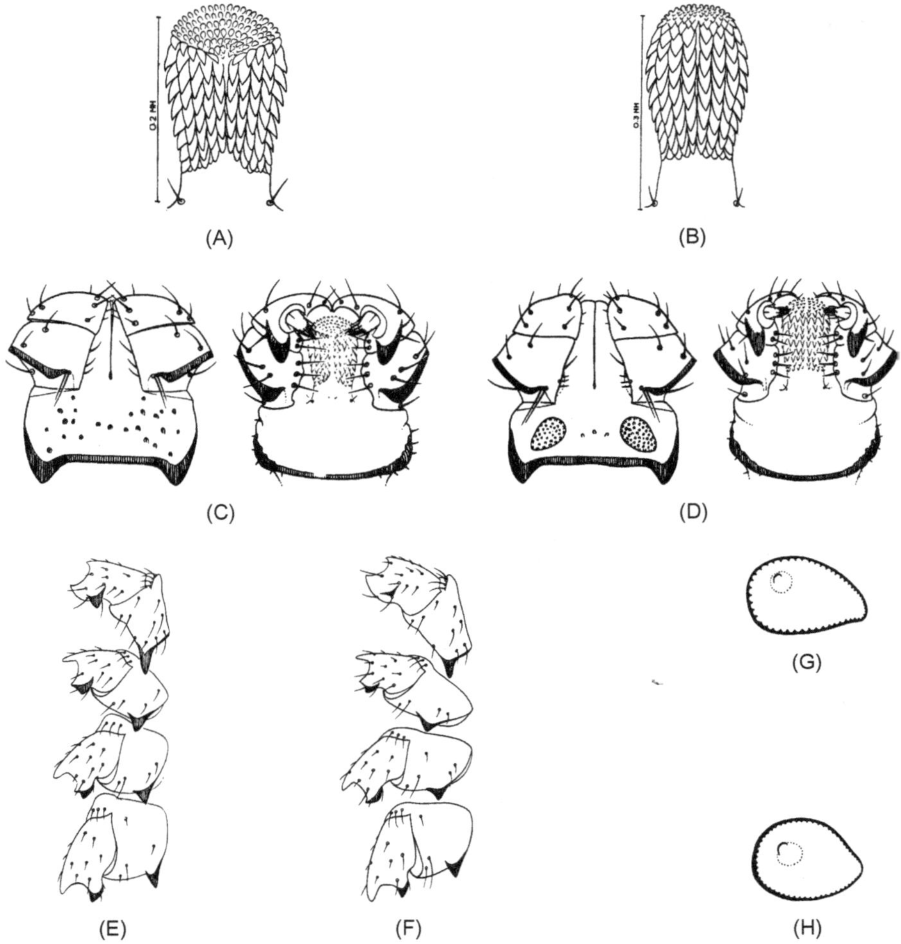

Figure 2.54 (A–B) Hypostome of male and female, respectively; (C–D) capitulum of male and female, respectively; (E–F) coxa of male and female, respectively; (G–H) spiracular plates of male and female, respectively.
Source: NIV. unpublished drawings.

central field of the scutum, definitely triangular dorsal projection of the spiracular plate, and seven ventrointernal setae of femur IV, etc., differ from those in *darjeeling*. In the *birmaniae* female, the presence of a ventral spur on trochanter I, broadly rectangular genital operculum, and nine ventrointernal setae of femur IV serve to differentiate it from *darjeeling*. *H. birmaniae* is generally smaller and paler, and more yellowish than this species. Two other species that have been recorded outside India, and related to *darjeeling*, are *H. capricornis* and *H. atherurus*. Male of *H. capricornis* lacks coxal spurs, and has a 4/4 dental formula and short, stout tarsi; thus, it is easily separated from the *darjeeling* male. The female of *capricornis* is unknown. *H. atherurus* is a small species easily distinguished from *darjeeling* by very short ventral spurs of palpal segment III, different palpal outlines, dorsointernal and ventrointernal setae each number two, semicircular spiracular plates, reduced spurs of female coxa, etc.

Ecology[69]

It occurs in West Bengal state of India. In West Bengal, in the Darjeeling district, it occurs at an altitude range of 1,800–3,200 ft.

Host[69]

Immature stages: Not available.

Adults: Goat antelope, serow, wild boar, barking deer, human, Artiodactyl mammal.

Distribution[69,120]

India (West Bengal, Assam), Thailand, Burma.

Disease Relationship

Not known.

2.7 Subgenus *Kaiseriana*

General Characteristics

It consists of 33 species and is the largest subgenus under the genus *Haemaphysalis*. Twenty-seven out of 33 *Kaiseriana* species are Oriental in distribution. Adults are structurally specialized, some extremely, for parasitizing Artiodactyla; most also feed on carnivores but other hosts are exceptional. Immatures parasitize small mammals, especially Rodentia. Only immatures of *H. longicornis* occasionally parasitize birds. The hallmark of *Kaiseriana* is a hair-hooking spur extending from the posterodorsal margin of adult palpal segment III. In *cornigera* group males, this spur is supplemented or replaced by a gap in the external surface of the palpus; working as an efficient hair-grasping gap. The Indian *Haemaphysalis* under the subgenus *Kaiseriana* can be grouped into at least four groups as per the classification of Hoogstraal and Kim: *cornigera* group, *bispinosa* group, *aculeate* group, and *hystricis* group.

Keys to Identify Species of Subgenus Kaiseriana

Males

1. Palpi are broadened; dorsal spur on palpal segment III is absent; all coxa bear long spurs but variable in length. ....[2]
Palpi are campanulate; dorsal spur is present or ridged dorsobasal on palpal segment III; coxa I shows large spur but others have small spurs or ridges................................[4]

2. Ventrobasal spur on palpal segment II is well developed; infrainternal setae number four; ventrobasal spur of palpal segment III reaches or overlaps to two-thirds of palpal segment II, presence of single elongate; lanceolate or dagger-like spur on coxa IV..[*H. spinigera*]
Ventrolateral spur on palpal segment II is low or obsolete; infrainternal setae number five; ventrobasal spur on palpal segment III reaches to half of palpal segment II, presence of two long spurs on coxa IV..[3]

3. Shallow and medium lateral groove is extending up to coxa III; evenly scattered punctations are evenly scattered; both spurs on coxa IV are posterolaterally directed and well spaced, the outer one is needlelike long while the inner is smaller; trochanters ventrally show small and triangular spur...................[*H. shimoga*]
Deep and long lateral grooves are extending up to coxa II; punctations are irregularly scattered; both the spurs on coxa IV are posteroexternally directed, closely spaced, and approximately equal in size; trochanters ventrally have no spurs...................[*H. anomala*]

4. Cornua are well developed with enlarged posterodorsal spur...................[5]
Cornua are very small with small to moderate posterodorsal spur on palpal segment III, in some cases posterodorsal spur is triangular or curved; coxal spurs are well developed to reduced...................[6]

5. Palpal segment II and III are nearly equal in length; dorsal and ventral spurs of palpal segment III are sharply triangular, tapering to acute apex, and extending beyond palpal segment II and externally directed; infrainternal setae number four; dental formula is 4/4; lateral grooves are moderate in size and extending up to anterior of coxa III without enclosing any festoon; spurs on coxa I and trochanter are pointed.................. [*H. cuspidata*]
Palpal segment II is larger than palpal segment III, dorsal and ventral spurs on palpal segment III are broadly triangular; reaching to two-thirds of palpal segment II dorsally and up to one-third ventrally; laterally directed; infrainternal setae are long, lanceolate, and number five; dental formula is 5/5; lateral grooves are small and extending up to only anterior of coxa IV and enclosing first festoon on both sides; spurs on coxa I and trochanter I are spatulate...[*H. aculeata*]

6. Spurs on palpal segment III are conventionally triangular; moderately large and overlapping about one-third of palpal segment II dorsally and two-thirds of palpal segment III ventrally and present in mid-line; infrainternal setae number six; dental formula is 6/6; punctations are numerous, small, and evenly distributed; coxal spurs are large, pointed; only trochanters I and II have short, triangular spurs ventrally...................[*H. davisi*]
Spurs on palpal segment III posterodorsally and posteroventrally are large, reaching almost half of segment II dorsally while two-thirds or more of segment II ventrally and present in the mid-line and curved outwardly; dental formula is 4/4; infrainternal setae number five; punctations are small, numerous and irregularly scattered, and shallow; lateral groove is medium length, not enclosing any festoon; coxal spurs are triangular and look blunt..[*H. ramachandrai*]

Posterodorsal and posteroventral spurs on palpal segment III are prominent and medium sized; present in mid-line and overlapping about one-third of segment II ventrally and up to mid-length dorsally; infrainternal setae are long, lanceolate, and number five; punctations are few, small, and irregularly scattered; coxal spurs are blunt or ridge-like except spur of coxa I where it is long..........[*H. bispinosa*]

7. Spurs on palpal segment III are either reduced/blunt or have been placed toward internal position; posteroventral spur when present is in middle or internal side; infrainternal setae are short and feathery; coxal spurs are much reduced and moderate to small in size..........[8]

8. Dorsobasal and dorsoventral spurs on palpal segment III are short and widely triangular; infrainternal setae are short, feathery, and number six; lateral grooves are very short, reaching to apical margin level of coxa IV, enclosing first festoon on both sides; coxal spurs are very short, broadly triangular, and sharply pointed..........[*H. kinneari*]

Spurs on palpal segment III are reduced and blunt, overlapping half of the palpal segment II ventrally and just crossing the suture dorsally; infrainternal setae are very short and number four; lateral grooves are absent; punctations are rare; coxal spurs are short and broadly triangular..........[*H. hystricis*]

The palpi are compact and give triangular shape to capitulum; the posterodorsal and posteroventral spur of palpal segment III are conventionally triangular and pointed but lying on inner side of the palpi; infrainternal setae number eight and are feathery, closely spaced; spurs on coxa are small; coxa I and IV spurs are triangular and pointed, while coxa II and III spurs are rounded or ridge-like..........[*H. wellingtoni*]

Posterodorsal spur of palpal segment III is short and toward internal margin, while the posteroventral spur is heavy, gradually tapered, and extending to the base of palpal segment II and in mid-line; punctations are numerous, large, and scattered all over the scutum; infrainternal setae number seven, are feathery, and closely placed; coxal spurs are robust or blunt; coxa I is with a long spur, while coxa II to IV are with ridge-like small spurs..........[*H. intermedia*]

Posterodorsal spur of palpal segment III is short, triangular, and toward inner margin, while the posteroventral spur is in mid-line but moderate and reaching to the mid-line of palpal segment II; punctations are small and medium size and relatively few; infrainternal setae number five, are long, lanceolate, and well spaced; coxal spurs are small; coxa I spur is moderate, while other spurs on coxa are small or obsolete.......... [*H. sambar*]

Dorsobasal spur is obsolete; ventrobasal spur extends to half or more of segment II; infrainternal setae number eight to nine and are very short, feathery, and closely spaced; coxal spurs are moderate; coxa I spur is moderate and conventionally triangular, while other coxa spurs are small..........[*H. kutchensis*]

9. Small ridge-like spur or elevation of dorsobasal side of palpal segment III; in some cases the posteroventral spur is large and elongate; ventrobasal spur is shorter; infrainternal setae are shorter and feathery or lanceolate; coxal spurs are much shorter..........[10]

10. Small ridge-like or crest like dorsobasal spur on palpal segment III; ventrobasal spur is slender, extending up to half of palpal segment II; infrainternal setae number five, short but thin and well spaced; punctations are shallow, numerous, and small; lateral grooves are obsolete or very short and not enclosing any festoon; coxal spurs are triangular and blunt; coxa I spur is narrowly elongate, while coxa II to IV spurs are broad and shorter consecutively; trochanters I to IV ventrally show triangular spurs..........[*H. turturis*]

Posterodorsal spur of palpal segment III is ridge-like; salience is moderate due to upcurved palpal segment II; infrainternal setae are seven to nine in number, closely placed,

and feathery; moderate lateral groove enclosing first festoon, extending to coxa III; trochanters show spurs ventrally..................................[*H. paraturturis*]
Dorsobasal spurs of palpal segment III are absent or crest-like; infrainternal setae are seven in number, widely spaced, lanceolate, or feathery; lateral grooves are long, enclosing first festoon and extending up to coxa II; punctations are numerous but are widely scattered; trochanter spurs are absent..................................[*H. silvafelis*]

Keys to Identify Females

1. Palpi are broadened; dorsal spur on palpal segment III is absent, all coxa bear long spurs but variable in length...................................[2]
   Palpi are campanulate, dorsal spur present or dorsobasal spur on palpal segment III is ridged; coxa I sometimes shows large spur but others have small spurs or ridges...[4]
2. Ventrolateral spur on palpal segment II is well developed; infrainternal setae number four; ventrobasal spur of palpal segment III reaches or overlaps to half of palpal segment II; porose areas are more or less circular and widely spaced; coxa I spur is moderate and thin, while other coxa bear small spur..[*H. spinigera*]
   Ventrolateral spur on palpal segment II is low or obsolete; infrainternal setae number five or more; coxa I spur is moderate to long in length and decreases in size from coxa II onward..[3]
3. Infrainternal setae are five in number but short; porose areas are oval but small; coxa I shows a very long, pointed spur, while other coxa bear small and more or less equal-sized spurs; trochanters ventrally show small, triangular spur...................[*H. shimoga*]
   No ventrolateral spur on palpal segment II; infrainternal setae are eight in number and are long; porose areas are oval and big; punctations are irregularly scattered; coxa I spur is long but blunt, while others are broadly triangular or ridge-like; trochanters ventrally have no or obsolete spurs..[*H. anomala*]
4. Cornua are well developed with enlarged posterodorsal spur...................[5]
   Cornua are very small or moderate with small to moderate posterodorsal spur on palpal segment III..[6]
5. Palpal segment II and III are nearly equal in length; dorsal and ventral spurs of palpal segment III are sharply triangular, tapering to acute apex, extending up to the base or beyond palpal segment II and externally directed; infrainternal setae number four; dental formula is 4/4; spurs on coxa I are pointed, while other coxa bear ridges; trochanter I bears a spatulate spur..[*H. cuspidata*]
   Palpal segment II is larger than palpal segment III; dorsal and ventral spurs on palpal segment III are broadly triangular, reaching to two-thirds of palpal segment II dorsally and up to one-third ventrally; laterally directed; infrainternal setae are long, lanceolate and number four; dental formula is 5/5; spurs on coxa I and trochanter I are spatulate...[*H. aculeata*]
6. Spurs on palpal segment III are conventionally triangular, moderately large and overlapping about one-third of palpal segment II dorsally and one-half of palpal segment II ventrally, and present in mid-line; infrainternal setae number five; dental formula is 5/5 punctations are numerous, small, and evenly distributed; coxal spurs are moderate, pointed on coxa I, and blunt on others..[*H. davisi*]
7. Spurs on palpal segment III posterodorsally and posteroventrally are large, reaching almost one-third of segment II dorsally while two-thirds or more of segment II ventrally and present in the mid-line and curved outwardly; dental formula is 4/4; porose areas

are oval, elongate, and well spaced; infrainternal setae are five in number, long, and lanceolate; punctations are small, numerous, and irregularly scattered; coxal spurs are moderate; coxa I shows a long and pointed spur while other coxa bear small, blunt spurs, or ridges..[*H. ramachandrai*]

8. Posterodorsal and posteroventral spurs on palpal segment III are prominent and medium sized, present in mid-line, and overlapping about one-third of segment II ventrally and up to mid-length dorsally; infrainternal setae are long and lanceolate and number five; dental formula is 4/4; porose areas are circular but spaced widely; punctations are few, small, and irregularly scattered; coxal spurs are blunt or ridge-like, except spur of coxa I where it is long..[*H. bispinosa*]

9. In some cases the posterodorsal and posteroventral spurs are completely on internal side of palpi, spurs on palpal segment III reduced except in *H. intermedia* and have been placed toward interior; infrainternal setae are short and feathery; coxal spurs are much reduced now to moderate or small in size..[10]

10. Dorsobasal and dorsoventral spurs on palpal segment III are short and widely triangular; infrainternal setae are short, feathery, and number seven; porose areas are oval and spaced from each other; dental formula is 4/4; coxal spurs are very short, broadly triangular, and sharply pointed..[*H. kinneari*]

Spurs on palpal segment III are reduced and blunt, overlapping half of the palpal segment II ventrally and just crossing the suture dorsally; infrainternal setae are very short and number four; porose areas are oval, elongate, and inclined toward interior; punctations are rare; coxal spurs are short and broadly triangular.................................. [*H. hystricis*]

11. The palpi are compact and give triangular shape to capitulum, the posterodorsal and posteroventral spur of palpal segment III is conventionally triangular and pointed but lying on inner side of the palpi overlapping one-half of palpal segment II; infrainternal setae are number 10 and are feathery and closely spaced; porose areas are more or less circular in shape and are widely spaced; spurs on coxa are small; coxa I and IV spurs are triangular and pointed, while coxa II and III spurs are rounded or ridge-like..................................[*H. wellingtoni*]

Posterodorsal spur of palpal segment III is short and toward internal margin, while the posteroventral spur is heavy, gradually tapered, and extending to the base of palpal segment II and in mid-line to interior; porose areas are circular, big, and are moderately spaced from each other; punctations are numerous, large, and scattered all over the scutum; infrainternal setae number seven, are feathery, and closely placed; coxal spurs are robust or blunt; coxa I is with a long spur, while coxa II to IV are with ridge-like small spurs..[*H. intermedia*]

Dorsobasal spur is obsolete; ventrobasal spur is extending to half or more of segment II; infrainternal setae number 10 and are very short, feathery, and closely spaced; porose areas are circular but small; coxal spurs are moderate; coxa I spur is moderate and conventionally triangular, while other coxa spurs are small............................. [*H. kutchensis*]

12. Dorsobasal spur of palpal segment III is small ridge-like or elevated; ventrobasal spur is shorter, in some cases it is loner; infrainternal setae are shorter and feathery or lanceolate; coxal spurs are much shorter..[13]

13. Dorsobasal spur on palpal segment III is small ridge-like or crest-like; ventrobasal spur is slender, extending up to half of palpal segment II; infrainternal setae are number five, short but thin, and well spaced; punctations are shallow, numerous, and small; porose areas are oval and elongate and are much inclined toward interior; coxal spurs are triangular and

blunt; coxa I spur is narrowly elongate, while coxa II to IV spurs are broad and shorter consecutively; trochanters I to IV ventrally show triangular spurs....................[*H. turturis*]
Posterodorsal spur of palpal segment III is ridge-like; ventrobasal spur extends to half of segment II; salience is moderate due to up-curved palpal segment II; infrainternal setae are seven in number, closely placed, and feathery; porose areas are very small and oval; trochanters show spurs ventrally..[*H. paraturturis*]
Dorsobasal spurs of palpal segment III are absent or crest-like; ventrobasal spur extends up to two-thirds of palpal segment II; infrainternal setae are seven in number, widely spaced, and feathery; porose areas are elongately oval and inclined interiorly; punctations are numerous but are widely scattered; coxa I spur is elongate while others are short; trochanter spurs are absent..[*H. silvafelis*]

2.7.1 Haemaphysalis spinigera

Neumann first described this species in 1897 on the basis of adults collected from Sri Lanka. Nuttall and Warburton repeated the same description and illustrations in 1915 and by Sharif in 1928. This is a tick species that has been studied more exhaustively than any other tick species in India as this is the main vector of KFD, the only tick-borne human viral disease in India (Figure 2.55).[47]

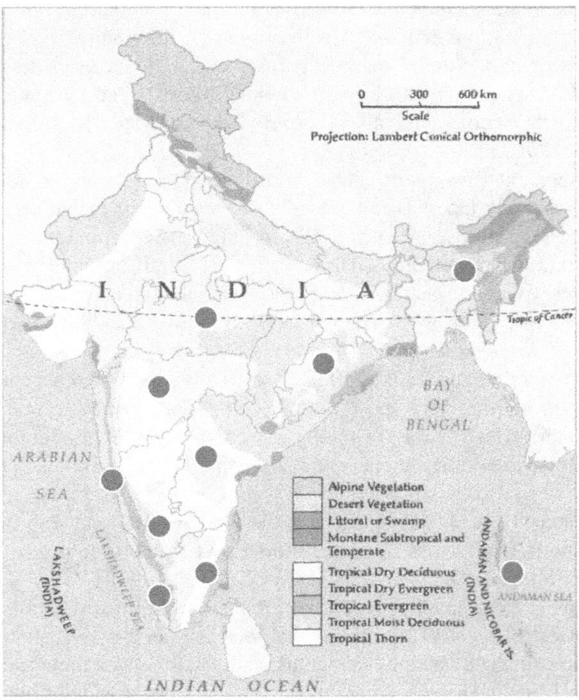

Figure 2.55 Distribution of *H. spinigera* in India. (●) Place of first record.
Map not to scale.

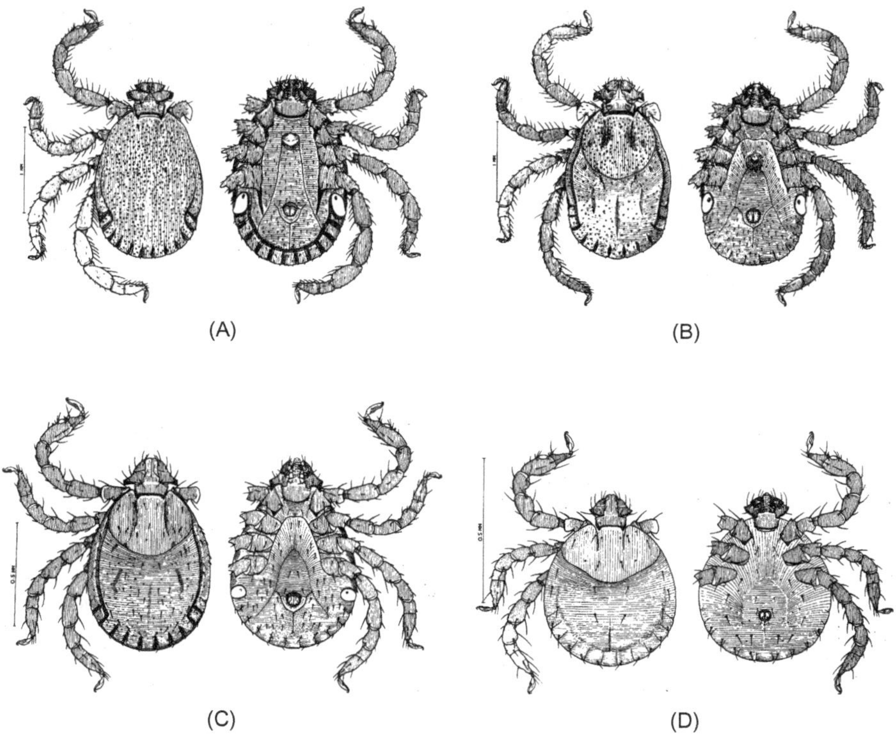

Figure 2.56 *Haemaphysalis spinigera.* (A) Male, dorsal and ventral side; (B) female, dorsal and ventral side; (C) nymph, dorsal and ventral side; (D) larva, dorsal and ventral side. *Source*: Ghalsasi and Dhanda, Oriental Insects 1974; 8(4): 505–20; NIV unpublished drawings.

Male (Figures 2.56 and 2.57)[14,18,47,104,127]

Overall body length is approximately 2.02–2.70 mm and 1.16–1.65 mm wide.

Capitulum: Basis capitulum is dorsally rectangular, approximately twice as wide as long; cornua are triangular, approximately one-third as long as basis capitulum. Palpi are short, widely salient, and almost as broad as long. Hypostome is clavate, as long as palpi, and dental formula is 5/5, with 9–11 denticles per file.

Scutum: Scutum is pyriform, approximately 1.4 times as long as wide; lateral margins are convex, widest at the level of coxa IV, and slightly depressed at the level of spiracles. Lateral grooves are shallow, extending to about mid-length of scutum anteriorly and enclosing first two festoons posteriorly. Cervical grooves are obsolete. Punctations are moderate in size, deep, and uniformly distributed all over scutum. Festoons number 11, somewhat rectangular in outline. Genital plate is at the level of coxa II, ovoid, and slightly narrow posteriorly.

Legs: Legs are long and stout. Coxa are each with characteristic spur. Spur of coxa I is long, tapering, sharply pointed, and extending half-length of coxa II; spurs of coxa II

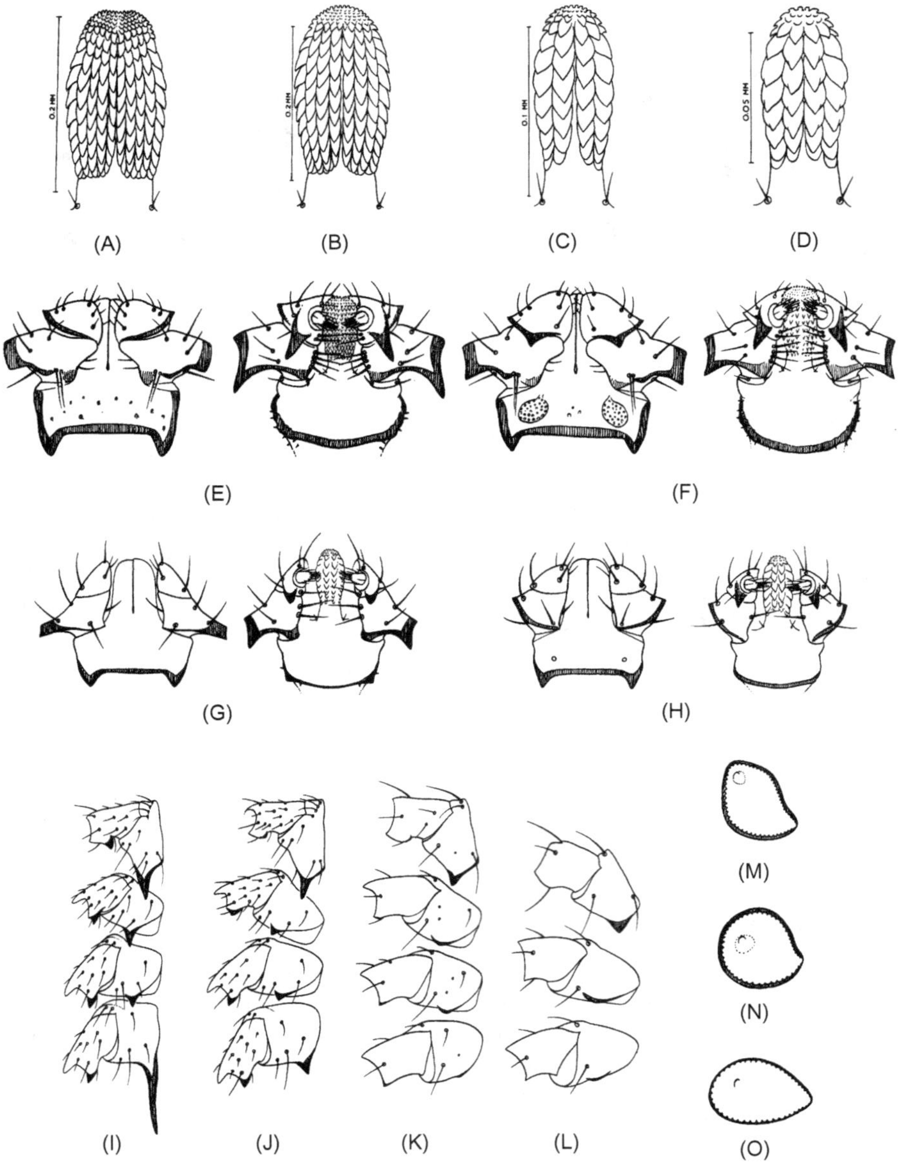

Figure 2.57 (A–D) Hypostome of male, female, nymph, and larva, respectively; (E–H) capitulum of male, female, nymph, and larva, respectively; (I–L) coxa of male, female, nymph, and larva, respectively; (M–O) spiracular plates of male, female, and nymph, respectively. *Source*: NIV. unpublished drawings.

and III are small, triangular, sharply pointed; spur of coxa IV is very long, dagger-like, directed posteriorly, and extending up to or beyond the level of anal opening.

Female (Figures 2.56 and 2.57)[47,104,127]

The female is 2.15–3.00 mm long, 1.35–1.80 mm wide.

Capitulum: Basis capitulum is dorsally almost twice as wide as long; margins are as in male; cornua are slightly shorter than that of male, one-third as long as the base of basis capitulum; porose areas are oval, wide apart by almost twice their largest diameter. Palpi are as in male except segment III dorsally with a well-developed, triangular, posteriorly directed spur at its posterior margin. Hypostome dental formula is 4/4, with 11–13 denticles per file.

Scutum: Scutum is sub-circular in outline. Cervical grooves are deep, narrow, sub-parallel, and extending up to mid-length of scutum. Punctations are moderate in size, deep, and uniformly distributed. Marginal grooves are long, narrow, and enclosing first two festoons. Festoons number 11. Genital area is circular in outline; operculum is broadly U-shaped.

Legs: Legs are slightly narrower than in males. Coxa I is with a long, bluntly pointed spur; II to IV are with small triangular, bluntly pointed spurs.

Nymph(Figures 2.56 and 2.57)[47]

Capitulum: Basis capitulum is approximately 2 times as wide as long. Cornua are triangular, approximately one-third as long as basis capitulum. Posterior margin between cornua is straight. Hypostome is clavate, extends slightly beyond the level of palpi, approximately 2.5 times as long as wide; dental formula is 2/2, sometimes 3/3 apically, with eight to ten denticles per file.

Scutum: Cervical grooves are deep anteriorly, parallel up to mid-length of scutum, then divergent and shallow posteriorly. Punctations are approximately 12 each with a minute seta in addition to six pairs of sensilli. Marginal grooves are long, deep, and enclosing first two festoons. Festoons number 11, each bearing a seta dorsally and ventrally except the median.

Legs: Legs are moderately long and stout. Coxa I is with a distinct triangular pointed spur; II and III are each with ridge-like spur.

Larva (Figures 2.56 and 2.57)[47]

Capitulum: Basis capitulum is dorsally approximately twice as broad as long; cornua are one-fourth as long as basis capituli. Each palpus is approximately 1.3 times as long as wide, conical in outline. Hypostome is with seven to eight denticles per file.

Scutum: Scutum is wider than long, outline trapezoid, with moderate anterior emargination, and posterior margin broadly rounded; cervical grooves are sub-parallel, reaching mid-length of scutum. Punctations are six in number, each with a single seta, in addition to three pairs of sensilli. Festoons number 11.

Legs: Legs are moderately long and stout. Coxa I has a short triangular spur, and II and III have small ridge-like projections; other leg characters are as in nymph.

Species Relationship[47]

H. spinigera is a member of group *cornigera*, subgroup *spinigera* in the subgenus *Kaiseriana*. The adults as well as the nymph of *H. spinigera* are characterized by broad flared salience on palpal segment II, which is produced into spur-like projection at its ventrolateral angle. Adults of other Indian species, which possess this characteristic, are *H. indica* and *H. canestrinii*. From both these species, the male of *H. spinigera* can be easily distinguished by the presence of an elongated dagger-like spur on coxa IV and its 5/5 dental formula instead of 4/4. The female can be separated by the presence of a median triangular spur on the dorsobasal margin of palpal segment III. Moreover, both the sexes of *H. spinigera* possess a long and pointed spur on coxa I, which in the case of *H. indica* and *H. canestrini* are rounded. Other Indian species in which the adults possess broad and flared salience on palpal segment II are *H. shimoga*, *H. anomala*, *H. doenitzi*, *H. megalaimae*, and *H. minuta*. Of these, *H. shimoga* and *H. anomala* are the closest. In both these species, coxa IV in the male possesses two long dagger-like spurs, while the female lacks ventrolateral spur on palpal segment II. The adults of the remaining species mentioned above can also be distinguished by the absence of ventrolateral spur on palpal segment II.

Among the nymphs, the species possessing a sharp and distinct spur on the ventrolateral angle of palpal segment II are *H. canestrini*, *H. indica*, *H. anomala*, and *H. shimoga*. The former two can be distinguished from *H. spinigera* by small spurs on the ventral side of palpal segment II, and on coxa I to IV, and by smaller cornua. *H. anomala* and *H. shimoga* nymphs are indeed difficult to distinguish from *H. spinigera*, but a careful examination reveals that the dorsolateral angle of palpal segment II is angular in *H. spinigera* and rounded in *H. anomala* and *H. shimoga*. Besides, the ventrolateral spur in *H. spinigera* is shorter and more pointed. The palpal segment II of the nymphs of *H. kinneari* and *H. kyasanurensis* also possess wide and flared salience, in which the ventrolateral margin is slightly angular but lacks a distinct spur.

In the larva of *H. spinigera*, unlike the adults and the nymph, the salience on palpal segment II is not broad and flared. It shows resemblance to *H. kyasanurensis*, *H. shimoga*, *H. anomala*, *H. kinneari*, and *H. turturis*. Trapido et al. (1964) have well described the distinguishing characters of *H. spinigera* from all these species, except *H. anomala*. *H. anomala* shows more resemblance to *H. shimoga* than to *H. spinigera*.

Life Cycle Under Laboratory Conditions[47,14]

Ghalsasi and Dhanda in 1974 and Bhat in 1979 studied biology of this species under laboratory conditions. In the laboratory, the life cycle was completed in

Table 2.2 Lifecycle of *H. spinigera* Under Laboratory Conditions at Ambient Temperature

Stage	Phase	Duration (in days)
Engorged females	Pre-oviposition	2–5
	Oviposition	18–23
	Post-oviposition	1–39
Egg	Incubation	29–33
Larva	Quiescent	3–10
	Life span (maximum)	320
	Parasitic	2–6
	Molting	10–15
Nymph	Quiescent	3–10
	Life span (maximum)	276
	Parasitic	2–6
	Molting	12–15
Adults	Quiescent	5–10
	Life span (maximum)	323 (M) and 339 (F)
	Parasitic	7–16

118–160 days at ambient temperature, which ranged from 18 to 35 °C and relative humidity from 80% to 90%. The unfed males and females from the colony readily fed on calf or rabbit and engorged within 8–13 days. The weight of a fully engorged female fed on laboratory rabbit ranged from 0.105 to 0.23 g and those fed on calf from 0.146 to 0.298 g. Oviposition usually commenced 2–5 days after detachment and continued for 17–25 days. The number of eggs laid by the females fed on rabbit ranged from 388 to 3,136 and those fed on calf from 1,804 to 3,536. The eggs hatched 25–30 days after laying. The larvae were not ready to feed until 5–7 days after hatching. They readily fed and engorged on chick, mice, or rabbit within 3–6 days and molted to nymphs in 13–16 days after detachment. The prefeeding, feeding, and the molting periods of the nymph were almost the same as for larvae. They fed well on chick or rabbit. The females were ready to feed 25–30 days after molting (Table 2.2).

Behavioral Studies of Different Developmental Stages Under Forest Conditions[17]

Engorged females: In one of his field experiments carried out by Bhat, 45 ticks were released in 9 lots, each consisting of 5 ticks, during 1964 and 1965. The engorged females, as soon as released on the ground cover, began to seek their way downward, crawling slowly and moving their forelegs. They dragged themselves under the litter and were out of sight within 5–10 min. Movement of these ticks ceased as soon as they came in contact with the underlying hard soil or the humus. Ticks released on the grazing ground also moved horizontally till they reached dark places under the grass cover. Ticks released during monsoon months came to rest just below the litter and laid eggs on the surface of humus. Those released during dry months from October to June showed a marked *burrowing* behavior. After reaching the humus they dug up small pits with their hypostome and forelegs and burrowed themselves, leaving only the posterior end exposed. The eggs were laid inside these pits. The lateral and downward progression, including burrowing, was confined to the first one or two days only. The direction of the movement was zig-zag and at random. The distance traveled between the release point and egg-laying site was measured for 23 ticks. All the other ticks were found lying eggs within a distance of 7.5–30 cm from the place of release.

Pre-oviposition phase: During dry months, when the temperature under the litter was low (17 ± 1 °C), pre-oviposition phase varied from 8 to 11 days. In open areas they began to lay eggs on the fifth day. During rainy season when the humus temperature was slightly higher (21 ± 1 °C), it varied from 5 to 8 days. No locomotion was observed during oviposition and each female laid all eggs in a single mass. Duration of incubation period varied from 30 to 52 days during dry months and 25–40 days during wet months. On the grazing field, development was observed only from October to December. During monsoon months females and their broods were washed away by rainwater, and from January to May the eggs shriveled up under the hot sun. *Oviposition* continued for 10–17 days.

Unfed larvae: Three broods consisting of 1- to 11-day-old eggs deposited under the litter in July 1963 completed hatching at the end of August of the same year. Larvae crawled up and clung in a compact mass to the undersurface of a dry leaf of the litter. The cluster remained stationary for about a month. All three broods had begun to dissociate into small clusters distributed within a radius of 7–10 cm at the end of September. During this process a few larvae crawled over to neighboring leaves. The process of *clustering and declustering* continued for a few more days. The brood had covered a radius of 30 cm. During this movement the area occupied by the brood increased until nearly all larvae settled under the leaves. Clusters settled on leaf-tips withdrew to the lower surface on rainy days and re-occupied the tips as soon as the environment dried. Larvae from these broods survived till December 10, 1963. Clusters from broods hatched during June and July stayed longer than those hatched during later months. Larvae hatched in October and November disappeared from vegetation by the end of March. The persistence was for 80–130 days. Larvae that hatched in rainy season and stayed dormant for a considerable period survived for 150–230 days. Irrespective of the time of hatching, all the broods settled on the vegetation after the monsoon disappeared before the end of March.

Engorged larvae: Twenty-one groups of freshly engorged larvae, each with 200 or 300 specimens, were studied in different microhabitats in different seasons from December 1963 to June 1965. Immediately after release, they crawled in all directions at random. On forest litter they crawled along the upper surface of the superficial leaves. After reaching the edges, some dropped to lower layers and some crawled along the lower surface. Subsequently, all of them penetrated the litter and settled in the humus after 4–6 h. They remained sensitive for 5–8 days and could be induced to move by slight disturbance. Later, they became inactive and passed on to the molting phase. Molting larvae were traced under deep layers of litter and loose humus, and were seen hidden inside crevices and cavities of mud lumps. They could penetrate as deep as 2 cm into the loose humus. Centripetal search in small sectors revealed that larvae had moved only 20–30 cm away in search of shelter.

Larvae released in the forest during monsoon months settled on the superficial layers of litter. Larvae released on grazing ground during dry months settled under grass and in earth crevices, of which only a small proportion was recovered as nymphs. Larvae released on grazing ground during rainy seasons were washed away and there was no recovery of nymphs. On the forest edge, where the ground was covered with a thin layer of litter mixed with grass, a small proportion of nymphs was recovered.

Larval molting: Larvae molted into nymphs 22–29 days after the release. Freshly emerged nymphs appeared on vegetation soon after molting. Larvae released on more exposed areas molted earlier than those on less exposed areas. Larvae released during monsoon months molted earlier than those released at other months. Nymphs appeared on vegetation and on the litter soon after molting, irrespective of season. Freshly molted nymphs containing traces of larval blood meal emerged on the surface. Emerging nymphs from some groups were recovered periodically to estimate the proportion of larvae molted into nymphs. They were re-released in other places to study their longevity. The recovery of nymphs inside the forest was 24–78%. The grazing field was most unfavorable niche for larval molting and only up to 4% recovery could be made, except in one group which yielded 25.5% nymphs. Nymphs survived on litter and on vegetation for a maximum span of 6 months. Though nymphs emerged early in the rainy months, they could survive for 3–6 months. The mortality rate was higher due to heavy rain.

Engorged nymphs: A total of 10,725 freshly fed nymphs, which were introduced into different microhabitats of the forest biotope in 36 groups each having 50–2,000 individuals, between December 1963 and June 1965, were studied for their behavioral pattern. Thirty-two groups were released during dry season from October to May, the period of nymphal activity, and four were released during rainy season. When released, engorged nymphs crawled at random in all directions seeking a way into the deeper layers of litter. Underneath the litter they settled at different depths in litter and humus. They could be traced as deep as 2.5 cm in the humus. The activity ceased 24–48 h after the release. When disturbed by lifting the litter cover and exposing to light, they resumed their activity, which ceased when they again found a shelter. The activity totally ceased when they entered the molting phase after 7–10 days.

During dry months the nymphs settled in the deeper layers of litter or humus. They were seen in the crevices, cavities in the lumps of mud, and between decomposing leaves. During rainy season they were traced in the superficial layers of the litter. On grazing land they settled in the crevices, but were either desiccated during dry months or were washed away during monsoon months. The dispersal was as far as 30 cm from the point of release. Forest ants, particularly from group nos. 5, 7, and 10, predated a large number of nymphs. Nymphs molted into adults 24–35 days after release. Each batch took 2–5 days to complete the molting.

Newly emerged ticks: Adults molted between December 1963 and May 1964 remained dormant during the entire dry season under the litter. Clusters of two to ten individuals were observed under individual leaves in the litter. However, a minor activity was observed during summer showers: ticks resting under deeper layers of litter and humus crawled up to the superficial layers. No significant lateral dispersal or emergence from the litter was observed. Similarly, adults from nymphs released from November 1964 to March 1965 spend the entire dry period in a state of dormancy under the litter. However, two to five individuals from each batch emerged from the litter and climbed on the plants after a heavy pre-monsoon shower. Adults that molted from June to mid-November invariably dispersed and climbed on the plants immediately after molting.

Dormant ticks spent the dry months in a state of repose with the legs folded and clinging to the undersurface of the litter leaves. When the leaves were turned upside down and exposed to light they became active within 1−3 min, crawled toward the undersurface, and again came to rest. Under direct sunlight, activation was immediate and the ticks immediately rushed toward darker places.

Invariably the dormant ticks became active after the first few monsoon rains. Ticks lying under thick forest canopy became active a few days earlier than those lying under thin canopy or under direct sunlight. Gradually they emerged from the litter and settled on the undergrowth vegetation. Total displacement was observed 10−40 days after the onset of activity. Initially ticks were found on plants of all heights up to 150 cm, but later the ticks settled on smaller plants (2−15 cm), descended, spread out, and settled on larger plants.

H. spinigera *as a Reservoir of KFD*[46,115]

H. spinigera is the principal vector of KFD, the nymphal stage being the one biting the humans and transmitting the virus. Both epizootics and epidemics coincide with the nymphal prevalence and activity. Studies have shown that virus can persist in the unfed stages through rainy seasons under the natural conditions. Survival of the nymphs through rainy season suggests the carryover of the virus by the infected nymphs through the rainy season, enabling the virus to start a new cycle in the following season. Field studies have shown that unfed ticks can carry the virus from one year to the next. Viral persistence in ticks suggests that the virus can be held in an unapparent state by infected ticks, a significant aspect of their role as reservoir of vertebrate infection.

Human Infestation with H. spinigera[15,46,128,129]

In KFD area, 95% of the isolations are from *Haemaphysalis* (see Table 2.3). *H. spinigera* is the most predominant species and has also yielded the maximum number of isolates. This species is considered to be the chief vector of the disease. A yearlong study, in which 4,668 persons were examined, had revealed that *H. spinigera* was the predominant species infecting man. There was an average of 2.5 ticks per infected person.

Table 2.3 KFD Virus Isolation from Questing Ticks Collected by Hand picking and Dragging (1961–1972)

Species	Number of Ticks			Number of Isolations		
	Larvae	Nymphs	Adults	Larvae	Nymphs	Adults
H. spinigera	19	253,946	88,624	0	479	359
H. turturis	41	123,738	53,645	1[a]	431	265
H. kinneari	81	12,597	26,646	2[a]	26	49
H. kyasanurensis	21	14,712	629	1[a]	41	4
H. wellingtoni	0	11,754	401	0	17	1
H. bispinosa	1	2,907	702	0	11	0
H. minuta[a]	0	2,113	104	0	6	0
H. cuspidata[a]	0	1,517	509	0	11	2
H. aculeata[a]	0	220	10	0	0	0
H. intermedia[a]	0	0	181	0	0	0
H. cornigera[a]	0	0	136	0	0	0
H. mixed species	18108	13,386	928	0	30	0
Total	18271	436,890	172,515	4	1,052	680

[a]KFD virus had been isolated from these species after 1972.

Disease Relationship[7,23,109,130,132,134,135,145]

This is the most predominant species among *Haemaphysalis* in the KFD area with the maximum number of KFD isolations. As many as 479 isolations from nymphs and 359 isolations from adults have been recorded during the period from 1961 to 1972 from the KFD area. There is no report of KFD isolation from larva.[132,135,145] Under laboratory studies, the transstadial as well as transovarial transmission of KFD virus by *H. spinigera* ticks has been proved.[130] In addition to KFD, another arbovirus called Kaisodi has also been isolated from this species. From 1965 to 1970, as many as 77 and 52 isolates of Kaisodi virus have been reported from nymphs and adults, respectively.[5,109,125]

In another laboratory study, unfed *H. spinigera* was infected by parenteral inoculation of Ganjam virus. *H. spinigera* distinctly supported the multiplication of the virus as progressively rising titers were observed in ticks at various intervals.

Changes in the Ecosystem and Spread of H. spinigera[16,20]

Boshell (1969) reported the possible emergence of KFD in the forest biotope in Shimoga district, Karnataka, India, due to ecologic changes. Formation of interphases due to clearance of forest patches is said to have triggered the multiplication of tick vector, *H. spinigera*, resulting in the outbreak of KFD for the first time in this forest. Out of thousands of tick specimens belonging to 15 species of *Haemaphysalis* collected from KFD area for two decades, i.e., 1957–1977 in connection with different ecological studies, *H. spinigera* has been found to be the predominant species. This species has been more prevalent in deciduous and semi-deciduous forest than in evergreen or semi-evergreen forests.

Hosts[16,47,93,114,117,119,120,139]

Immature stages: Indian gray mongoose, mouse deer, sambar deer, black-naped hare, crested porcupine, Asiatic jackal, toddy cat, peacock, white throated ground thrush bird, jungle fowl, jungle myna, buffalo, shrew, jungle babbler, crow pheasant, red vented bulbul, rat (1, 5, 6, 9), sheep, tiger, jungle cat, langur monkey, cattle, bonnet monkey, panther, wild dog, squirrel, man, and from flag dragging and handpicking.

Adults: Asiatic jackal, Sambar deer, cattle, wild animals, buffalo, tiger, leopard, bear, Indian bison, jungle cat, small Indian civet, black-naped hare, mouse deer, wild dog, bullock, rat, jungle fowl, jungle myna, peacock, white throated ground thrush bird, and from flag dragging and handpicking under stones and bushes.

Distribution[47,93,113,114,117,120]

India (Karnataka, Andaman and Nicobar islands, Andhra Pradesh, Goa, Kerala, Madhya Pradesh, Maharashtra, West Bengal, Meghalaya, Orissa, Tamil Nadu), Sri Lanka.

Ecology[47,118]

H. spinigera is essentially a forest-inhabiting species, where it parasitizes on a variety of mammals and birds. *H. spinigera* is the commonest of all ticks found in ground drags and on vegetation in the KFD area. All the areas where the species has been reported are forests and the vegetation ranges from wet evergreen to moist deciduous. The rainfall is from heavy to moderate. Boshell (1969) has given a detailed ecologic description of the KFD area in Karnataka state. Other areas where the species have been reported also appear to have a more or less similar ecology.

2.7.2 Haemaphysalis shimoga

This species, collected from KFD area, was first described as a new subspecies of *Haemaphysalis cornigera* by Hoogstraal and Trapido in 1964. Later on, the subspecies status was raised to species level (Figure 2.58).[142]

Male (Figures 2.59 and 2.60)[142]

Mean body length measures 3.1 mm, width 1.8 mm.

Capitulum: Basis capituli are rectangular, 1.8–2.0 times as wide as long; cornua are three-fourths as long as the base of basis capitulum, distinctly longer than their own basal width, sharply tapering. Palpi are short and widely salient. Hypostome is

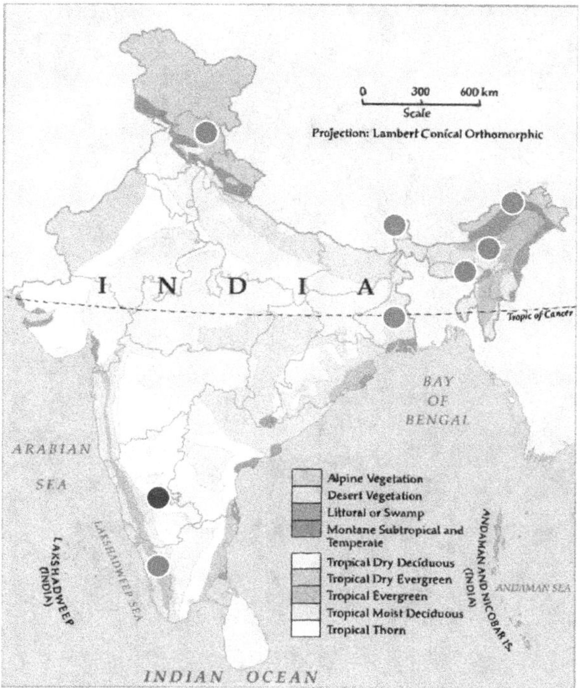

Figure 2.58 Distribution of *H. shimoga* in India. (●) Place of first record.
Map not to scale.

as long as palpi, lateral margins almost straight (or very slightly convex); dental formula is 5/5, with approximately 8 (inner)–14 (outer) denticles in a file.

Scutum: Scutum approximately 1.6 times as long as wide, lateral margins convex, widest at the level of coxa IV. Lateral grooves are very shallow, wide to the level between coxa III and IV, enclose one or no festoons. Cervical grooves are deep, narrow, parallel, and short. Punctations mostly small, numerous, discrete, noncontiguous, and scattered over entire surface including lateral ridges and festoons. Festoons are eleven in number are long, and distinctly separated.

Legs: Coxa are each (I to III) with one or (IV) two spurs.

Female (Figures 2.59 and 2.60)[142]

Overall mean body length is approximately 3.2 mm, width 1.8 mm.

Capitulum: Basis capitulum is with outline similar to that of male but with cornua slightly shorter than those of the male; porose areas are ovate, widely spaced, and of moderate size. Palpi widely salient basally, each palpus approximately 1.5 times as long as basal width. Hypostome similar to that of male except that it is approximately 3 times as long as wide; dental formula is 4/4, with 13–17 denticles in a file.

Scutum: Scutum is sub-circular, margins are rounded. Cervical grooves are deep, narrow, and parallel, on the apical half of scutum. Punctations are small, rare

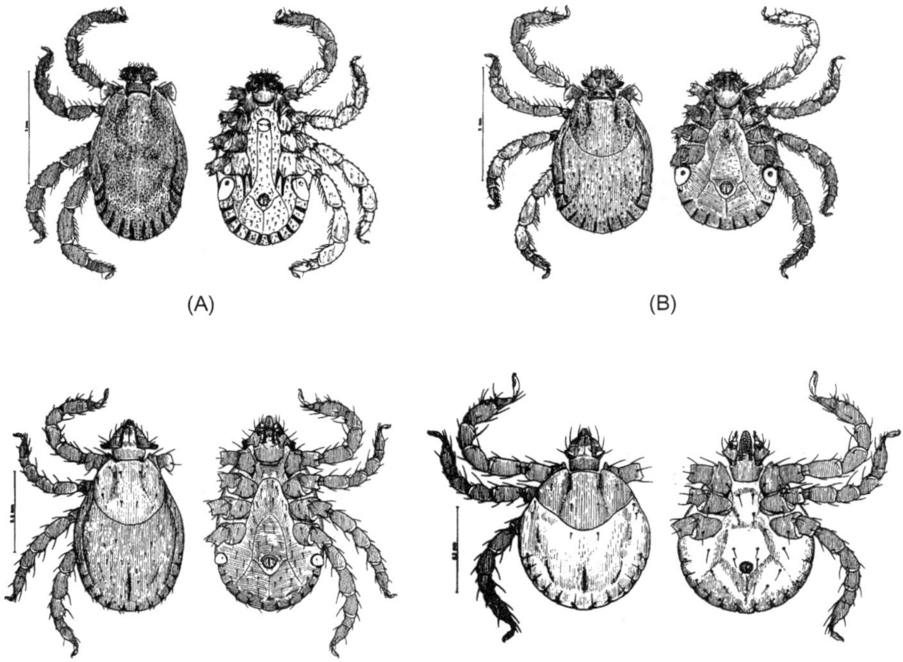

(A) (B) (C) (D)

Figure 2.59 *Haemaphysalis shimoga.* (A) Male, dorsal and ventral side; (B) female, dorsal and ventral side; (C) nymph, dorsal and ventral side; (D) larva, dorsal and dentral side. *Source*: Trapido and Hoogstraal (1964), Journal of Parasitology (vol 50(2)), Allen press Publishing Services.

in central field, and mixed small and medium size and moderately numerous in lateral fields.

Nymph (Figures 2.59 and 2.60)[142]

Total body length is approximately 1.3 mm.

Capitulum: Basis capituli are approximately 2.5 times as wide as long; cornua (dorsally) are very small. Suprainternal seta is single; infrainternal setae number two, long, slender, undifferentiated. Hypostome is longer than palpi; lateral margins are slightly convex; dental formula is 2/2, with 9–12 broad denticles in a file.

Scutum: Scutum is sub-circular, margins rounded. Cervical grooves are anteriorly deep, narrow, parallel pits; posteriorly shallow. Punctations number approximately 12.

Larva (Figures 2.59 and 2.60)[142]

Total body length is approximately 0.60 mm, width 0.45 mm.

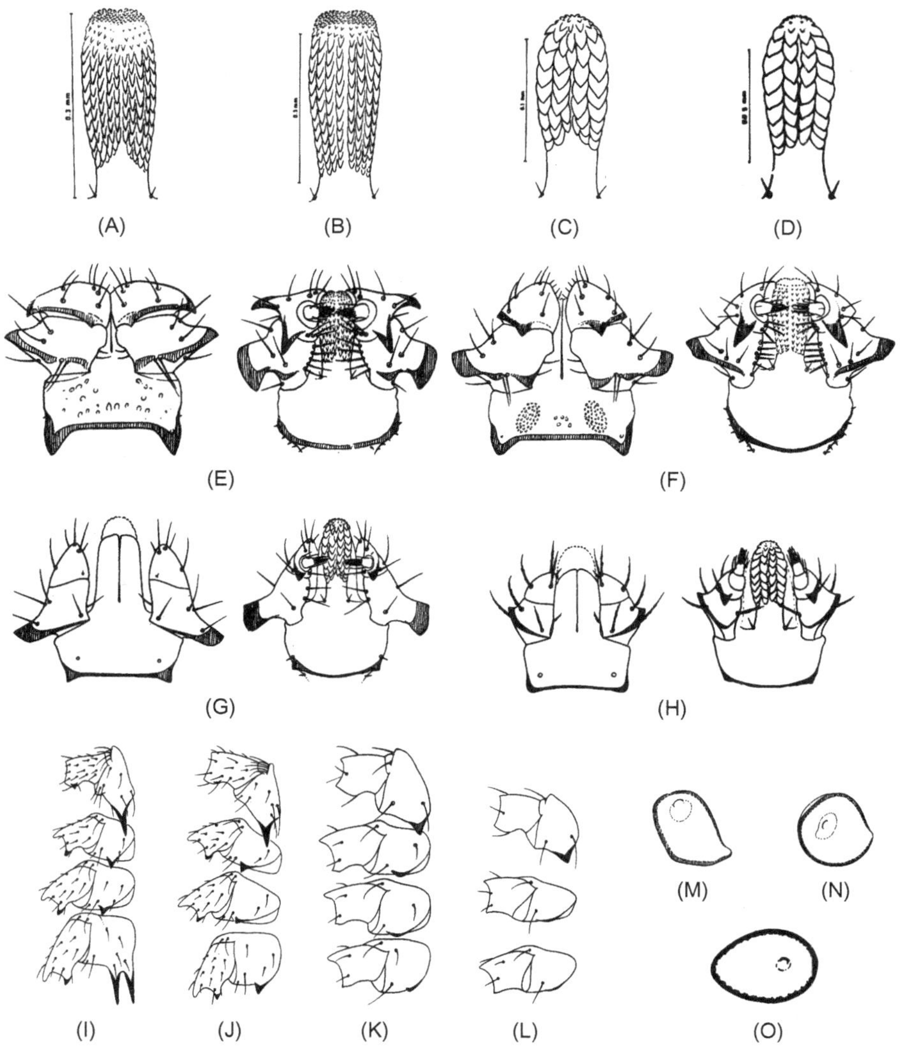

Figure 2.60 (A–D) Hypostome of male, female, nymph, and larva, respectively; (E–H) capitulum of male, female, nymph, and larva, respectively; (I–L) coxa of male, female, nymph, and larva, respectively; (M–O) spiracular plates of male, female, and nymph, respectively.
Source: NIV. unpublished drawings.

Capitulum: Basis capituli are with straight sides, twice as wide as long; dorsal cornua are very small (may be reduced to rounded margins in mounted specimens); ventral cornua in each basolateral corner are very small; a pair of small posthypostomal bristles is present. Palpi are approximately 1.3 times as long as wide.

Related Species[142]

Males of this species are differentiated from *H. cornigera* by the long, sub-equal spurs of coxa IV (inner spur much longer than outer), the form of the basoventral palpal margin, and the form of the lateral wings of palpal segment III. The Indian species is also readily separated from the Formosan *H. cornigera taivana* Sugimoto, which has a definite median dorsobasal spur on palpal segment III; long, distinct lateral grooves on the scutum; a larger spur on coxa III; and the two long spurs that meet apically from coxa IV.

Life History[9]

The maximum life span of adults observed under laboratory conditions (90–100% RH and at 17–28 °C) was 335 days. During laboratory rearing experiments, most of the adults attached within 24 h after release on the host cow calf. The average amount of blood imbibed by the tick was approximately 170 times that of their average unengorged body weight (2.5 mg). The repleted females had a weight range of 272–500 mg. The oviposition commenced 4–7 days after the engorgement and lasted for 18–31 days. Egg laying commenced abruptly, reached a peak between the third and the seventh day, and tapered gradually with slight fluctuations. The spent females died 6–15 days after the completion of oviposition. The loss in the initial weight after the completion of oviposition ranged from 56% to 76%. Hatching commenced 35–42 days after the initiation of oviposition. The process started with the oldest eggs laid on the first day and ended with those laid on the last day. The duration for complete hatching of a brood ranged from 13 to 31 days, and was more or less equal to the period taken for oviposition. In 10 broods observed, the maximum longevity of unfed larvae was 338 days. The larvae could be fed successfully as early as 5–6 days after hatching. The engorgement took place in two well-defined phases. The larvae stayed on the chicks in the first preparatory phase for approximately 24 h, before passing on to the second "repletion" phase prior to dropping. The larvae, which did not pass on to the second phase in the second night were carried on for the subsequent nights for repletion followed by dropping on the subsequent days. The maximum longevity of unengorged nymphs was 219 days. During the parasitic phase the nymphs followed a similar sequence of events as the larvae did. The age of the nymphs fed on the chicks ranged from 28 to 142 days. Engorged nymphs molted into adults 25–29 days after dropping (Table 2.4).

Host[26,114,142]

Immature stages: Information not known.

Adults: Sambar deer, cattle, goat, sheep, cow, bison.

Table 2.4 *Haemaphysalis shimoga* Life Cycle Under Laboratory Conditions

Stage	Phase	Number of Days	Number of Observations	Host Animal
Egg	Incubation period	35–42	30 broods	–
Larvae	Quiescent phase	5–7	2 broods	–
	Life span (maximum)	338	10 broods	–
	Parasitic phase	3–7	14 broods	Chick and *Rattus blanfordi*
	Molting phase	10–19	12 groups, 100 each	–
Nymph	Quiescent phase	1–2	2 groups, 100 each	–
	Life span (maximum)	219	6 groups, 100 each	–
	Parasitic phase	3–6	15 groups, 100–200 each	Chick
	Molting phase	25–29	8 groups, 100 each	–
Adult	Quiescent phase	20–25	2 groups, 25 each	–
	Life span (maximum)	335	2 groups, 25 each	–
Female	Parasitic phase	8–15	32 specimens	Calf
	Pre-oviposition phase	5–8	32 specimens	–
	Oviposition phase	18–31	32 specimens	–
	Post-oviposition phase	6–15	32 specimens	–

Distribution[26,113,114,142]

India (Karnataka, Kerala, Meghalaya, Himachal Pradesh, Assam, West Bengal, Arunachal Pradesh, Sikkim).

Disease Relationship

So far no KFD virus isolations have been recorded from this species of ticks, though this species has been collected repeatedly from KFD area.

2.7.3 Haemaphysalis anomala

This species was described by Warburton in 1913 from a single male from wolf in Chota Nagpur plateau, Jharkhand state (previously part of Bihar state), India. This species has an unspecialized/anomalous male palpal outline, which deviates from general characters of *cornigera* group; hence, perhaps the name *anomala* (Figure 2.61).

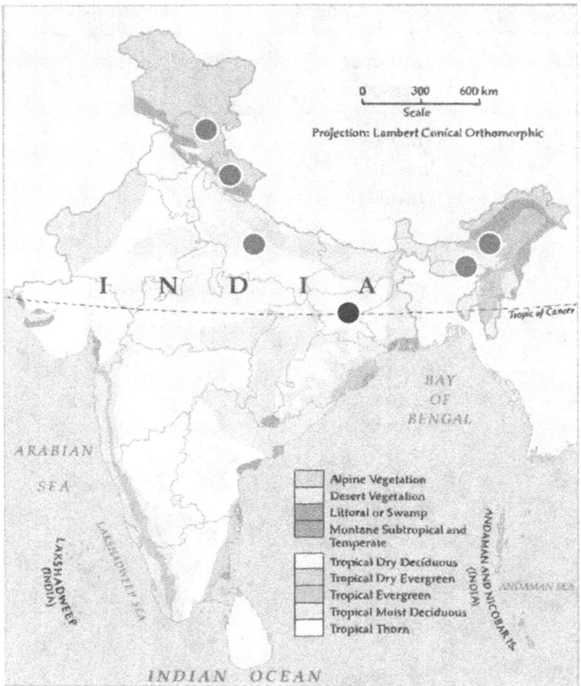

Figure 2.61 Distribution of *H. anomala* in India. (●) Place of first record.
Map not to scale.

Male (Figures 2.62 and 2.63)[67,127]

The mean body length is approximately 2.70 mm (2.50–2.96), width 1.60 mm (1.52–1.70).

Capitulum: Basis capitulum is dorsally approximately 1.8 times as wide as long; external margins are sub-parallel (slightly indented at mid-length); posterior margin is between cornua straight; cornua are triangular, as wide as long, with pointed apices, and one-half as long as the base of basis capituli. Infrainternal setae number five. Hypostome is not as long as palpi. Dental formula is 5/5, with 8 (internal)–11 (external) narrowly elongate denticles in a file.

Scutum: Cervical grooves are deep, comma-shaped pits, and shallow. Lateral grooves are narrow, deep, and long; extend to anterior one-fourth or one-fifth of scutal length and enclose anterior pair of festoons. Punctations are numerous, small, and discrete, tending to be slightly larger and more widely spaced anteriorly. Festoons number 11. Spiracular plates sub-quadrate; dorsal projection triangular, large, elongate, and pointed dorsally.

Legs: Coxa are each with characteristic spurs. Spur of I needlelike, overlapping anterior one-half of coxa II.

Female (Figures 2.62 and 2.63)[67,127]

Overall body length is approximately 3.0 mm, width 1.6 mm.

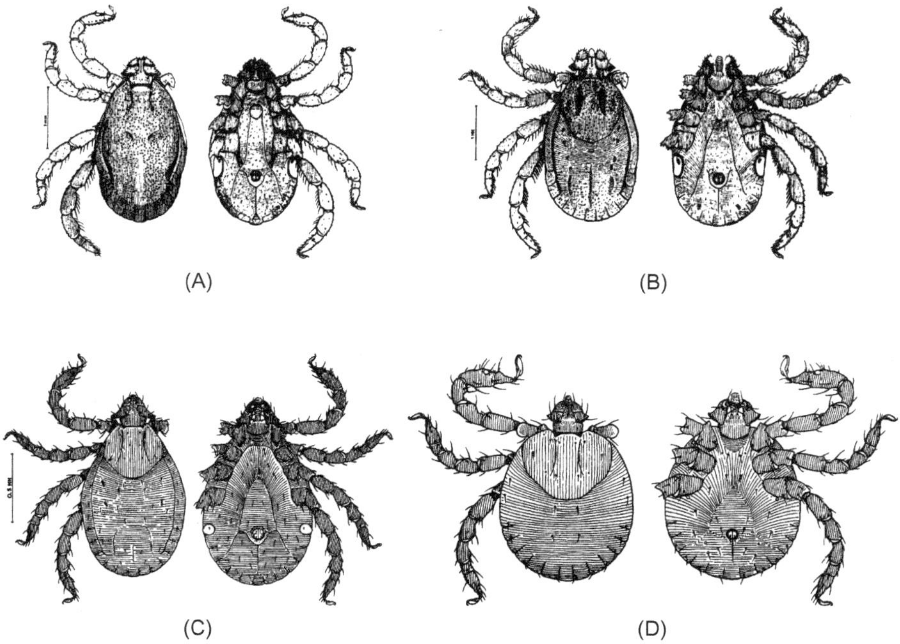

Figure 2.62 *Haemaphysalis anomala*. (A) Male, dorsal and ventral side; (B) female, dorsal and ventral side; (C) nymph, dorsal and ventral side; (D) larva, dorsal and ventral side. *Source*: Hoogstraal et al. (1967), Journal of Parasitology (vol 53(1)), Allen press Publishing Services.

Capitulum: Basis capitulum is dorsally twice as wide as long; cornua are very slightly less than one-half as long as the base of basis capituli; porose areas are elongate, ovate, tilted, and widely spaced. Palpi are each 1.4 times as long as wide. Infrainternal setae number eight. Hypostome is with 4/4 dental formula, with approximately 11 (internal) to 13 (external) denticles in a file.

Scutum: Scutum is approximately 1.1 times as long as wide. Cervical grooves are deep, narrow, sub-parallel, extending beyond scutal mid-length. Punctations are slightly larger but otherwise similar to those of male anteriorly. Genital operculum is short, wide, and margins gradually rounded. Spiracular plates are large and subcircular, and dorsal projection is small.

Nymph (Figures 2.62 and 2.63)[80]

Total body length is approximately 1.3 mm, breadth 0.7 mm.

Capitulum: Basis capitulum is dorsally approximately 2 times as broad as long; cornua are triangular, approximately one-third as long as the base of basis capitulum. Palpi are broadly salient. Hypostome is slightly longer than palpi; dental formula is 2/2, with seven or eight denticles in a file.

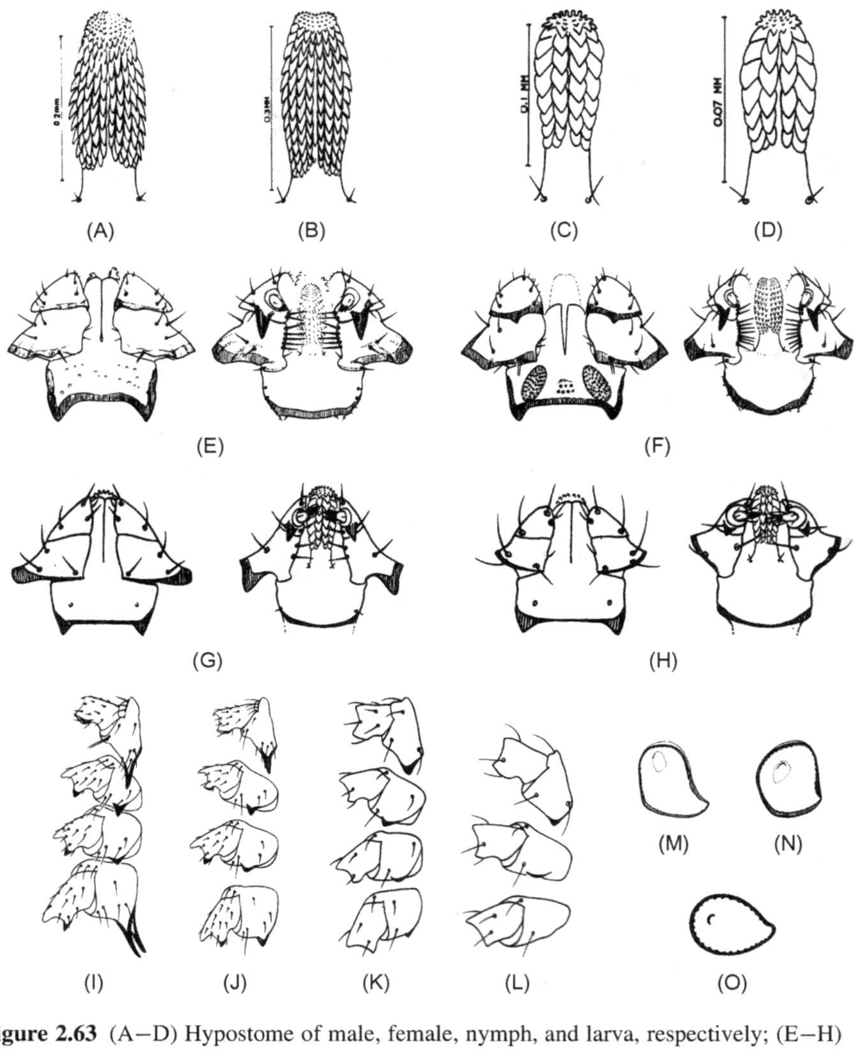

Figure 2.63 (A–D) Hypostome of male, female, nymph, and larva, respectively; (E–H) capitulum of male, female, nymph, and larva, respectively; (I–L) coxa of male, female, nymph, and larva, respectively; (M–O) spiracular plates of male, female, and nymph, respectively.
Source: NIV. unpublished drawings.

Scutum: Scutum is 1.10–1.25 times as broad as long. Cervical grooves are narrow, extending to posterolateral margins. Punctations are obscure, five or six pairs.

Legs: Legs are moderately long and narrow. Coxa are each with triangular or sub-triangular spur.

Larva (Figures 2.62 and 2.63)[80]

Overall body length is approximately 0.65 mm, breadth 0.45 mm.

Capitulum: Basis capitulum is dorsally approximately 2 times as broad as long; external margins are straight; cornua are broadly triangular, one-third as long as the base of basis capitulum. Palpi are broadly campanulate, each palpus about 1.3 times as long as broad. Hypostome is as in nymph, except slightly broader and with one or two fewer denticles in each file.

Scutum: Scutum is about 1.3 times as broad as long. Cervical grooves are narrow and extend beyond mid-length. Punctations are obscure.

Related Species[80]

This species belongs to *cornigera* group. The *H. cornigera* group consists of 10 taxa occurring in certain areas of the Oriental, Palearctic, Australian, and Ethiopian faunal regions. Adults are characterized by broadly salient palpi, mostly with unusual modifications in the outline and posterodorsal spur of segment III, and by the presence of one or two lanceolate spurs of the male coxa IV. Spurs in coxa IV in *H. anomala* are closely spaced and approximately equal in size, whereas in *H. shimoga* it is well spaced, the outer one is needlelike and long while the inner one is smaller.

The related species among the Indian *Haemaphysalis* is *H. shimoga*. The immature stages of *shimoga* and *anomala* are much alike, but the cornua of *anomala* are at least twice as large as in *shimoga* and the scutum of each stage of *anomala* is more elongate. Adults of these species differ more markedly than nymphs and larvae.

Ecology[80]

The species has been reported from moist deciduous forests between 800 and 3,400 ft altitude in western Himalayan foothills and valleys. Temperature reaches 40 °C in summer and 1 °C, or rarely slightly less, in winter. Much of the total annual rain (125–300 cm) falls during the monsoons, which begin in late June or early July and continue to mid-September. Most adult ticks were taken during the monsoons; all immatures were collected in pre-monsoon and post-monsoon months. *H. anomala* survived the faunal changeover in these foothill and valley forests. Adult and immature ticks apparently thrive on the large and small vertebrate animals now inhabiting this environment.

Life Cycle[80]

The three-host type of life cycle was completed in 129 days (parent female feeding to F_1 female feeding) under the laboratory conditions. The original parent female was taken fully engorged from a cow at Satyanarayan, Uttar Pradesh.

Hosts[67,80,120]

Immature stages: Rat, coucal, crow pheasant, and from flag dragging through vegetation.

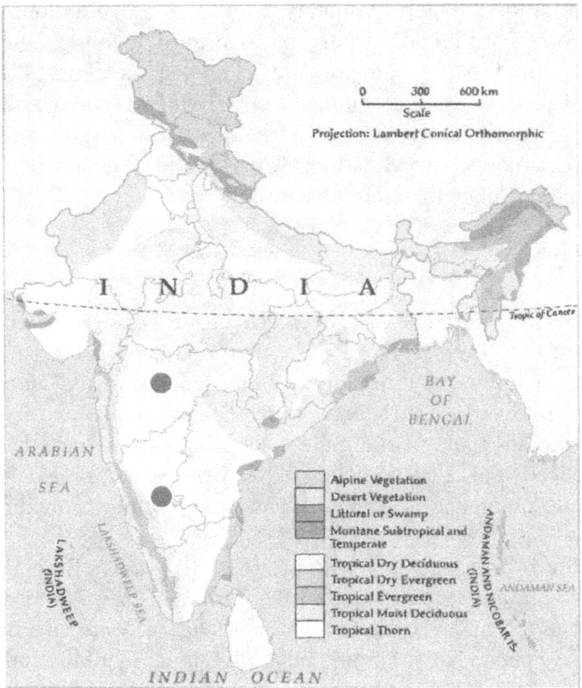

Figure 2.64 Distribution of *H. cuspidata* in India. (●) Place of first record.
Map not to scale.

Adults: Cow, calf, buffalo, coucal, barking deer, leopard, rats, wolf, man, cattle, goat, rodents, sambar deer, wild ox, and from flag dragging through vegetation.

Distribution[67,80,120]

India (Assam, Jharkhand, Himachal Pradesh, Uttar Pradesh, Uttarakhand, Jammu and Kashmir, Meghalaya), Nepal, Sri Lanka, Thailand, Cambodia, Vietnam.

2.7.4 Haemaphysalis cuspidata

Haemaphysalis cuspidata was first described by Warburton in 1910 from the specimens collected from mouse deer and palm civet in Sri Lanka (Figure 2.64).

Male (Figures 2.65 and 2.66)[104]

Mean length is about 1.2–1.3 mm.

Capitulum: Cornua are very long and sharply pointed, and triangular. Palpal segment II is with moderate salience and without spur on both dorsal and ventral side. Infrainternal setae number four. Palpal segment III shows a prominent, elongately triangular, sharply pointed spur on dorsal side extending beyond palpal segment II, as well as a retroverted, sharply pointed spur, extending beyond palpal

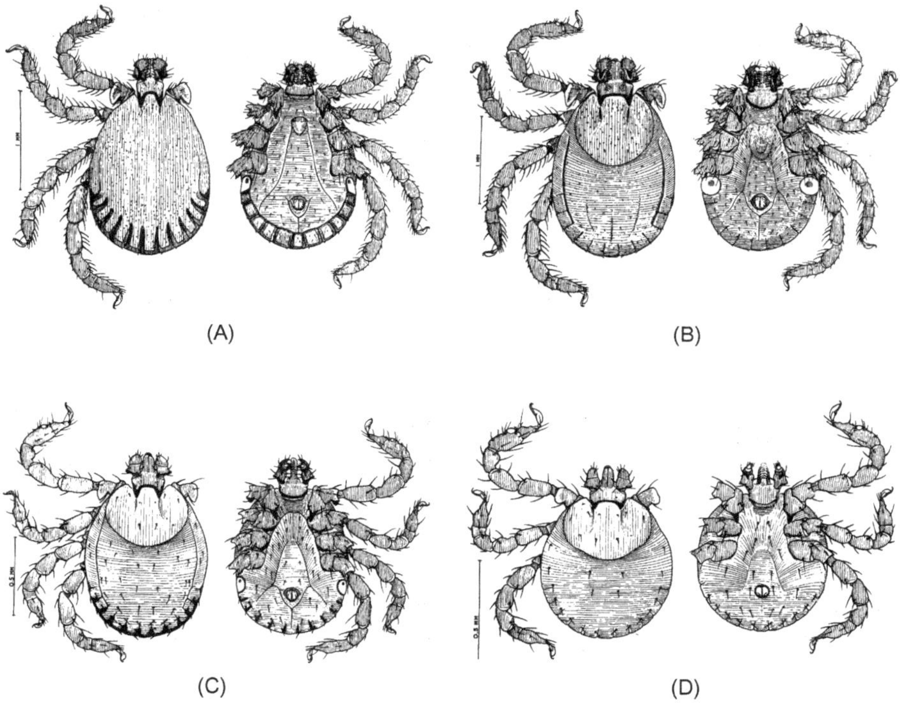

Figure 2.65 *Haemaphysalis cuspidata.* (A) Male, dorsal and ventral side; (B) female, dorsal and ventral side; (C) nymph, dorsal and ventral side; (D) larva, dorsal and ventral side. *Source*: NIV. unpublished drawings.

segment II on ventral side. The dental formula is 4/4, with seven or eight denticles in each row.

Scutum: Scutum is ovoid in shape and is about 1 mm in length. Punctations are very small and are not deep, scattered all over the scutum. Festoons are very deeply marked, measuring 11 in number. Lateral grooves are short not enclosing any festoons on either side and extending up to third coxa on each side.

Legs: Coxa I bears a very prominent, sharp, and pointed spur, overlapping anterior portion of coxa II. Coxa II, III, and IV also show spur but are smaller than coxa I and diminish from II to IV. Trochanter I bears a blunt spatulate spur, while trochanters II, III, and IV show crest-like outgrowth.

Female (Figures 2.65 and 2.66)[104,Redescribed]

Average body length is about 1.3–1.5 mm.

Capitulum: Cornua are triangular and pointed but not as big as in male, length of cornua is approximately half the length of base of basis. Palpal segment II shows low salience and no spur on either side. Infrainternal setae number four. Palpal segment III shows a big spur overlapping more than half of the palpal segment II on

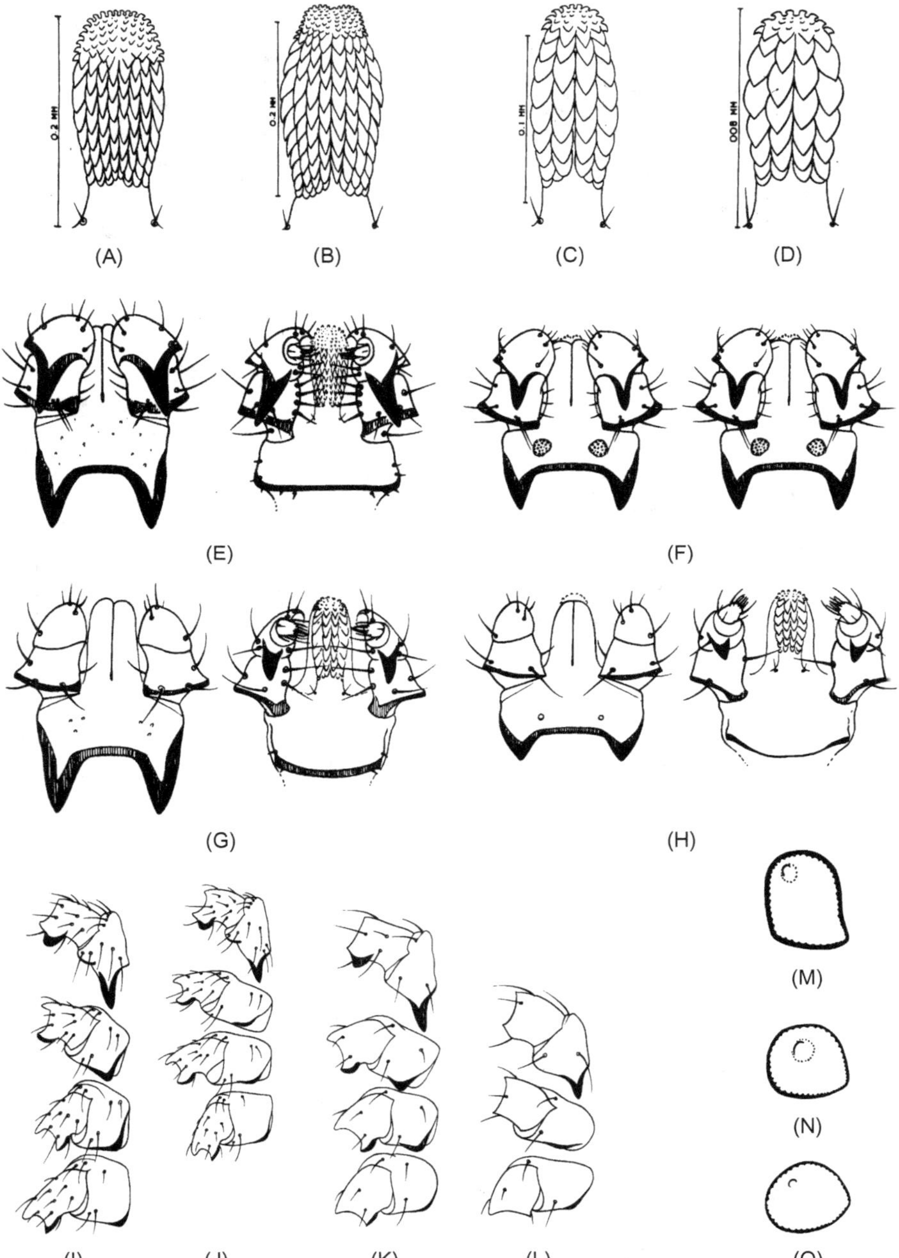

Figure 2.66 (A–D) Hypostome of male, female, nymph, and larva, respectively; (E–H) capitulum of male, female, nymph, and larva, respectively; (I–L) coxa of male, female, nymph, and larva, respectively; (M–O) spiracular plates of male, female, and nymph, respectively. *Source*: NIV. unpublished drawings.

dorsal as well as huge and pointed spur, protruding over the palpal segment II reaching to base of palpal segment I on ventral side. Dental formula is 4/4, containing eight or nine denticles in each row.

Legs: Coxa I as in male shows an elongated, pointed sharp spur, extending along half of the coxa II. Coxa II to IV show no spur but a crest-like outgrowth toward the posterior side. Trochanter I shows a spatulate-shaped spur but on trochanters II to IV there are a crest-like outgrowth like that of the coxa.

Nymph (Figures 2.65 and 2.66)[104]

The average length is about 1 mm.

Capitulum: Cornua are about the same length of that of basis capitulum. Basis capitulum shows few punctations. Palpal segment II shows very less salience and no spur on either side. Two infrainternal setae are present on ventrointernal side of the palpal segment II. Palpal segment III shows no spur on dorsal side but a small, triangular, pointed spur is present on ventral side extending to half of palpal segment II. Dental formula is 2/2, containing eight or nine denticles in each row.

Legs: Coxa I has a big, bluntly pointed triangular spur, extending up to half of coxa II. Coxa II and III show small spur-like outgrowths. Coxa IV shows a crest-like structure. Trochanter I shows a small spur, while trochanters II to IV show only crest-like outgrowths.

Larva (Figures 2.65 and 2.66)[104, Redescribed]

It is about 1 mm in length.

Capitulum: Cornua are very short and with triangular pointed apices, about half the length of basis capitulum. Palps are quite well spaced or widely arranged. Palpal segment II is with one infrainternal seta and shows very low salience. Palpal segment III on dorsal side shows no spur but on ventral side shows a short, triangular spur extending just beyond the intersegmental margin between segment II and III. Dental formula is 2/2, with six denticles in each row.

Legs: Coxa I shows a small, triangular spur extending just anterior to coxa II, while coxa II and III show no spur but a crest-like structure. Trochanters show no spur or crest-like structure.

Life History of H. cuspidata[11]

Bhat et al. in 1974 studied the life history of this species. The minimum duration of a total life cycle under the laboratory condition was observed to be about 120–784 days at 90–100% RH and at room temperature between 17 and 27 °C.

The maximum life span of adults was 317 days. Most of the hungry (questing) adults attached themselves to the experimental hosts within 48 h after putting them on the respective hosts. Copulating pairs were observed on the fifth day after

Table 2.5 Life Cycle of *Haemaphysalis cuspidata*

Stage	Phase	Duration (in days)
Female[a]	Parasitic	9–15
	Pre-oviposition	8–18
	Oviposition	25–46
	Post-oviposition	1–8
Egg	Incubation	40–55
Larva[b]	Quiescent	3–5
	Life span (maximum)	156
	Parasitic	2–5
	Molting	17–21
Nymph[b]	Quiescent	3–5
	Life span (maximum)	121
	Parasitic	2–6
	Molting	19–24
Adult	Quiescent	5–10
	Life span (maximum)	317

[a]Fed on man.
[b]Fed on chick.

release. Females repleted and dropped between the 9th and 15th day (from 10 to 22 h). Two rather arbitrarily defined phases of repletion could be recognized. The first "preparatory phase" lasted until the last but one day of parasitic phase. The ticks changed their original brown color into cream. The copulation was generally observed at the end of this phase. The second "repletion phase," which terminated with detachment, was of short duration, ranging from 12 to 36 h. The average weight of blood ingested by the females was approximately 150 times that of their average unengorged body weight (1 mg). The fully repleted females had a weight range of 105–190 mg. Males fed exclusively on tissue fluid. The detachment of females resulted in moderate hemorrhage at the site of attachment. The oviposition commenced 8–18 days after detachment from the host, and lasted for 25–46 days. Oviposition commenced suddenly, reached a peak between the 2nd and the 15th day, and terminated gradually. The loss in weight after oviposition ranged from 60% to 75% with an average of 67%; 53% of the entire weight loss was accounted for the weight of the eggs. The total number of eggs laid ranged from 1,014 to 2,127 with an average of 1,590 eggs/female. The approximate weight of eggs laid by one female was 50 mg. The females died within 8 days after termination of oviposition. Hatching started 40–55 days after the commencement of oviposition. It commenced with the oldest eggs laid on the first day and sequentially ended with those laid on the last day. The duration for hatching of a brood ranged from 25 to 46 days, and was more or less equal to the period taken for oviposition. The maximum longevity of unfed larvae was about 156 days. Larvae could be fed successfully as early as 5–6 days after hatching. The time taken for engorgement was

2–5 days. Engorged larvae molted into nymphs 17–21 days after dropping. The observed maximum longevity of unengorged nymphs was 121 days. During their parasitic phase of 2–6 days they followed a similar sequence of events as the larvae (Table 2.5).

Hosts[11,114,119,139,143]

Immature stages: Mouse deer, crested porcupine, palm civet, gray jungle fowl, magpie robin, house shrew, Malabar gray hornbill, bonnet monkey, langur monkey, crow pheasant, cattle, buffalo, from vegetation of KFD area of Karnataka (India).

Adults: Rat, shrew, jackal, mouse deer, wild mammals, leopard, black-naped hare, bird species, palm civet, toddy cat, gray mongoose, crow pheasant, from vegetation in KFD area of Karnataka (India).

Distribution[11,113,114]

India (Karnataka, Maharashtra), Sri Lanka.

Disease Relationship[11]

KFD virus isolations, two from adults, twelve from nymphs off vegetation, and two from nymphs attached to langur monkey, have been recorded from the KFD field collections during the period of 1961 to 1972.

2.7.5 Haemaphysalis aculeata

This species is characterized by palpae with almost parallel borders, very well-developed cornua that are nearly as long as the basis capitulum. This species was first described by Lavarra in 1904 from a male specimen taken from mouse deer in the East Indies (Southeast Asian islands). Later on, female was described by Warburton in 1909 as *H. longipalpis* specimens taken from mouse deer at Colombo (Figure 2.67).

Male (Figures 2.68 and 2.69)[104,127]

It is about 1.5–1.8 mm in length and is oval in shape.

Capitulum: Capitulum is very elongated and is about 0.5 mm in length with huge, prominent, triangular, and pointed cornua. The length of cornua is more than the length of the basis capitulum. Palpal segment II shows four infrainternal setae on ventrointernal side. Palpal segment III is with huge spurs on both dorsal and ventral sides; the spurs are huge, heavy, bluntly triangular, and extending up to the base of palpal segment II. Hypostome is about 0.2 mm in length, and dental formula is 5/5, with eight or nine denticles per file.

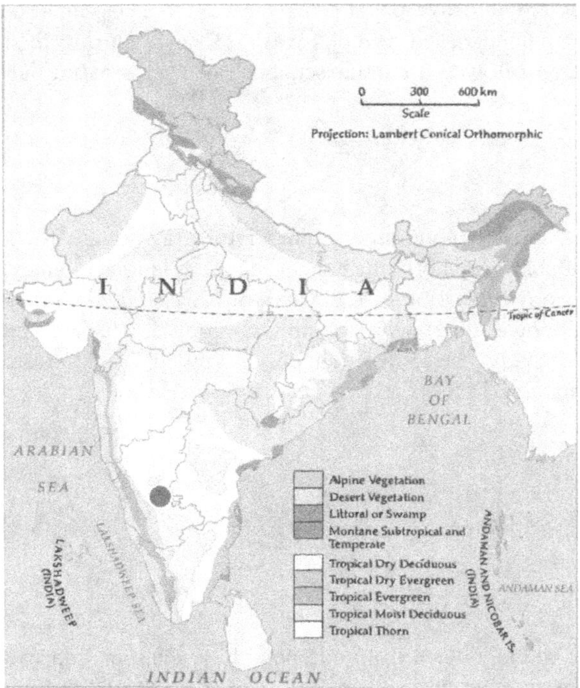

Figure 2.67 Distribution of *H. aculeata* in India. (●) Place of first record. Map not to scale.

Scutum: Cervical grooves are small and short, enclosing one festoon on each side. Festoons are deeply marked and are about equal in length and breadth.

Legs: Coxa I is showing a spatulate spur while coxa II to IV show no spur or crest-like structure. Trochanter I is having a large spatulate spur, while the trochanters II to IV show no spur or crest-like structure.

Female (Figures 2.68 and 2.69)[104]

It is about 1.5–1.8 mm in length and is ovoid in shape.

Capitulum: Porose areas are oval and are small in area. Infrainternal setae are lanceolate and long and four in number. Palpal segment III is with short spur on dorsal side and a moderate spur on ventral side. On ventral side the spur is broadly triangular and is just extending beyond the intersegmental margin between palpal segment II and III. On dorsal side, the spur is moderate, triangular, and extends up to half of the palpal segment II. The dental formula is 4/4 and is with eight or nine denticles per file.

Legs: Coxa I is with a moderate, triangular, and pointed spur, extending up to the anterior part of the coxa II. No spur or crest-like structure is present on coxa II to IV. Trochanter I shows spatulate spur, while trochanters II to IV show no spur or crest-like structure.

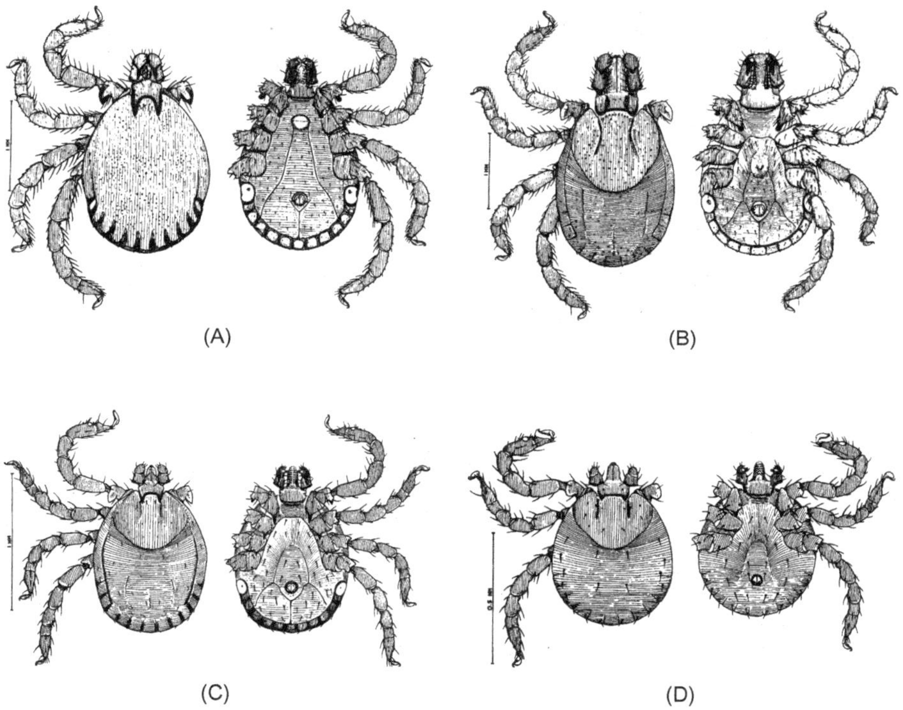

Figure 2.68 *Haemaphysalis aculeate*. (A) Male, dorsal and ventral side; (B) female, dorsal and ventral side; (C) nymph, dorsal and ventral side; (D) larva, dorsal and ventral side. *Source*: NIV. unpublished drawings.

Nymph (Figures 2.68 and 2.69)

Total length is about 1 mm and is oval in shape.

Capitulum: Remarkable cornua are present at the posterior end of the basis capitulum. Length of the cornua is about equal to that of the basis capitulum. Two long and lanceolate infrainternal setae are present. Palpal segment III is with a moderate, bluntly pointed, triangular spur on ventral side, extending up to mid-length of the palpal segment II. Hypostome is about 0.1 mm in length and is extending a little beyond the apex of the palps. The dental formula is 2/2, with seven or eight denticles per file.

Legs: Coxa I is with a moderately large, triangular, and bluntly pointed spur, extending up to anterior half of coxa II. Coxa II and III are with crest-like outgrowths while coxa IV shows no spur or crest-like outgrowth. Trochanter I has a broadly triangular short spur whereas trochanters II to IV show crest-like outgrowths.

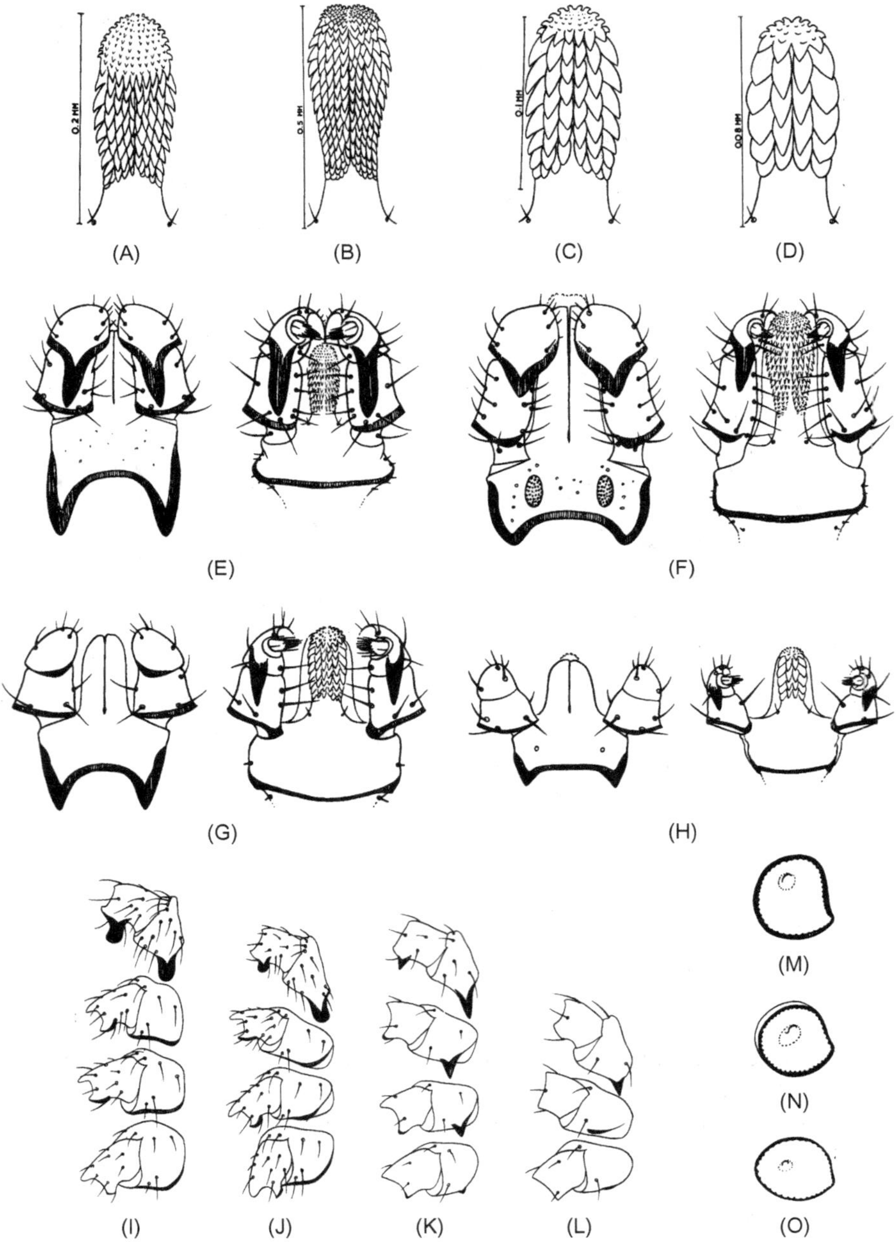

Figure 2.69 (A–D) Hypostome of male, female, nymph, and larva, respectively; (E–H) capitulum of male, female, nymph, and larva, respectively; (I–L) coxa of male, female, nymph, and larva, respectively; (M–O) spiracular plates of male, female, and nymph, respectively.
Source: NIV. unpublished drawings.

Larva (Figures 2.68 and 2.69)[Redescribed]

Larva is small and is about 0.7–0.9 mm in length.

Capitulum: The cornua are also very short, broadly triangular, and about half the length of the basis capitulum. Palps are very loosely arranged, palpal segment II is with no or very less salience. One long infrainternal seta is present on ventrointernal side of the palpal segment II. Palpal segment III shows no spur on dorsal side but a short, broadly triangular spur on ventral side, reaching up to anterior of the palpal segment II. Hypostome is short, about 0.08 mm in length, with 2/2 dental formula and with six or seven denticles per file.

Legs: Coxa I shows small, triangular spur, just reaching to anterior of the coxa II. Coxa II and III show no spur or crest-like structure. Trochanters also show no spur or crest-like structure.

Immature stages: Mouse deer, crested porcupine, langur monkey, buffalo, rat, bonnet monkey, and birds like babbler, bulbul, Myna, Indian robin, and crow pheasant.

Adults: Mouse deer, man, monkey, bird species, cow, leopard, mongoose.

Distribution[113,114,127]

India (Karnataka), Sri Lanka.

Disease Relationship

No KFD virus isolation has been reported from this species of ticks.

2.7.6 Haemaphysalis davisi

Hoogstraal et al. first described this species in 1970 from tick samples taken from domestic and wild mammals and in vegetations of Arunachal Pradesh and Sikkim in India and from Burma. This species is named after Dr. Gordon E. Davis who collected most of the material recorded from Burma (Figure 2.70).

Male (Figures 2.71 and 2.72)[70]

Length from palpal apices to posterior scutal margin is 2.55–2.93 mm (average 2.80), breadth 1.58–1.88 mm.

Capitulum: Basis capitulum is dorsally approximately 1.5 times as broad as long; external margins essentially straight (slightly indented at mid-length); cornua are broadly triangular, approximately one-third as long as the base of basis capitulum. Palpi are moderately salient posteriorly, combined breadth is approximately 1.4 times that of basis capitulum. Hypostome is slightly shorter than palpi. Dental formula is chiefly 6/6 (7/7 anteriorly), with seven to eleven denticles in a file.

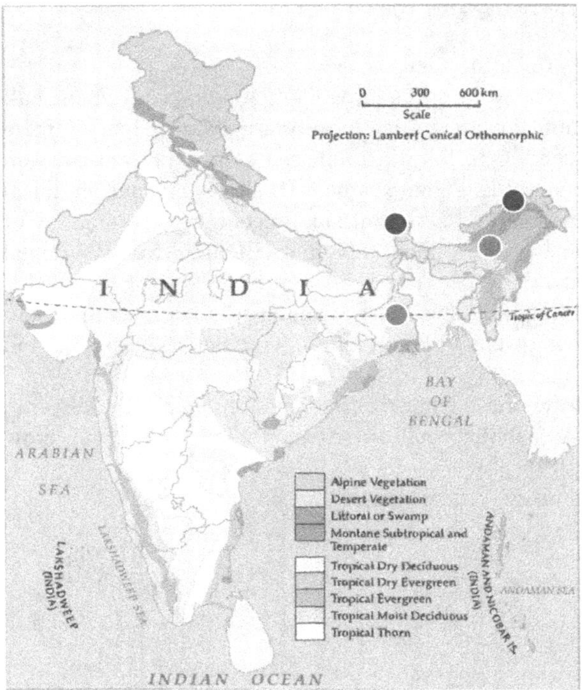

Figure 2.70 Distribution of *H. davisi* in India. (●) Place of first record.
Map not to scale.

Scutum: Scutum is approximately 1.5 times as long as broad, margins broadly rounded, broadest in the middle. Lateral grooves are narrow, defined, moderately long, extend somewhat beyond scutal mid-length, and do not enclose festoons. Cervical pits are small and converging. Cervical grooves are short and obscure. Punctations are discrete, small, relatively numerous, and regularly scattered over entire surface except in a smooth narrow posteromedian line. Festoons are narrowly elongate, and number 11. Legs are long and thin.

Legs: Coxa are each with a prominent triangular spur; spur of I is lanceolate, extending beyond anterior margin of coxa II; spurs of II to IV are broadly triangular.

Female (Figures 2.71 and 2.72)[70]

Length is about 2.40–3.30 mm, breadth 1.35–1.95 mm.

Capitulum: Basis capitulum is dorsally approximately twice as broad as length; cornua are broadly triangular, rounded apically, and approximately one-third as long as the base of basis capitulum; porose areas are large, irregularly oval, and widely spaced. Palpi outlines are quite like those of male. Hypostome is as long as palpi; dental formula is 5/5, with nine or ten denticles in a file.

Scutum: Scutum is as broad as long; greatest breadth slightly anterior of mid-length. Cervical grooves are deep, linear, slightly converging, extending beyond

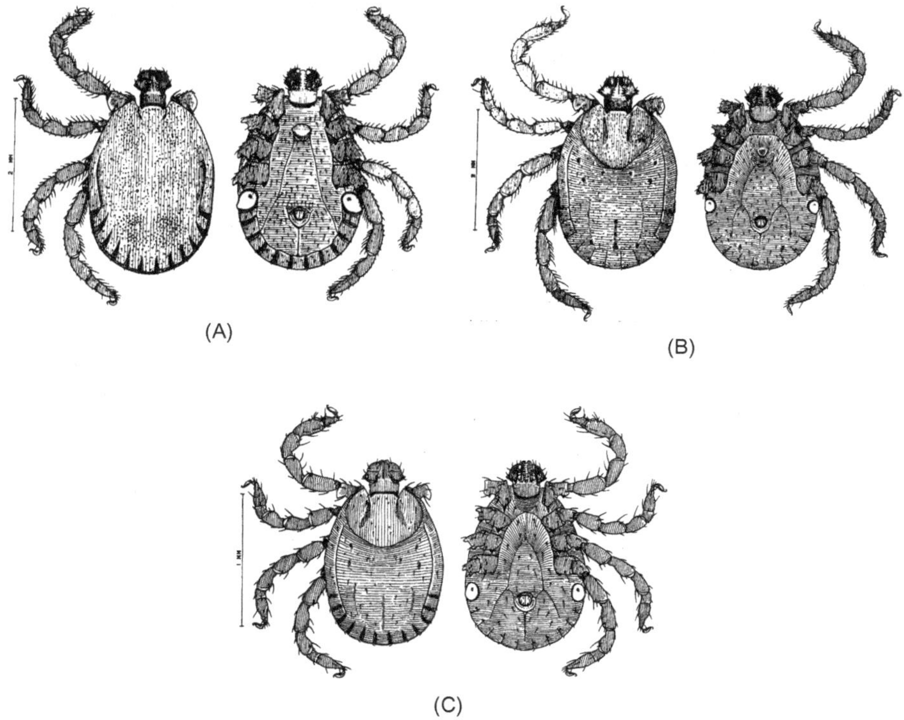

Figure 2.71 *Haemaphysalis davisi*. (A) Male, dorsal and ventral side; (B) female, dorsal and ventral side; (C) nymph, dorsal and ventral side.
Source: Hoogstraal et al. (1970), Journal of Parasitology (vol 56(3)), Allen press Publishing Services.

scutal mid-length. Punctations are as in male. Genital operculum is about four-fifths as long as anterior breadth.

Legs: Legs are approximately as in male. Coxa I is with spur somewhat shorter than in male; spurs of II and III are broad, sub-equal in length, and extending slightly beyond posterior margin of coxa; spur of II is pointed apically; spurs of III and IV are rounded apically.

Nymph (Figures 2.71 and 2.72)[70]

Length is approximately 1.6 mm, breadth 1.1 mm.

Capitulum: Basis capitulum is dorsally approximately 2 times as broad as long, margins straight; cornua are triangular, approximately one-half as long as the base of basis capituli. Palpi are campanulate, each palpus approximately 1.5 times as long as broad. Hypostome is as long as palpi; dental formula is 3/3, with seven to nine denticles in a file.

Scutum: Scutum is approximately 1.5 times as broad as long, greatest breadth at mid-length. Cervical grooves are shallow. Punctations are few and obscure.

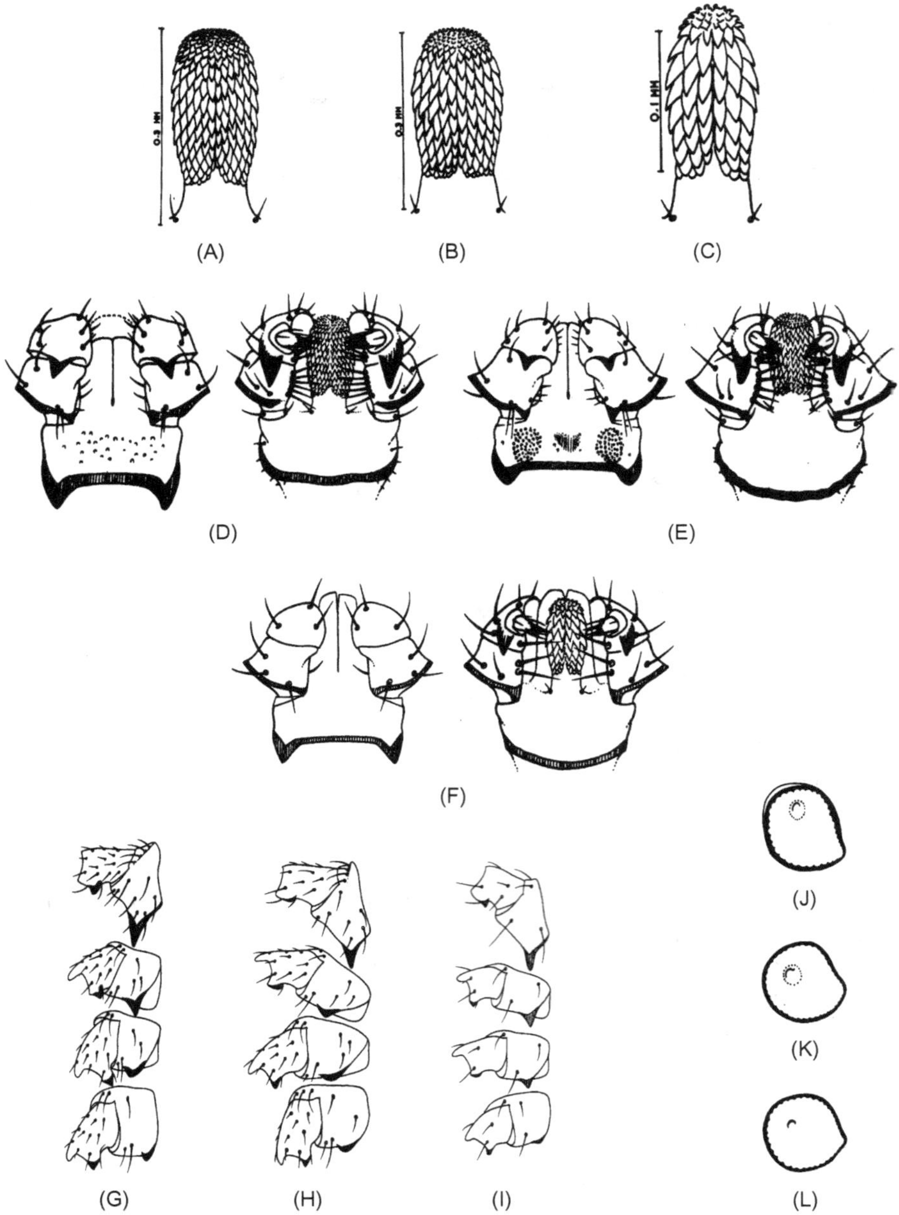

Figure 2.72 (A–C) Hypostome of male, female, and nymph, respectively; (D–F) capitulum of male, female, and nymph, respectively; (G–I) coxa of male, female, and nymph, respectively; (J–L) spiracular plates of male, female, and nymph, respectively.
Source: NIV. unpublished drawings.

Related Species[70]

H. davisi, a member of the "bispined" *H. bispinosa* group, is most closely related to *H. ramachandrai* and *H. bispinosa*. *H. ramachandrai* is also a distinctly smaller species (average male length 2.2 mm) with 4/4 (rarely 5/5 male) adult and 2/2 nymphal dental formula; obscure, short lateral grooves; and fewer ventrointernal setae on femur IV. Other features of *ramachandrai*, such as somewhat smaller coxal spurs, more broadly rounded posterior margins of female scutum and genital operculum, different tarsal outlines, absence of a ventral spur on nymphal trochanter I, etc., also serve to differentiate this species from *H. davisi*. *H. bispinosa* is a much smaller and narrower parasite with 4/4 adult and 2/2 nymphal dental formula, deeper lateral grooves, fewer and larger scutal punctations, much reduced spurs on coxa II to IV, half the number of ventrointernal setae on femur IV, etc. Thus, it is easily distinguished from *H. davisi*.

Host[70,120]

Immature stages: Sambar deer, barking deer, cattle, sheep, Indian bison, goats, gyacha (hybrid of cattle and gaur), mules, and from flag dragging in vegetation.

Adults: Cattle, horse, Mithan or South Asian ox, mule, sheep, goat, gaur, Indian muntjac, sambar deer, tiger, hog badger, domestic cow, gyacha (hybrid of cattle and gaur), and from flag dragging in vegetation.

Distribution[70]

India (Sikkim, Arunachal Pradesh, Assam, West Bengal), Burma.

Disease Relationship

Not recorded.

Ecology[118]

H. (K.) davisi is recorded from tropical and temperate zones between approximately 470 and 9,000 ft altitude, contrast with those of a closely related species *H. ramachandrai*, which has been taken only in tropical localities between 200 and 3,200 ft altitude in West Bengal and Himachal Pradesh states of India and from Nepal. These ticks have been recorded from domestic and wild mammals and from vegetation.

2.7.7 Haemaphysalis ramachandrai

This species was first described by Dhanda et al. in 1970 from lab-reared and field-collected specimens from lowlands of Himalayan foothills of India and Nepal. The species is named after Ramchandra Rao, the then director of Virus Research

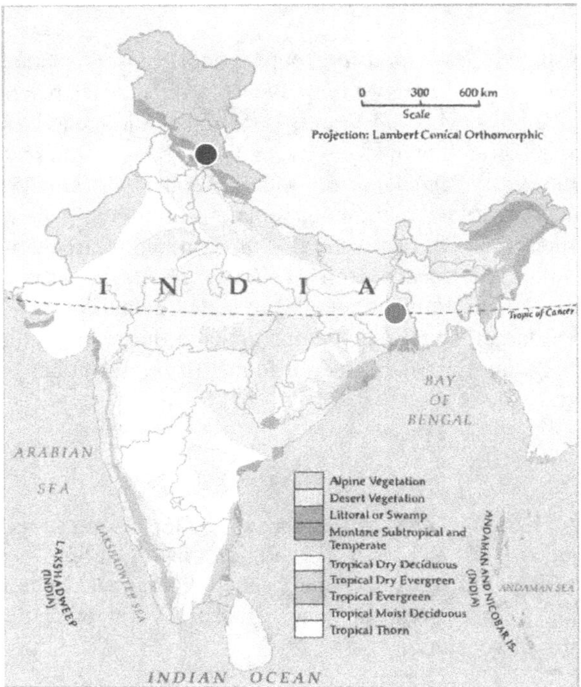

Figure 2.73 Distribution of *H. ramachandrai* in India. (●) Place of first record. Map not to scale.

Center, Pune (India) (now National Institute of Virology), to acknowledge his many years of contribution to the study of vectors and vector-borne diseases in India (Figure 2.73).

Male (Figures 2.74 and 2.75)[31]

Length is about 2.03–2.29 mm, breadth 1.33–1.46 mm.

Capitulum: Basis capitulum is dorsally approximately 1.5 times as broad as long; cornua are broadly triangular, one-half to three-fifths as long as the base of basis capitulum. Palpi are moderately salient posteriorly, combined breadth is approximately 1.25 times the breadth of basis capitulum. Hypostome is as long as palpi; dental formula is 4/4 or 5/5, with seven (internal) to ten (external) denticles in a file.

Scutum: Scutum is broadly oval, 1.5–1.7 times as long as broad; margins are broadly rounded, greatest breadth is immediately anterior of spiracular plates. Lateral grooves are shallow, obscure, short, do not reach scutal mid-length or enclose festoons. Cervical pits are small and rounded. Cervical grooves are short and obscure. Punctations are shallow, inconspicuous, small, moderately numerous, irregularly scattered, and absent in a smooth, narrow posteromedian line. Festoons are narrowly elongate, and number 11.

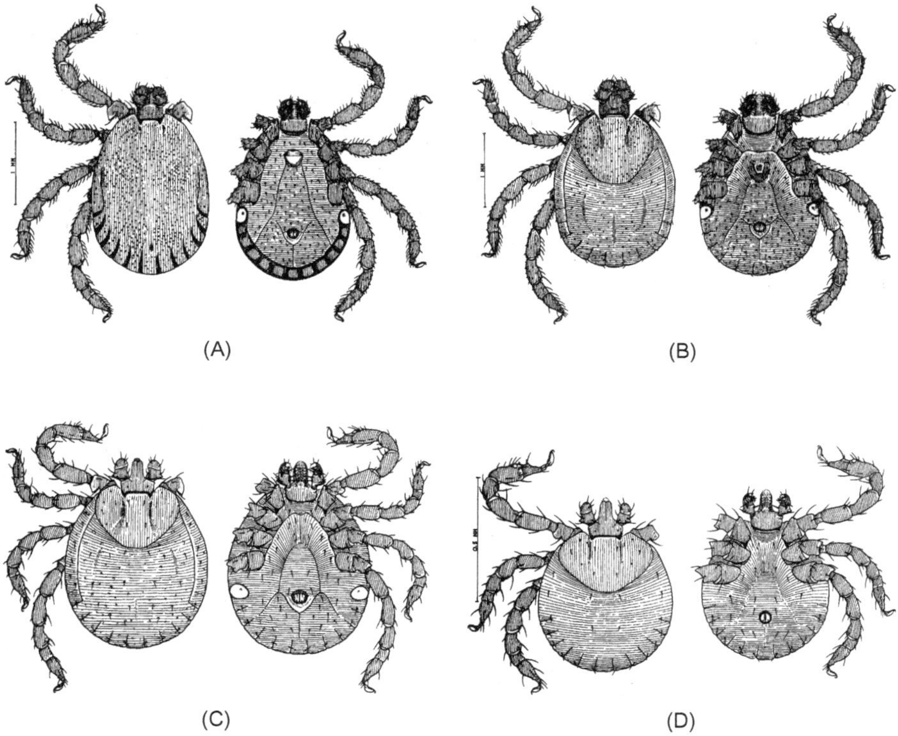

Figure 2.74 *Haemaphysalis ramachandrai*. (A) Male, dorsal and ventral side; (B) female, dorsal and ventral side; (C) nymph, dorsal and ventral side; (D) larva, dorsal and ventral side. *Source*: Dhanda et al. (1970), Journal of Parasitology (vol 56(4)), Allen press Publishing Services.

Legs: Coxa are each with a triangular spur; spur of I is lanceolate, extending to or slightly beyond anterior margin of coxa II; spurs of II to IV are broadly triangular, successively decreasing in size; those of II and III are extending well beyond posterior margin of coxa; spur of IV is quite small.

Female (Figures 2.74 and 2.75)[31]

Length is about 2.22–2.77 mm, breadth 1.44–1.66 mm.

Capitulum: Basis capituli are dorsally 2.2 times as broad as long; cornua are broadly triangular, almost one-third as long as the base of basis capituli; porose areas are moderately large, oval, widely spaced. Palpi are quite like those of male, posterior margin of segment II is somewhat broader than in male and with a slight posteroexternal protrusion. Hypostome is similar to that of male.

Scutum: Cervical grooves are shallow, extend from small sub-anterior pits to scutal mid-length. Punctations are as in male. Genital operculum is slightly longer than anterior breadth.

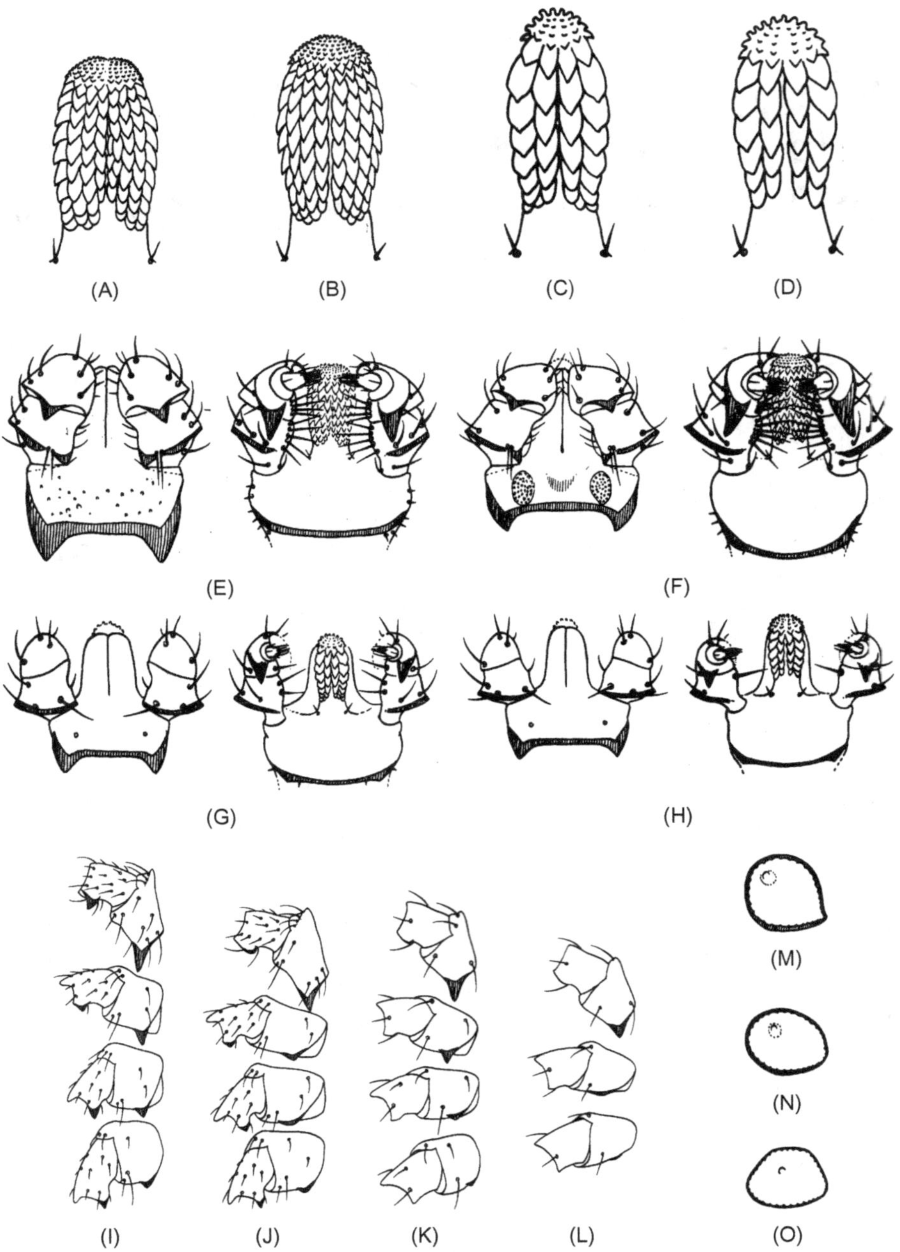

Figure 2.75 (A–D) Hypostome of male, female, nymph, and larva, respectively; (E–H) capitulum of male, female, nymph, and larva, respectively; (I–L) coxa of male, female, nymph, and larva, respectively; (M–O) spiracular plates of male, female, and nymph, respectively. *Source*: NIV. unpublished drawings.

Legs: Legs are approximately as in male. Coxa are with spur outline as in male on I; spurs of II to IV are reduced to short, broadly triangular or rounded ridges, extending only slightly beyond posterior margins.

Nymph (Figures 2.74 and 2.75)[31]

Length is approximately 1.25 mm, breadth 0.80 mm.

Capitulum: Basis capitulum is dorsally approximately 2.1 times as broad as long; cornua are broadly triangular, approximately one-half as long as the base of basis capitulum. Palpi are campanulate. Hypostome is as long as palpi; dental formula is 2/2, with six or seven denticles in a file.

Scutum: Scutum is approximately 1.3 times as broad as long, greatest breadth is at the level of anterior two-fifths of length. Cervical grooves are linear, parallel to scutal mid-length, thence shallow and diverging to margins. Punctations are few and obscure.

Legs: Coxa are each with spur quite similar to that of female except that spur of II is broadly triangular and extends well beyond coxal margin.

Larva (Figures 2.74 and 2.75)[31]

Length is approximately 0.75 mm, breadth 0.50 mm.

Capitulum: Basis capitulum is dorsally approximately 2.2 times as broad as long; cornua are broadly triangular, approximately one-third as long as the base of basis capitulum. Palpi are quite similar to those of nymph except that internodorsal margin of segment II is mildly convex, and ventral spur of III extends to mid-length of segment II. Hypostome is with large corona of five rows of hooklets; dental formula is 2/2, with five or six denticles in a file.

Scutum: Scutum is 1.5 times as broad as long; cervical grooves are linear, extend to scutal mid-length; other characters are quite like those of nymph.

Legs: Coxa I is with spur broadly triangular, extending well beyond coxal margin; II is with a broadly rounded ridge on coxal margin; III is with a very small, rounded ridge on coxal margin.

Related Species[31]

Haemaphysalis ramachandrai is a member of the *H. bispinosa* group of tropical southern and temperate northeastern Asia and outlying islands. At least seven distinctive species comprise this group. The species *ramachandrai* differs from other members of the "bispined" group by its small size and in scutal characters, reduced spurs of female coxa II to IV, absence of a ventral spur on female trochanter II, etc. Both *bispinosa* and *ramachandrai* are small species, but in *bispinosa* the long, distinct lateral grooves and punctations, and reduced coxal II to IV spurs, easily distinguish it from *ramachandrai.*

Host[31,120]

Immature stages: Leopard, sambar deer, chital deer.

Adults: Wild mammals, domestic cattle, buffalo, goat, Indian muntjac, leopard, tiger, sambar deer, chital deer, human (crawling).

Distribution[31,120]

India (Himachal Pradesh, West Bengal), Nepal.

2.7.8 Haemaphysalis bispinosa

This species was described from a single female collected in southern Chennai of Tamil Nadu, India, by Neumann in 1897. This species is characterized by the absence of well-developed cornua. The lateral borders of the palpae are not salient and the palpal dorsomedial elevated retroverted spur is on its posterior or basal border (Figure 2.76).

Male (Figures 2.77 and 2.78)[1,66,104,127,146]

Its size varies from 1.5 mm × 1.1 mm to 2.1 mm × 1.46 mm.

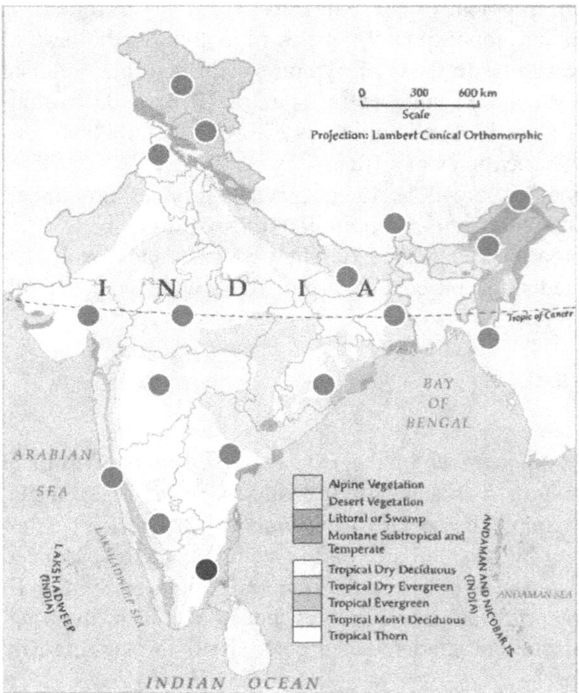

Figure 2.76 Distribution of *H. bispinosa* in India. (●) Place of first record. Map not to scale.

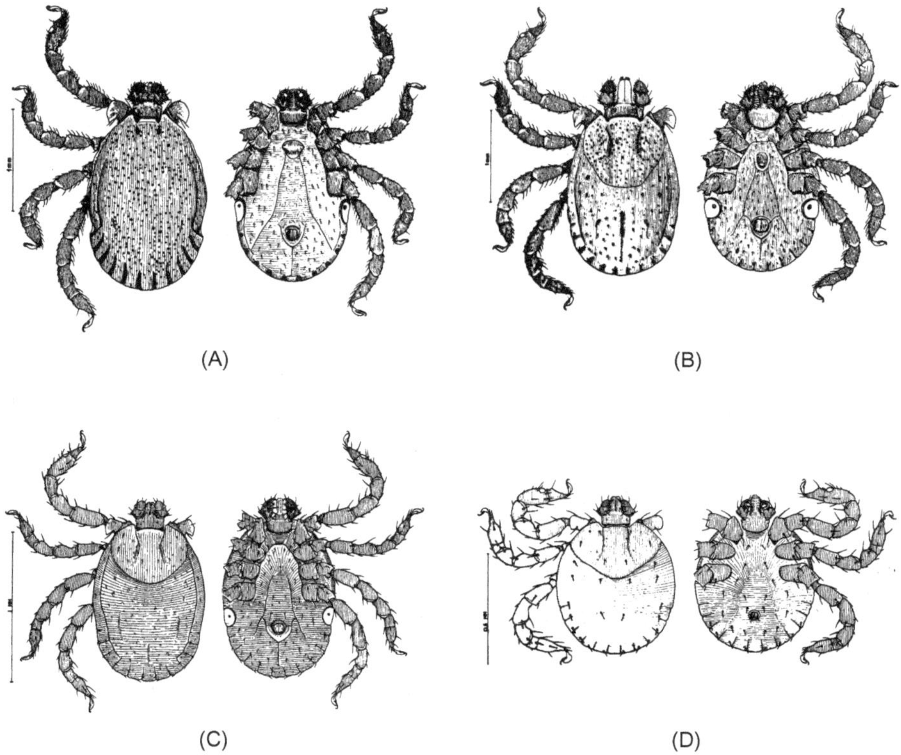

Figure 2.77 *Haemaphysalis bispinosa*. (A) Male, dorsal and ventral side; (B) female, dorsal and ventral side; (C) nymph, dorsal and ventral side; (D) larva, dorsal and ventral side. *Source*: Hoogstraal and Trapido (1966), Journal of Parasitology (vol 52(6)), Allen press Publishing Services; NIV unpublished drawings.

Capitulum: The capitulum is nearly as broad as long; its length varies from 0.35 to 0.4 mm. The base is twice as broad as long. The palps are half as broad as long. The hypostome has 4/4 rows of teeth with about nine teeth in each row.

Scutum: The punctations are fine and irregularly scattered all over the scutum. The cervical grooves are represented by small, posteriorly convergent depressions. The lateral grooves are shallow and commence opposite the second pair of legs and each includes the extreme festoon of the side. The festoons are twice as long as broad. The spiracle is sub-oval, longer than broad with its dorsal and ventral sides parallel.

Female (Figures 2.77 and 2.78)[1,66,104,127,146]

The unfed female measures 2.25 mm × 1.7 mm and is elongate-oval in shape but when replete measures 7.0 mm × 4.0 mm. The scutum is sub-cordiform.

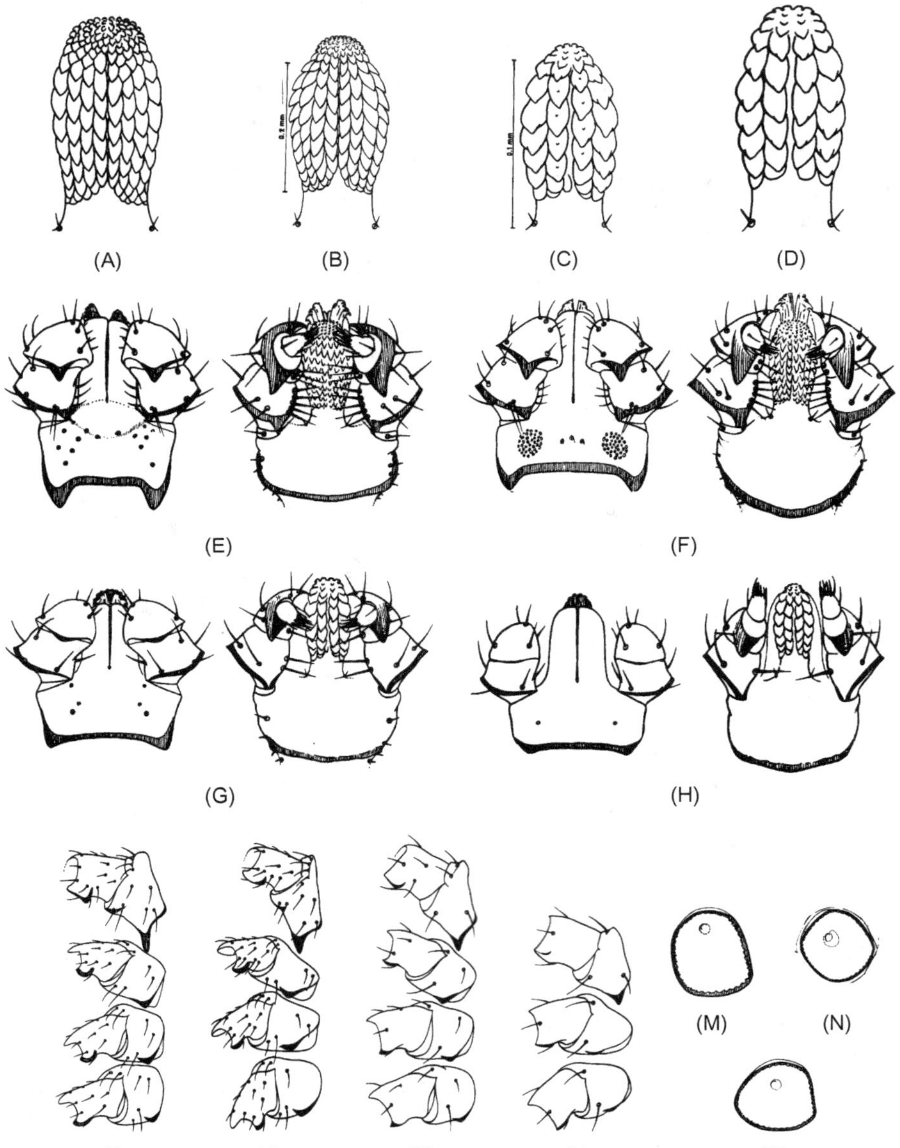

Figure 2.78 (A–D) Hypostome of male, female, nymph, and larva, respectively; (E–H) capitulum of male, female, nymph, and larva, respectively; (I–L) coxa of male, female, nymph, and larva, respectively; (M–O) spiracular plates of male, female, and nymph, respectively. *Source*: NIV. unpublished drawings.

Capitulum: The capitulum is comparatively strong and is 0.44 mm in length. The base is 3 times as broad as long. The porose areas are oval, and the interval between them is equal to twice their shortest diameter.

Scutum: The punctations are a little larger than those of the male, but are sparsely scattered except on the lateral fields, where they are comparatively numerous. The marginal grooves commence opposite the second pair of legs and each includes two festoons of the side. There are unequal punctations all over the general body. The spiracle is as broad as long.

Nymph (Figures 2.77 and 2.78)[104]

Capitulum: The capitulum is longer than broad and is 0.18 mm in length. The cornua are poorly developed. The hypostome is armed with 3/3 rows of teeth with about six teeth in each row; their number is fewer in the innermost row.

Scutum: The scutum is sub-pentagonal in shape with the posterior angle broadly rounded. It is broader than long and the surface has a granulated appearance. The scutum measures 0.40 mm in length and 0.49 mm in breadth approximately.

Larva (Figures 2.77 and 2.78)[104]

Capitulum: Without cornua, salience rounded, palpal segment II only slightly broader than palpal segment III. Palpal segment III is with distinct ventral spur, ventral spur of palpal segment III is short, thick, and blunt, reaching about one-fourth the distance to the basal margin of palpal segment II.

Legs: Coxa I is with spur reduced and blunt and sub-equal to or shorter than ventral spur of palpal segment III.

Life History[8,118]

Biology of this species was studied by Bhat in 1971 under the natural conditions of cattle sheds in KFD area where this species is known to mass breed. Freshly dropped and ovipositing females were seen in clusters of 50–60 along with egg masses inside the crevices of walls, up to 3 ft above the floor. The population of engorged adults observed increased in July. Engorged adults continued to appear till the end of August. From September onward their number gradually decreased and from December to May, they were almost absent. In the next rainy season from June to September, they reappeared and a similar sequence of events followed.

About a month after the appearance of engorged adults in large numbers, larvae appeared in clusters of a few to several hundred individuals all over the walls and pillars up to 5 ft above the floor. Clusters of engorged larvae and nymphs were observed in grooves and crevices of wooden pillars, walls, and ropes. Engorged larvae were found up to 5 ft and nymphs up to 8 ft above the floor. Engorged larvae, nymphs, and adults dropped in large number from cattle during early morning hours between 6 and 8 a.m., and the shed was teeming with crawling ticks.[10]

Studies carried out on the seasonal prevalence of this species on goats by Bhat et al. in 1986 indicated that maximum number of adult population was in October to December in Shimoga district of Karnataka state, India.

Studies on the biology of this species conducted at 22 °C by Sapre in 1949 indicated that the life cycle appeared to be completed in about 102 days. According to Sapre, the oviposition period was 7–15 days and the period of survival of female after oviposition had ceased to 3–6 days; the duration of the egg stage was about 30 days, the larvae were engorged in about 7 days; the nymphs were attached for about 9 days and the parasitic period of the adult was about 10 days.

The life of *H. bispinosa* was studied by Nakamura and Yajima in 1969 in Japan. Rearing and observation were conducted in the laboratory; rabbits were used as a host. The ticks were fed during their feeding period on the ears of rabbits. The average duration of each stage was 24 days in the egg, and 31.4 days in the larva. In the nymphal stage, 42 days were required for nymphs that become adults within the year and 211 days for nymphs that hibernate. Duration for adults was 31.4 days for adults that die within the year and 139 days for adults that hibernate in the pre-feeding period, 106 days for adults that hibernate in the post-feeding period. Generations are repeated twice a year for adults that hibernate, once a year for nymphs that hibernate.

An interesting feature about this tick appears to be the presence of obligatory parthenogenesis and the absence of spermatozoa in a fair proportion of the *H. bispinosa* males in Australia. In India, so far there is no report of parthenogenesis in this species.

Related Species

It is related with *H. ramachandrai*; the relationship has been discussed under *H. ramachandrai*.

Host[26,40,93,114,117,119,120,127]

Immature stages: Dog, jungle cat, black-naped hare, crested porcupine, Indian roller bird, cattle, goat, sheep, jackal, leopard, bonnet monkey, spotted deer, horses, langur monkey, buffalo.

Adults: Sheep, goat, jackal, mongoose, mule, barking deer, buffalo, cattle, cat, dog, donkey, horse, pony, rabbit, rat, spotted deer, tiger, common wood shrike, white handed babbler, black-naped hare, wild mammals, rodents.

Distribution[1,8,26,40,93,114,117,120,127,143]

India (Himachal Pradesh, Jammu and Kashmir, Punjab, Assam, West Bengal, Bihar, Madhya Pradesh, Maharashtra, Tamil Nadu, Arunachal Pradesh, Andhra Pradesh, Goa, Gujarat, Karnataka, Mizoram, Orissa, Sikkim, Andaman islands), Sri Lanka, lowlands of Nepal, Pakistan, Burma, China, USSR, Japan, Australia, New Zealand, Indonesia, Oceania, Siberia, Thailand, Malaysia, Borneo, East Indies (Southeast Asian islands), Vietnam.

Disease Relationship

A total of 11 KFD virus isolations have been reported from the nymphal stages of this species from KFD area during the period of 1961–1972. However, its role in the KFD transmission cycle is still not known.

Parasitism of H. bispinosa *Nymphs by a Hymenopteran Parasite –* Hunterullus sagarensis[39,40,41]

Parasitism of *H. bispinosa* nymphs by a chalcid parasite, *H. sagarensis*, has been reported from many villages in Karnataka and Maharashtra states. Mass breeding of *H. bispinosa* in cattle sheds in India is well known and the parasite had been recorded from fed nymphs collected from these cattle sheds.

Out of a total of 59 villages searched, fed nymphs of *H. bispinosa* were collected from 49 villages of Karnataka state. Of the 49 villages, 71.4% were positive for the parasites. Rate of infestation of ticks ranged from 0.3% to as high as 48.9%.

The female chalcid alighted on the nymph and oviposited inside the body cavity by piercing through the dorsal body wall with its stylets. Development of the parasite starts only when the parasitized nymphs take a full blood meal. None of the eggs showed any sign of development in unfed nymphs. Parasite larvae could be dissected out from fed nymphs on the first day of their detachment from the host. Five to ten days after detachment, the external symptom of parasitism could be noticed in parasitized nymphs. The inner content could be seen eaten up partly in the initial period and completely later on. The nymphs started dying from the seventh day. Total developmental period from egg to adult, excluding the dormant period in the unfed nymph, varied from 30 to 36 days under laboratory conditions. The adults emerged from the nymphs by gnawing holes on the posterior aspects. The immature stages completely destroyed the inner content of the tick and only the exoskeleton was left. Usually, one to three parasites emerged from a single nymph but in some cases as many as five were observed. The life span of adults was very short, ranging from 2 to 4 days. Sex ratio of 1,020 chalcid parasites emerged from 350 nymphs collected from the field showed that 384 were males and 636 females, giving an approximate male-to-female ratio of 1: 1.6.

These chalcids have been collected in all the months of the year. Though there is some difference in the percentage parasitism from month to month and between two localities, it is not possible to attach any seasonal significance. However, there was an apparent decline in the number of nymphs parasitized in the dry month of April. The presence of the parasite throughout the year is apparently due to the availability of a suitable host in all the seasons inside the cattle sheds.[8]

Laboratory experiments on host range showed that in addition to the nymphs of *H. bispinosa*, those of *H. spinigera*, *H. turturis*, and *H. kinneari* could be also parasitized. However, it did not oviposit on the nymphs of *B. microplus*, which is also a common cattle tick of this area. The parasitized unfed nymphs were as active as the nonparasitized ones until they were fully fed when the development of the chalcid flies starts inside the body. Only the unfed nymphs were susceptible for the parasitism.

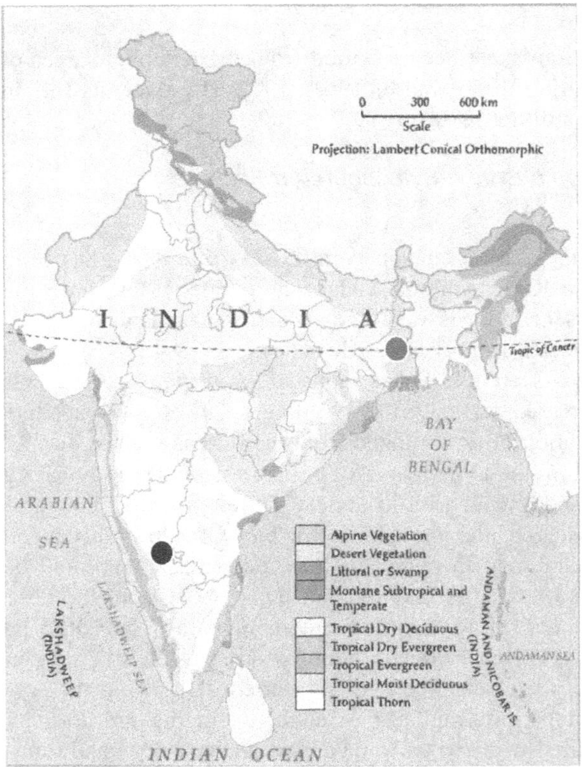

Figure 2.79 Distribution of *H. kinneari* in India. (●) Place of first record.
Map not to scale.

2.7.9 Haemaphysalis kinneari

Warburton first described this species in 1913 from a single female collected on a tiger in the Mysore area of Karnataka state in India, which was collected by N.B. Kinnear in 1910. Later, in 1913, Nuttall and Warburton together gave the illustrations (Figure 2.79).[140]

Male (Figures 2.80 and 2.81)[140]

Mean length is measuring 3.5 mm, width 2.4 mm.

Capitulum: Basis capituli are rectangular, slightly less than half as long as wide; cornua are short, widely triangular, approximately one-third as long as the base of basis capituli; surface with several deep punctations. Palpi are compact and subquadrate, lacking basolateral salience. Hypostome equals palpi in length, stout, lateral margins slightly convex, and dental formula is 4/4, with 9–11 denticles in a file.

Scutum: Scutum is flat, more or less depressed between lateral ridges; outline rather broadly oval, two-thirds as wide as long, widest at the level of coxa IV. Lateral grooves are very short, little longer than width of one festoon, enclosing

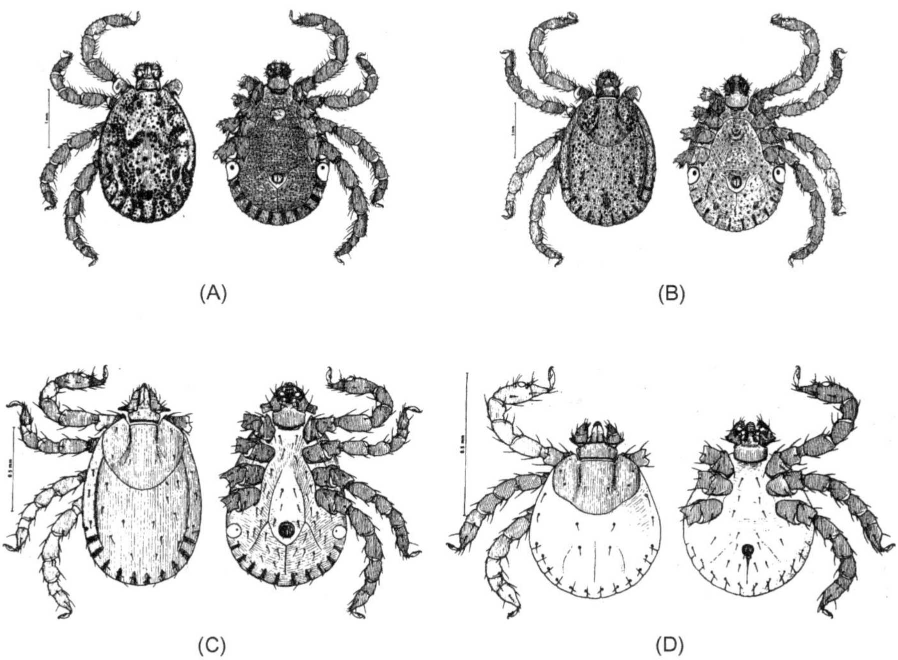

Figure 2.80 *Haemaphysalis kinneari*. (A) Male, dorsal and ventral side; (B) female, dorsal and ventral side; (C) nymph, dorsal and ventral side; (D) larva, dorsal and ventral side. *Source*: Trapoda et al. (1964), Journal of Parasitology (vol 50(1)), Allen press Publishing Services.

first festoon. Cervical grooves are shallow, with sub-circular pits at the level of coxa II; grooves are obsolete or almost so. Punctations are mostly deep, large and medium size, and confined to depression and certain elevated areas. Festoons number 11, and are clearly separated.

Legs: Coxa are each with a short, triangular, rather sharply pointed spur; spur on I is longest, on IV slightly shorter, those on II and III sub-equal and slightly shorter than that of IV; spurs on I, II, and III are arising midway between outer and inner margins, and on IV closer to inner margin.

Female (Figures 2.80 and 2.81)[140,104]

Mean length is about 3.7 mm, width 2.3 mm.

Capitulum: Basis capituli are slightly over twice as wide as long; cornua are widely triangular and short; porose areas are small, sub-ovate, and widely spaced. Palpi are similar to those of male except for somewhat greater length ratio and more angular apex of segment III; with group of three rather than two dorsal setae near posteromedial juncture of segment II. Hypostome is with two or three more denticles in each file than in male.

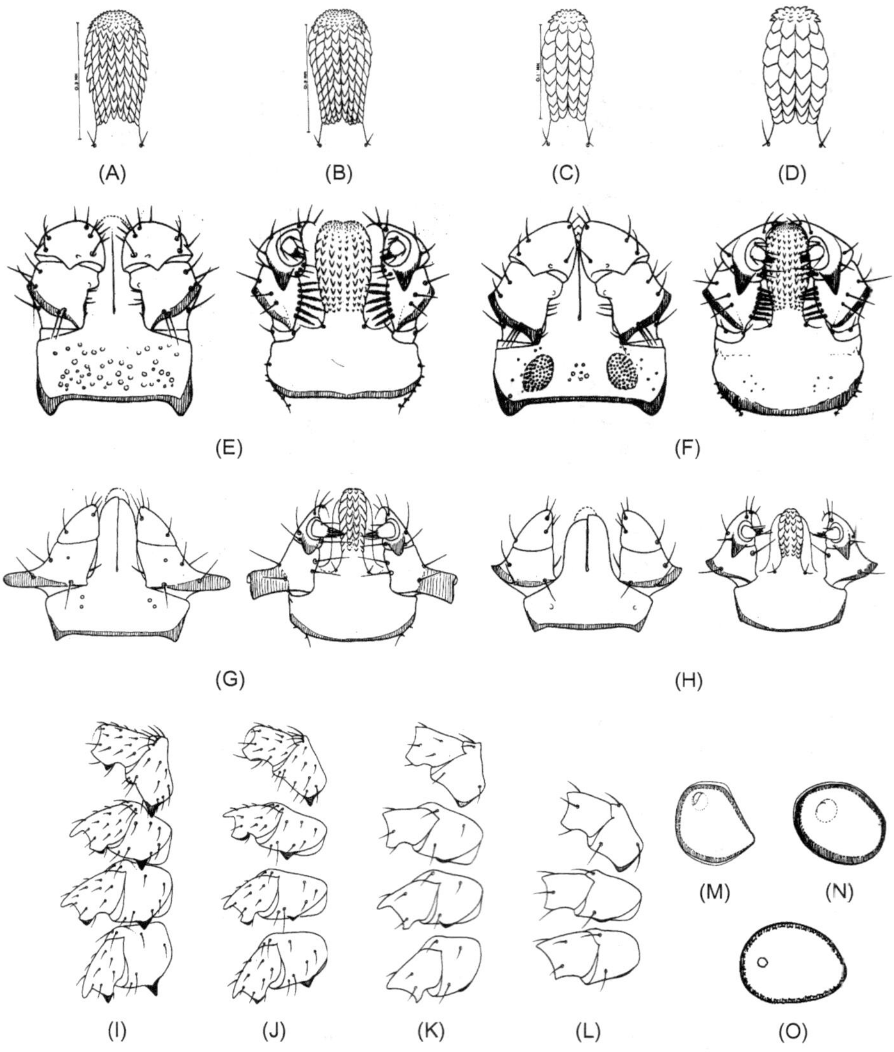

Figure 2.81 (A–D) Hypostome of male, female, nymph, and larva, respectively; (E–H) capitulum of male, female, nymph, and larva, respectively; (I–L) coxa of male, female, nymph, and larva, respectively; (M–O) spiracular plates of male, female, and nymph, respectively.
Source: NIV. unpublished drawings.

Scutum: Scutum is slightly wider than long. Lateral margins straight and parallel. Cervical grooves are arising slightly posterior of anterior scutal margin as narrow, short, parallel grooves, and continuing toward posterolateral scutal margin as shallow, wider, diverging grooves. Punctations are large and medium size, larger and more numerous in lateral fields, and fewer in median field.

Legs: Legs are similar to those of male in characters of coxal spurs, femoral setae, claws, and pulvilli.

Nymph (Figures 2.80 and 2.81)[140]

This stage differs greatly from adults in that the palpi are very widely salient basolaterally (a dorsobasal spur is lacking on segment III). Total length is approximately 1.24 mm, width 0.71 mm.

Capitulum: Basis capituli are almost 3 times as wide as long; cornua are small and pointed (or obsolete). Palpi are with wide basolateral salience. Hypostome is longer than palpi, lateral margins straight; dental formula is 2/2, with denticles in files of seven to nine on anterior two-thirds of shaft.

Scutum: Scutum is with length-width ratio equal, widest at the level of anterior fourth of scutal length. Cervical grooves are deep, narrow, parallel anteriorly, and shallow, wider, diverging posteriorly. Punctations number approximately six.

Legs: Legs are moderately stout. Coxa I has a moderate, triangular spur; II and III are each with a very short, widely triangular spur hardly if at all overlapping basal margin of coxa; IV is with spur obsolete or with a very small spur at inner margin basally.

Larva (Figures 2.80 and 2.81)[140]

Mean length is approximately 0.7 mm, width 0.4 mm.

Capitulum: Basis capituli are quite similar to that of nymph; a pair of posthypostomal bristles is present. Palpi are with segment I similar to that of adults but lacking setae. Hypostome is slightly longer than palpi; approximately 2½ times as long as wide; lateral margins are slightly convex; dental formula is 2/2, with denticles in files of eight or nine on apical two-thirds of shaft.

Scutum: Scutum is seven-tenths as long as wide. Cervical grooves are shallow. Punctations number six to eight, some bearing a small seta.

Legs: Coxa I is with a moderately long, triangular spur at inner margin; II is with a short, more or less widely triangular spur extending slightly beyond basal margin; III is with a very small (or obsolete) basal spur at inner margin.

Life History of Haemaphysalis kinneari

P.V.M. Mahadev has studied the life history of this species (personal communication). *H. kinneari* behaved as three-host tick under natural as well as laboratory conditions (at 20–22 °C and RH 90–100%). In the laboratory, life cycle completes within a minimum period of 172 days. The unfed adults from laboratory colony readily feed on rabbit or calf, the feeding period ranges from 7 to 19 days. The weight of fully fed female in laboratory ranges from 255 to 685 mg on calf and 60 to 563 mg on rabbit. Oviposition commences after 4–11 days of detachment and reaches the peak in 4–8 days with a subsequent decrease up to 45–60 days. Number of eggs laid range from 2,564 to 7,795 (fed on calf). In general, larvae take 2–5 days, nymphs take 3–7 days to feed on white leghorn chickens, and

Table 2.6 Life Cycle of *Haemaphysalis kinneari* Under Laboratory Conditions

Stage	Event	Days for Completion	
		Room Temperature (24–30 °C)	**Controlled Temperature (20 ± 1 °C)**
Adult	Pre-feeding	20–25	20–25
	Feeding	6–19	7–11
	Pre-oviposition	4–9	8–11
	Oviposition	22–45	30–40
Egg	Incubation	35–42	44–62
Larva	Pre-feeding	3–4	3–4
	Feeding	3–6	3–5
	Molting	13–22	22–29
Nymph	Pre-feeding	5–7	5–7
	Feeding	3–6	4–7
	Molting	12–15	25–28

adults take 5–12 days on rabbits. These hosts are considered to be suitable for maintenance of temporary laboratory colonies. Duration of various events increased with decrease in ambient temperature. Under laboratory conditions, engorgement and dropping of these stages seems to coincide with the activity of hosts. The duration of parasitic phase of larvae, nymph, and adults differs with the host offered under laboratory conditions (Table 2.6).

Related Species[140]

H. kinneari is closely related to *papuana*. *H. papuana* is distinctly smaller, 2.8 by 1.8 mm, than *H. kinneari*, 3.5 by 2.4 mm. Ridge-like development on dorsobasal margin of palpal segment III in *papuana* distinguishes it from *kinneari*, which has a small triangular spur at the same locus. The sexually dimorphic palpal basolateral salience of the female of *H. papuana* is absent in *H. kinneari*.

H. kinneari belongs to a group in which the palpi of immature stages are broadly salient while in adults they are not salient. This character differs it from closely related *H. bispinosa* group in which immature palpi are not salient; also from *H. doenitzi* group in which all palpi are broadly salient basolaterally.

Host[114,117,139]

Immature stages: Wild boar, mouse deer, crested porcupine, Indian gray mongoose, langur monkey, cattle, buffalo, different rat species, squirrel, mice, shrew, Asiatic long-tailed mouse, leopard, wild hare, wild dog, bonnet monkey.

Adults: Asiatic jackal, wild boar, tiger, monkey, rodents, buffalo, leopard, wild hare, mongoose (gray), wild dog.

Distribution[113,114,117]

India (Karnataka, West Bengal).

Disease Relationship[131,135]

This species is third in number with regard to the abundance and number of KFD virus isolations in the KFD area. As many as 26 isolations from nymph, 49 isolations from adults, and 2 isolations from larvae have been reported during the period of 1961–1972. Studies have shown that this species can transmit the KFD virus from infected to clean animals under the laboratory conditions.

Ecology[118]

It is one of the common species found on the ground tracks and forest vegetation in KFD area, next in abundance only to *H. spinigera* and *H. tuturis*.

2.7.10 Haemaphysalis hystricis

Supino first described *Haemaphysalis hystricis* in 1897. He described male and female without illustrations, collected from *Hystrix bengalensis* in Burma. The species is perhaps named after the host species. This species has long been confused with *bispinosa*, *papuana nadchatrami*, *birmaniae*, *semermis*, and several other species of *Haemaphysalis*. Later on in 1965, Hoogstraal et al. cleared the identity of this species (Figure 2.82).

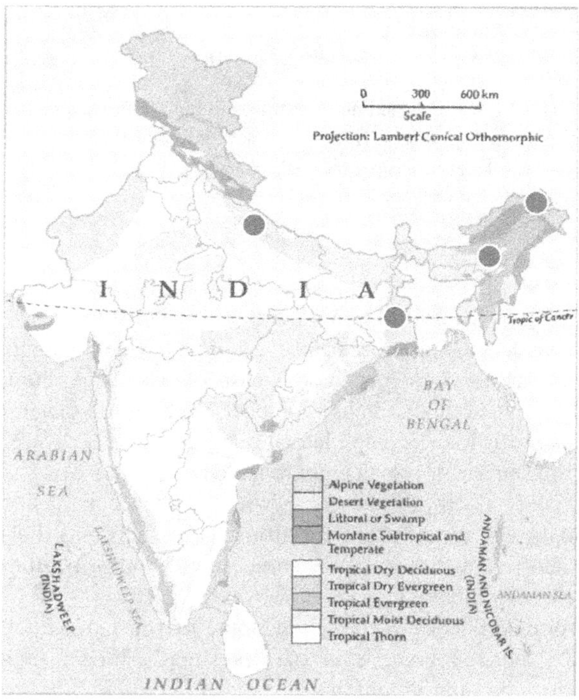

Figure 2.82 Distribution of *H. hystricis* in India. (●) Place of first record. Map not to scale.

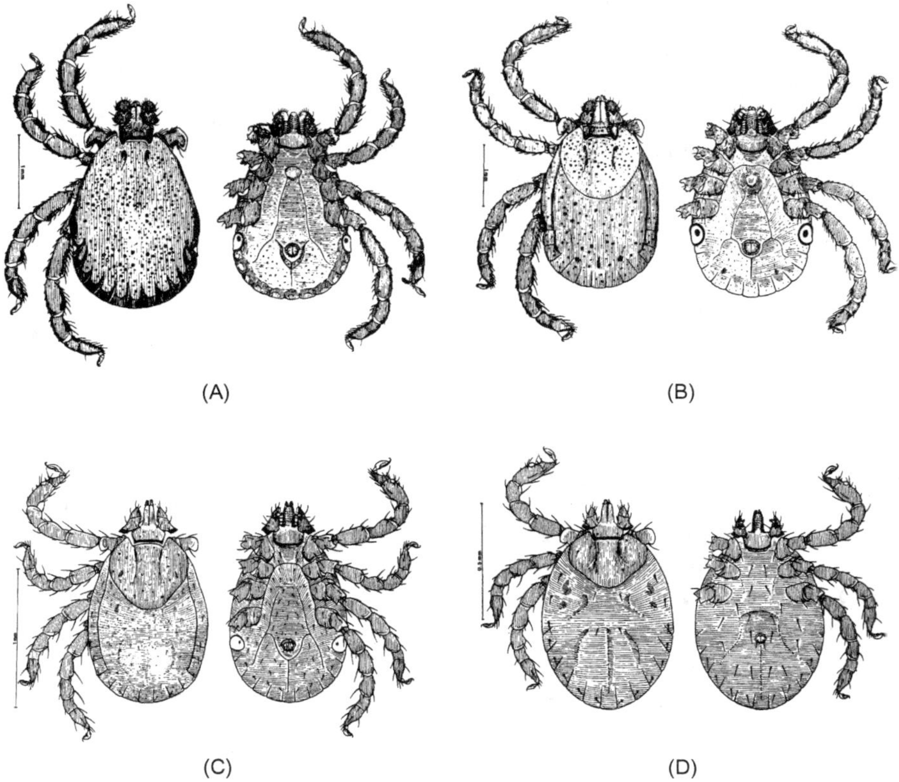

Figure 2.83 *Haemaphysalis hystricis.* (A) Male, dorsal and ventral side; (B) female, dorsal and ventral side; (C) nymph, dorsal and ventral side; (D) larva, dorsal and ventral side. *Source*: Hoogstraal et al. (1965), Journal of Parasitology (vol 51(2)), Allen press Publishing Services.

Male (Figures 2.83 and 2.84)[1,59,104,127]

Mean length is approximately 3.2 mm, width 2.0 mm.

Capitulum: Basis capitulum is dorsally approximately 1.7 times wider than its length; cornua are approximately one-third as long as the base of basis capitulum, widely triangular, with roundly pointed apex. Palpi compact, lacking basolateral salience. Hypostome is approximately as long as palpi; lateral outline is slightly diverging from the base to apex; dental formula is 4/4, with eight to ten denticles in a file.

Scutum: Scutum is approximately 1.4 times as long as wide; posterior margin is gradually rounded. Lateral grooves are absent. Cervical grooves are short and shallow; pits are deep, oval, and short. Punctations rare, some small, some medium size, and widely scattered. Festoons number 11.

Legs: Coxa each with a prominent, widely triangular spur, margins of which gradually converge to narrowly rounded apex; spur of I is slightly longer than others, spurs of II, III, and IV are sub-equal in length.

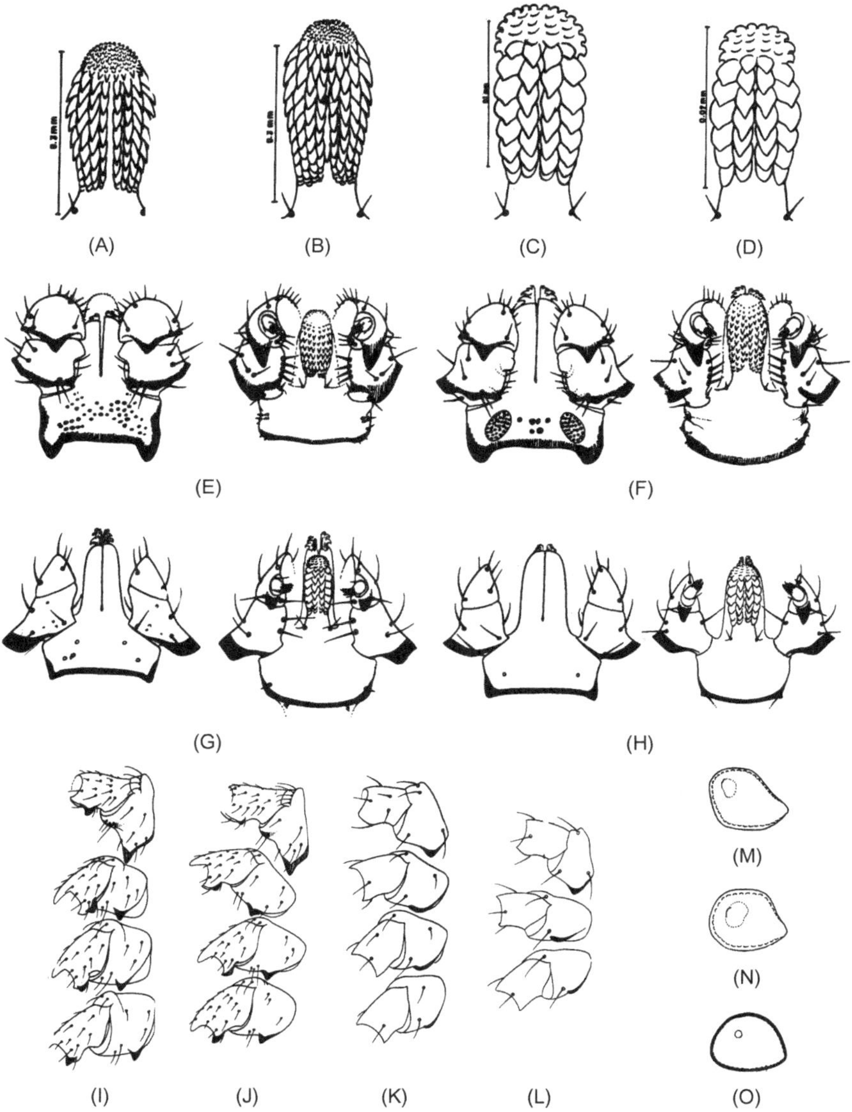

Figure 2.84 (A–D) Hypostome of male, female, nymph, and larva, respectively; (E–H) capitulum of male, female, nymph, and larva, respectively; (I–L) coxa of male, female, nymph, and larva, respectively; (M–O) spiracular plates of male, female, and nymph, respectively. *Source*: NIV. unpublished drawings.

Female (Figures 2.83 and 2.84)[1,59,104,127,146]

Mean length is approximately 3.4 mm.

Capitulum: Basis capituli are approximately 2.2 times as wide as long; cornua are approximately one-third as long as the base of basis capitulum; porose areas are small, oval, deep, tilted, and separated from each other by a distance equal to approximately twice the diameter of one pit. Palpi are quite similar to those of male except that each segment is comparatively slightly more elongate, and baso-ventral spur of segment II is blunter and usually somewhat more lateral than that of male. Hypostome is similar to that of male though slightly more elongate and with a few more denticles (11–13) in each file.

Scutum: Scutum is from 1.1 to 1.3 times as wide as long, and margins are gradually rounded. Cervical grooves and punctations are similar to those of male. Genital operculum is sub-quadrate.

Legs: Legs are essentially as in male except that ventral spurs of trochanters are reduced, those of I and II extending only slightly beyond basal margin and those of III and IV usually smaller or obsolete.

Nymph (Figures 2.83 and 2.84)[59]

This stage differs notably from adults in that the palpi are widely salient basolaterally and the basoventral spur of segment II is comparatively reduced. Mean body length is approximately 1.5 mm.

Capitulum: Basis capitulum is dorsally about 2.7 times as wide as long; cornua are reduced to very short, widely triangular spurs no more than one-fourth as long as the base of basis capituli. Palpi are widely salient (*leachi*-type). Hypostome is almost as long as palpi and stout; lateral outlines are parallel; dental formula is 2/2, with seven or eight denticles in a file.

Scutum: Cervical grooves are narrow, deep, elongate, slightly converging pits on anterior one-third and narrow, shallow, and slightly diverging grooves on central one-third of scutum. Punctations are rare.

Larva (Figures 2.83 and 2.84)[59]

Capitulum: Basis capitulum is 3 times as wide as long; lateral margins and posterior margin between cornua are straight; cornua are reduced to minute spurs or rounded angles.

Scutum: Scutum is 1.3 times as wide as long; outline, cervical grooves, and punctations are as illustrated.

Ecology[59]

Available data suggest that this is a forest-inhabiting species. *H. hystricis* appears to be a tick of low to intermediate altitudes (near sea level to 7,000 ft elevation), which in these latitudes are subtropical or comparatively temperate in climatic character. All carefully documented continental records, as well as that from Sumatra, are from mountainous areas. In the Ryukyu Islands, authentic data are from lowlands, and on Taiwan this tick has been recorded from sea level to 7,000 ft altitude. The altitudinal distribution of *H. hystricis* presumably includes lower altitudinal zones chiefly in the more northern islands of its geographic range.

Disease Relationship

Disease relationship of this species is not recorded yet.

Related Species

This species resembles *H. bispinosa* superficially but it can be separated by the absence of lateral grooves and the presence of a medial lobe-like projection on the dorsomedial border of palpal segment II in the adult. In the immature stages the palpal segments are more salient in *hystricis*.

Host[1,26,59]

Immature stages: Rat, human, Formosan reeves muntjac (deer), wild boar, tree shrew, and from flag dragging in Arunachal Pradesh (India).

Adults: Tree shrew, dog, man (unattached), rat (*R. brunneusculus* Hodgson, *R. fulvescens* Gray), tiger, wild boar, spinous tortoise, large scimitar babbler bird, hog badger, stink badger, buffalo, Himalayan black bear, cattle, pig, horse, Bengal porcupine, and from flag dragging in Arunachal Pradesh (India).

Distribution[1,26,59,120]

India (Uttar Pradesh, Assam, West Bengal, Arunachal Pradesh), Burma, Thailand, Laos, Vietnam, China, Taiwan, Indonesia, Ryukyu islands, Okinawa islands, Malaysia, Borneo, East Indies (Southeast Asian islands), Sri Lanka.

2.7.11 Haemaphysalis wellingtoni

H. wellingtoni was first described by Warburton from the specimen collected on domestic fowl in Borneo in 1907. In India, it has been recorded from a number of states, mainly from KFD area of Shimoga district in Karnataka state of India. It is considered as a bird tick (Figure 2.85).

Male (Figures 2.86 and 2.87)[1,104,127]

The male is 1.8 mm × 1.5 mm in size.

Capitulum: The capitulum is broader than long and is 0.38 mm in length. The base is sub-rectangular with convex lateral sides and is twice as broad as long. The cornua are short and blunt. The palps are 1½ times as long as broad and articles II and III are sub-equal. The hypostome has 4/4 or 5/5 rows of teeth, the innermost row of each side having five or six teeth and the others about ten in each row.

Scutum: The punctations are slightly unequal; the larger ones are mostly found on the lateral fields of the pseudo-scutum, which is slightly more raised than the

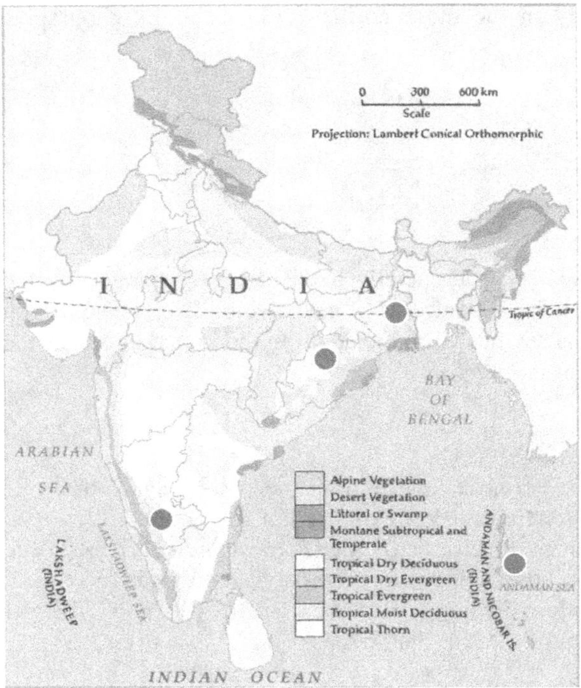

Figure 2.85 Distribution of *H. wellingtoni* in India. (●) Place of first record. Map not to scale.

scutum. The nonpunctate depressions are on either side of the center. The festoons are longer than broad.

Female (Figures 2.86 and 2.87)[1,104,127]

The scutum is 0.83 mm × 0.87 mm–1.1 mm × 1.0 mm in size.

Capitulum: The capitulum is comparatively large; its length being 0.56 mm. The base is 3 times as broad as long and is without cornua. The porose areas are oval, their long axes converging anteriorly; the interval between them is equal to their long diameter. The palp is longer and the lateral salience is more pronounced than in the male. The hypostome has 5/5 rows with about 13 teeth in each row.

Scutum: The punctations are fine, deep, and numerous. The cervical grooves extend over more than half the length of the scutum. The spiracle is pear shaped.

Nymph (Figures 2.86 and 2.87)[104]

Capitulum: Salience is less well developed, with recurvature in some species. Ventral retroverted spur of palpal segment III extends less than three-fourths the distance to the basal margin of palpal segment II; dorsal cornua are present but less than half-length of basis; infrainternal setae are four in number, feathery and closely set.

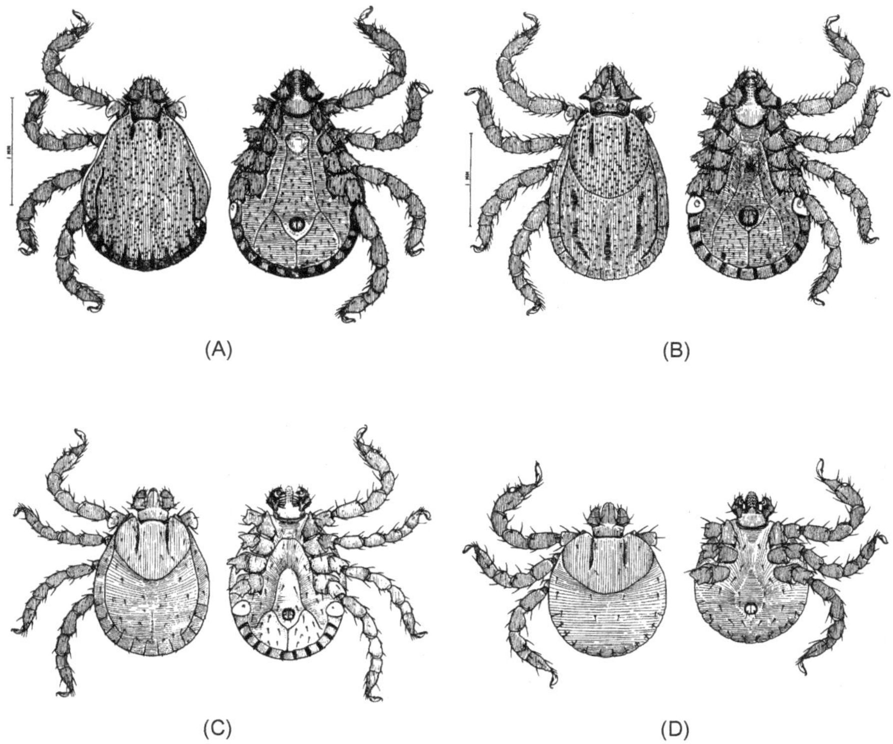

Figure 2.86 *Haemaphysalis wellingtoni*. (A) Male, dorsal and ventral side; (B) female, dorsal and ventral side; (C) nymph, dorsal and ventral side; (D) larva, dorsal and ventral side.
Source: NIV. unpublished drawings.

In ventral aspect, external profile of palps is without distinct recurvature; dorsobasal margin of palpal segment III reaching internal margin of palpal segment II only.

Legs: Coxa III and IV are with distinct sharp spurs, and ventral trochantal spurs are absent or obscure.

Larva (Figures 2.86 and 2.87)[104]

Capitulum: Without cornua, salience reduced, palpal segment II only slightly broader than III. Palpal segment III has a distinct, short, blunt, thick ventral spur reaching about one-fourth distance to basal margin of palpal segment II, ventral spur of palpal segment III shorter and more slender than the following.

Legs: Coxa I has a spur strong and sharp.

Host[12,13,93,114,116,117,119,121,127]

Immature stages: Toddy cat, mouse deer, barking deer, buffalo, langur monkey, dog, crow pheasant, shrew, cattle, red spur fowl, and from flag dragging through forest vegetation.

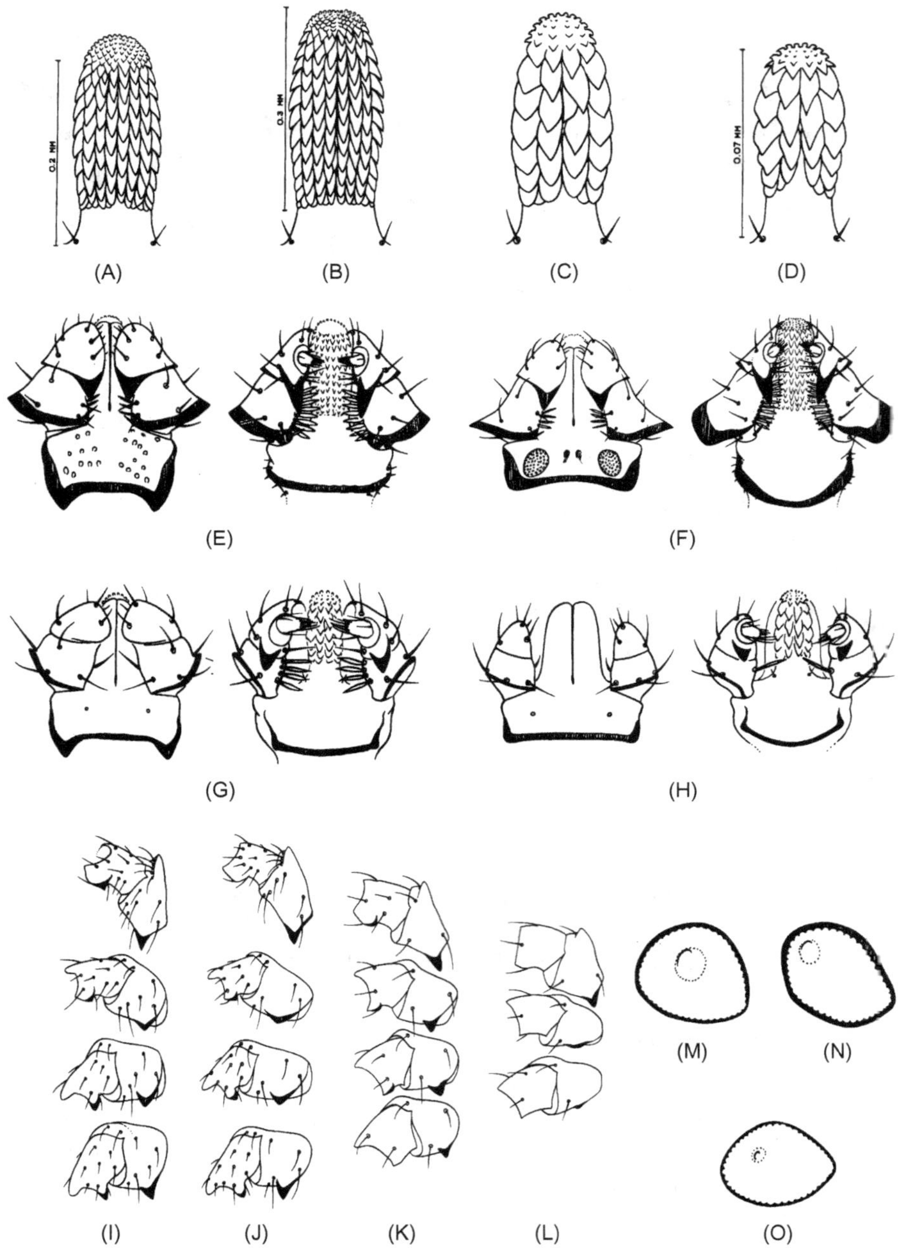

Figure 2.87 (A–D) Hypostome of male, female, nymph, and larva, respectively; (E–H) capitulum of male, female, nymph, and larva, respectively; (I–L) coxa of male, female, nymph, and larva, respectively; (M–O) spiracular plates of male, female, and nymph, respectively. *Source*: NIV. unpublished drawings.

Adults: Mouse deer, crested porcupine, monkey, small mammals, Gallinaceous and other bird species, crow pheasant, buffalo, dog, and from flag dragging through forest vegetation.

Distribution[1,12,93,113,127,143]

India (Andaman and Nicobar islands, Karnataka, Orissa, Assam, West Bengal), Sumatra, Siam, Borneo, Indonesia, Sri Lanka, Burma, Thailand, Malaysia, Vietnam, Myitkyina, East Indies (Southeast Asian islands), Siam, New Guinea.

Life History[12]

Biology of this species has been reported by Bhat and George in 1977. The life cycle of the species could be completed under ambient temperature (RH 90–100% and at 17–27 °C) within 98 days when all the stages were fed immediately after they started questing. The maximum life span of unfed adults was 308 days. They could attach within 24 h after release on the host (chick) and feed at any time during their lifetime from a few days after hatching. Copulating pairs were observed after the fourth day of release. Replete females dropped between the fifth and fifteenth day from 8 to 12 h. Two rather arbitrarily defined phases of repletion in females could be observed. The first "preparatory phase" lasted until the last but one day of parasitic phase. During this phase the rate of increase in size was very slow. The ticks changed their color from original dark brown to cream. Copulation was generally observed at the end of this phase. The second "repletion phase," which terminated with detachment, was of short duration lasting for 12–18 h. The average amount of blood imbibed by the females was approximately 3.35 times that of their average unengorged weight (0–98 mg). The fully repleted females weighed 175–454 mg. Oviposition commenced 4–7 days after the dropping and lasted for 13–24 days. The loss in weight after oviposition ranged from 68% to 84%. Of this, 50–70% accounted for the weight of the egg mass. The total number of eggs laid by the fully engorged females ranged from 2,104 to 5,853 (average 4,508 eggs). The number of eggs laid by each individual was directly proportional to its fed weight. The spent females died within 19 days after the completion of oviposition. Hatching of broods commenced 26–35 days after the initiation of oviposition and continued for as many days as the duration of oviposition. The maximum longevity of unfed larvae was 118 days. As adults, they could attach themselves and feed at any time during their lifetime within a few days after hatching. The larvae stayed on the chicks in the "preparatory phase" for at least 30 h before passing on to the "repletion phase," which lasted for not more than 10 h. Larvae, which did not pass on to the repletion phase in the later part of the second night were carried on to the subsequent night for repletion, followed by dropping in the following days. Most of the engorged larvae dropped from 08 to 18 h. Engorged larvae molted into nymphs 16–20 days after dropping. The maximum longevity of unengorged nymphs was 146 days. During their parasitic phase they followed a sequence of events similar to that of the larvae. Engorged nymphs molted into adults 24–31 days after dropping (Table 2.7).

Table 2.7 Life Cycle of *H. wellingtoni* Under Laboratory Conditions

Stage	Phase	Duration (in days)	No. of Observations
Female[a]	Parasitic	5–15 (mean 10.75)	20 specimens
	Pre-oviposition	4–7 (mean 4.8)	20 specimens
	Oviposition	13–24 (mean 17.7)	20 specimens
	Post-oviposition	1–19 (mean 4.85)	20 specimens
Egg	Incubation	26–35 (mean 29.8)	20 broods
Larva[a]	Quiescent	5–7	20 broods
	Life span (maximum)	118	2 vials of 2500 larvae each
	Parasitic	2–6	4 broods
	Molting	16–20	4 broods
Nymph[a]	Quiescent	3–5	Many
	Life span (maximum)	146	4 batches, 100 each
	Parasitic	3–7	11 batches, 200 each
	Molting	24–31	7 batches, 25 each
Adult	Quiescent	10–15	Many
	Life span (maximum)	308	5 batches, 50 each

[a]Fed on chicks.

Disease Relationship[12,13,23,116,134]

As many as 20 KFD virus isolations have been recorded from this species, out of which 17 have been recorded from nymphs during the period of 1961–1972, the remaining isolations being from adults and from larval stages.

Bhat and Naik (1978) have reported that this species can transmit the KFD virus under laboratory conditions.

Two more viruses, i.e., Ganjam and Kaisodi viruses, have been isolated from *H. wellingtoni* ticks.

2.7.12 Haemaphysalis intermedia

Nuttall and Warburton first described this species in 1909 and since then it has remained in a long taxonomic confusion. Hoogstraal and Trapido cleared up the confusion in 1963 by redescribing all the stages of *Haemaphysalis intermedia* (Figure 2.88).

Male (Figures 2.89 and 2.90)[57,127]

It is an exceptionally small haemaphysalid; body length measures approximately 1.25 mm, width approximately 0.6 mm.

Capitulum: Basis capitulum is rectangular; length is two-thirds as long as width; cornua are moderately large, sharply tapered, two-thirds as long as the base of basis capitulum; surface is with a few small punctations. Palpi of conical type are lacking basolateral salience. Hypostome is small, not reaching apex of palpi, lateral margins convex; dental formula is 4/4 in files of approximately seven denticles.

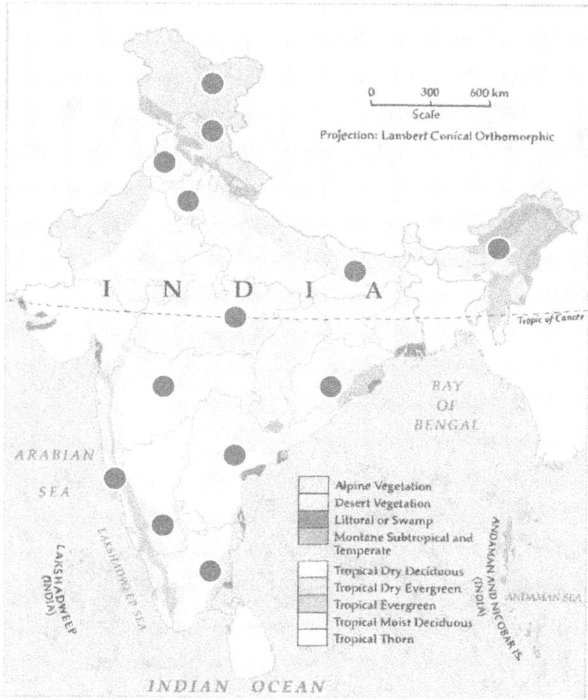

Figure 2.88 Distribution of *H. intermedia* in India. (●) Map not to scale.

Scutum: Scutum outline is broadly oval, width approximately two-thirds of length, widest at the level of coxa IV; lateral margins are gradually rounded. Lateral grooves are of medium length, extending to the level of coxa III and not enclosing festoons. Cervical grooves are formed by moderate pits and shallow, indistinct grooves. Punctations are numerous, moderate and large size, shallow to deep, and closely spaced to confluent. Festoons number 11 and are clearly separated.

Legs: Coxa I is with long, spine-like spur; II, III, and IV are each with a short, broadly triangular spur; that on III is largest, II is intermediate, and IV is smallest.

Female (Figures 2.89 and 2.90)[57,127]

Mean body length is approximately 1.5 mm.

Capitulum: Basis capitulum is half as long as wide; cornua are widely triangular, short, approximately one-third as long as the base of basis capitulum; porose areas are fairly large, shallow, appear to be partially confluent in some specimens. Palpi are similar to those of male; ventral spur of segment III does not quite reach basal margin of segment II. Hypostome is similar to that of the male but more elongate and extending to or very slightly beyond apex of palpi; dental formula is 4/4, with nine or ten denticles in a file.

Scutum: Scutum is with width and length sub-equal. Cervical grooves are distinct, convex, extending to or slightly beyond mid-length of scutum. Punctations

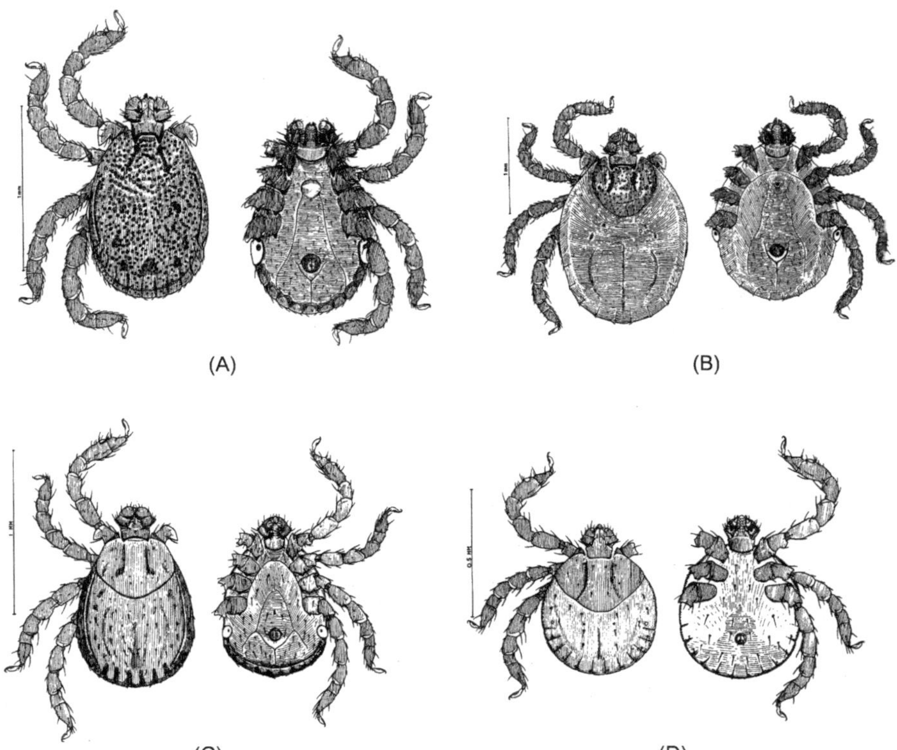

Figure 2.89 *Haemaphysalis intermedia*. (A) Male, dorsal and ventral side; (B) female, dorsal and ventral side; (C) nymph, dorsal and ventral side; (D) larva, dorsal and ventral side. *Source*: Hoogstraal and Trapido (1963), Journal of Parasitology (vol 49(5)), Allen press Publishing Services.

are moderately numerous and discrete; mostly small and medium size, a few large; and noncontiguous.

Legs: Legs are similar to those of male except that ridges and spurs on trochanters ventrally are reduced or obsolete; spur of coxa II does or does not extend beyond the basal margin of coxa; and tarsi are moderately elongate and lack an apical-ventral hook.

Nymph (Figures 2.89 and 2.90)[57]

Length is measuring approximately 0.9 mm, width 0.6 mm.

Capitulum: Basis capitulum is approximately half as long as wide; cornua are widely triangular, approximately half as long as the base of basis capituli. Palpi are with outline similar to that of female except that inner margin of segment III dorsally is only very slightly convex (almost straight). Hypostome is extending to or

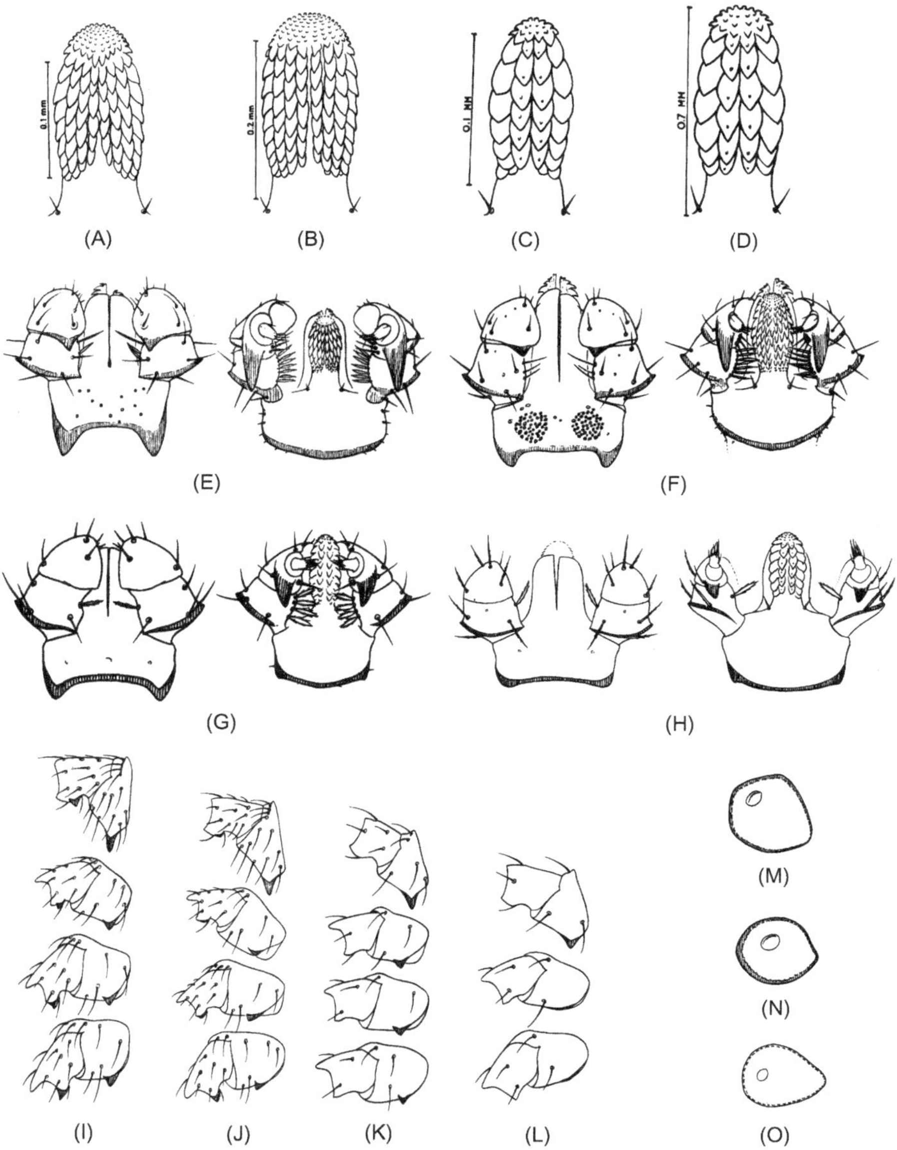

Figure 2.90 (A–D) Hypostome of male, female, nymph, and larva, respectively; (E–H) capitulum of male, female, nymph, and larva, respectively; (I–L) coxa of male, female, nymph, and larva, respectively; (M–O) spiracular plates of male, female, and nymph, respectively.
Source: NIV. unpublished drawings.

slightly beyond apex of palpi, lateral margins are slightly convex; dental formula is 2/2, with approximately eight denticles in a file.

Scutum: Scutum is approximately four-fifths as long as wide. Cervical grooves are straight, extending slightly beyond mid-length of scutum. Punctations are rare.

Legs: Coxa are with spurs similar to those of female.

Larva (Figures 2.89 and 2.90)[57]

Larva measures approximately 0.6 mm in length and 0.5 mm in width.

Capitulum: Basis capitulum is approximately 2½ times as wide as long; cornua are obsolete, suggested by rounded basolateral junctures; lateral and posterior margins are straight. Palpi with basal and outer margins are quite similar to those of nymph while inner margin is straight. Hypostome is extending slightly beyond the apex of palpi; lateral margins are slightly convex; dental formula is 2/2, with six or seven denticles in a file.

Scutum: Cervical grooves are concave, extend to posterior margin of scutum. Punctations are rare.

Legs: Coxa I is with a small, triangular, inner spur; coxa II and III are each with a very slightly elevated ridge basally (spurs obsolete).

Related Species[57]

H. intermedia, although closely related to *H. bispinosa*, is easily distinguished by a combination of characters including very small size; inner situation of the widely triangular, flat, dorsobasal spur of palpal segment III, and the widely convex inner margin numerous large, often contiguous scutal punctations; and medium length of lateral grooves. Specialists have confused Indian material of *intermedia* with *bispinosa* through a failure to appreciate the structured difference in the dorsobasal spur of palpal segment III in the two species, particularly in specimens in which the mouthparts are occluded with flesh of the host and the palps are spread outwardly and rotated, distorting its apparent position. Further contributing to the misidentification of these species is the fact that while this spur is more inwardly situated in *intermedia* there is some variation in its location and some specimens have it approaching the median position of *bispinosa*. However, the spur in intermedia is a thin, flat structure with its internal margin contiguous with the margin of the palpal segment, while in *bispinosa* the spur is stout, thickened, elevated, and set slightly anterior to the basal margin of the palp. The adults also resemble *H. kutchensis* in general fades and such morphological details as the ventral spur of palpal segment III, form of the spurs of the coxa and trochanters, and number and appearance of the infrainternal setae. But the species is readily separated from *kutchensis* by the distinctive appearance of the dorsobasal and internobasal margin of palpal segment III; in *kutchensis* the internobasal margin extends medially well beyond the internal margin of segment II and a basal spine is lacking, in contrast with *intermedia* which has a flat, basal, pointed, and widely triangular spur.

Life History[88]

Dhanda and Ilkal reported the life history in 1981 at ambient temperature ranging from 22.7 to 32.2 °C and at relative humidity from 80% to 90%. The entire life cycle took between 85 and 142 days. Most of the adults released on the host (chick) attached within 24 h. Ticks commenced feeding from the third day of their release. Copulating pairs were found during feeding. The females completed engorgement 7–9 days after release. Two distinct phases, viz., "preparatory" and "repletion," were observed at the time of feeding. The average amount of blood taken by a female was approximately 80 times its average unfed weight (0.77 mg). The weight of fully repleted female ranged from 39.0 to 91.7 mg. Oviposition commenced 3–5 days after detachment of the fed female, and lasted between 10 and 19 days. The female laid 263–884 eggs. The number of eggs laid was maximum on the third or the fourth day after commencement of oviposition, declined gradually subsequently. The number of eggs laid by the fed females was directly proportional to their body weight. The correlation coefficient was $r = 0.93$. The number of eggs laid per milligram increase in body weight of fed female was 12.00. The average weight of an egg ranged from 0.0408 to 0.0511 mg. The incubation period of eggs ranged between 20 and 40 days. The sequence of hatching of the eggs was the same as the sequence of laying, i.e., those laid early hatched first and those laid subsequently hatched later. The minimum pre-feeding period for the larvae was 4 days. The maximum number of the fed larvae was recovered with 10-day-old larvae. The feeding period ranged from 2 to 4 days. It took 7–12 days for the engorged larvae to molt into nymphs after dropping. The nymphs were ready to feed within a few hours of emergence. After release, the nymphs remained attached for 3–7 days to complete the blood meal. The time taken for the emergence of adults from the fed nymphs was 8–11 days (Table 2.8).

Studies carried out by Bhat et al. in 1986 have shown that the goats are the main host of this species in KFD area.

Table 2.8 Duration of Various Events in the Life Cycle of *H. intermedia*

Events	Duration (in days)	Host
Female feeding	7–9	Rabbits
Pre-oviposition	3–5	–
Oviposition	10–19	–
Egg hatching (from commencement of oviposition)	20–40	–
Pre-larval feeding (optimum period)	10–15	–
Larval feeding	2–4	Chickens
Larva–nymph molt	7–12	–
Pre-nymphal feeding (optimum period)	15–20	–
Nymphal feeding	3–7	Chickens
Nymph–adult molt	8–11	–

Host[57,93,114,117,121]

Immature stages: Jungle cat, mouse deer, Asiatic jackal, black-naped hare, goat, sheep, leopard, squirrel, birds (gray partridge, painted spur fowl, Indian plaintive cuckoo, Indian pitta, bush lark, brown shrike, gray drongo, brahminy myna, common myna, common wood shrike, red vented bulbul, white headed babbler, tailor bird, Blyth's red warbler, magpie robin, Indian robin, coucal), and by flag dragging through forest undergrowth.

Adults: Goat, sheep, buffalo, dog, squirrel, Asiatic jackal, bonnet monkey, hare, jungle cat, wild boar, black-naped hare, cattle, red jungle fowl, domestic chicken, and by flag dragging through forest vegetation.

Distribution[57,93,113,114,121]

India (Haryana, Himachal Pradesh, Jammu and Kashmir, Andhra Pradesh, Bihar, Assam, Delhi, Goa, Punjab, Maharashtra, Madhya Pradesh, Karnataka, Orissa, Tamil Nadu), Sri Lanka.

Disease Relationship[21,22,88]

A number of KFD virus isolations have been made from this species. In addition, Ganjam related to Nairobi sheep disease virus, Bhanja, and Barur have been isolated in India from this species.

2.7.13 Haemaphysalis sambar

Hoogstraal first described this species in 1971. He described it from a single specimen collected on Sambar deer from the mountains of South Western India. This single male of *H. sambar* was collected in 1911 by C.R. Howlett at about 3,000 ft altitude in Travancore district of Kerala, India (Figure 2.91).[75]

Male (Figure 2.92)[75]

Overall body length is approximately 2 mm, breadth 1 mm.

Capitulum: Basis capitulum is dorsally approximately 1.6 times as broad as long; cornua are broad anteriorly, elongately triangular, almost one-half as long as the base of basis capitulum. Palpi are broadly campanulate; combined breadth is approximately 1.5 times the breadth of basis capitulum. Infrainternal setae number five. Hypostome is longer than palpi; dental formula is 5/5, with eight to ten denticles in a file.

Scutum: Scutum is 1.5 times as long as broad. Cervical pits are short and relatively broad. Cervical grooves are obsolete. Lateral grooves are very short and narrow (all but obsolete). Punctations are small and medium size, inconspicuous, shallow, and relatively few. Festoons number 11, first pair is closed.

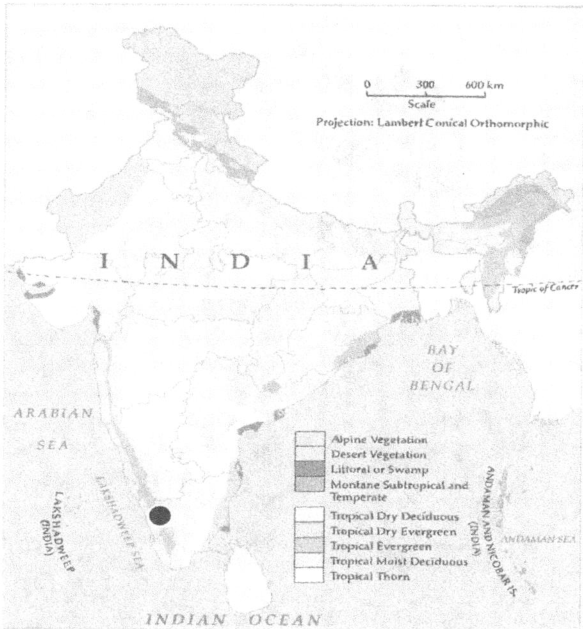

Figure 2.91 Distribution of *H. sambar* in India. (●) Place of first record.
Map not to scale.

Legs: Coxa I spur is elongately triangular, not reaching anterior margin of coxa II; II to IV spur are each broadly rounded, extending somewhat beyond the coxal margin; II and III spur are sub-equal; IV spur is smaller.

Related Species[75]

This species appears to be related morphologically to *H. intermedia* group. The *intermedia* group is characterized by a peculiarly developed projection of the posterodorsal margin of the adult palpal segment III. This projection differs distinctly from species to species but appears to be within reasonable limits as a group pattern. Each *intermedia* group species recorded has distinctive lateral grooves and coarse punctations, thus differing markedly from *sambar* in scutal appearance. *H. intermedia* species also has a 4/4 dental formula (5/5 in *sambar*).

Host[75]

Immature stages: Not recorded.

Adults: Sambar deer (single male).

Distribution[75]

India (Kerala).

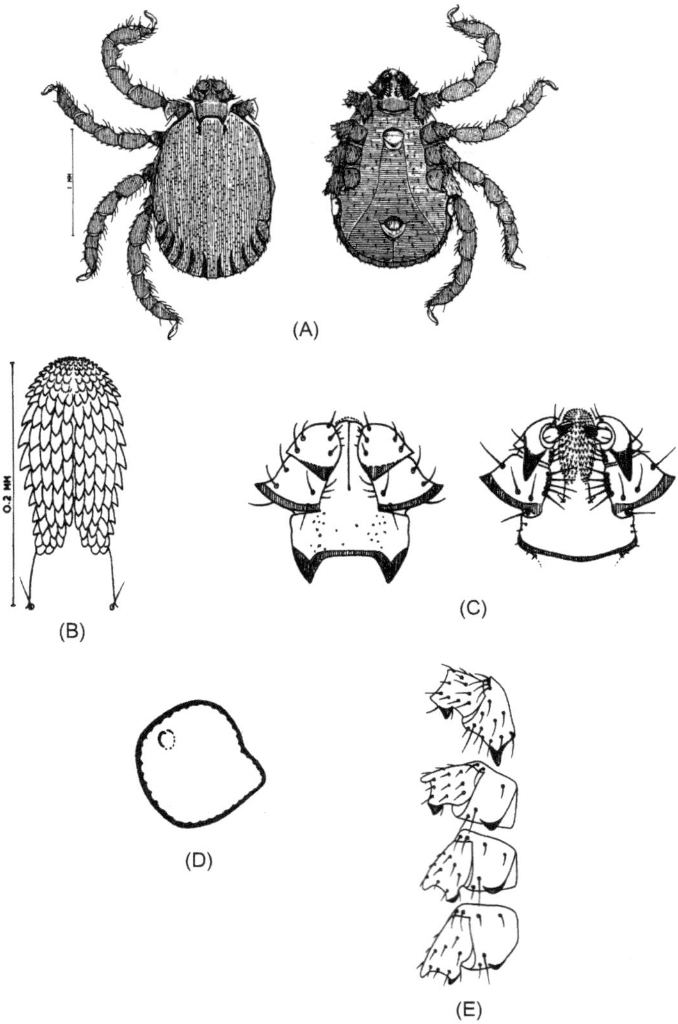

Figure 2.92 *Haemaphysalis sambar*. (A) Male, dorsal and ventral side; (B) hypostome; (C) capitulum, dorsal and ventral side; (D) spiracular plate; (E) coxa.
Source: Hoogstraal (1971), Journal of Parasitology (vol 57(1)), Allen press Publishing Services; NIV unpublished drawings.

2.7.14 Haemaphysalis kutchensis

Hoogstraal and Trapido first described this species in 1963 from Kutch area, Gujarat state of India. The name is derived from the place of its first collection (Figure 2.93).

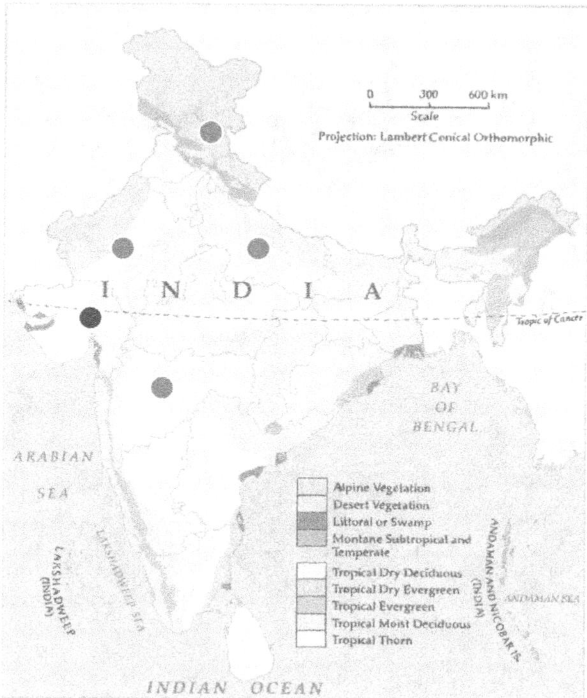

Figure 2.93 Distribution of *H. kutchensis* in India. (●) Place of first record. Map not to scale.

Male (Figures 2.94 and 2.95)[55]

Mean body length measures approximately 2.0 mm, width 1.1 mm.

Capitulum: Basis capitulum is dorsally approximately 1.4 as wide as long; surface has a few weak punctations; cornua are narrowly pointed apically and approximately one-fourth as long as base. Palpi are triangular in outline. Hypostome is moderately stout; lateral margins are very slightly convex; dental formula is 4/4, with seven to nine denticles in a file.

Scutum: Scutum outline is pyriform, with length measuring approximately 1.7 mm and width approximately 1.2 mm. Punctations are fairly numerous, shallow, more or less distinct, of variable size, noncontiguous, scattered over entire surface. Lateral grooves are deep, distinct, extending anteriorly to the level of coxa III, posteriorly enclosing first one or two festoons. Cervical grooves are short, moderately deep, straight or very slightly concave. Festoons number 11 and are distinct. Genital operculum is large and is situated between coxa II.

Legs: Coxa each is with a large, rather widely triangular spur; spurs of I and II are sub-equal in length but I is narrower; spurs of II, III, and IV are sub-equal in width but those of III and IV are slightly shorter than II.

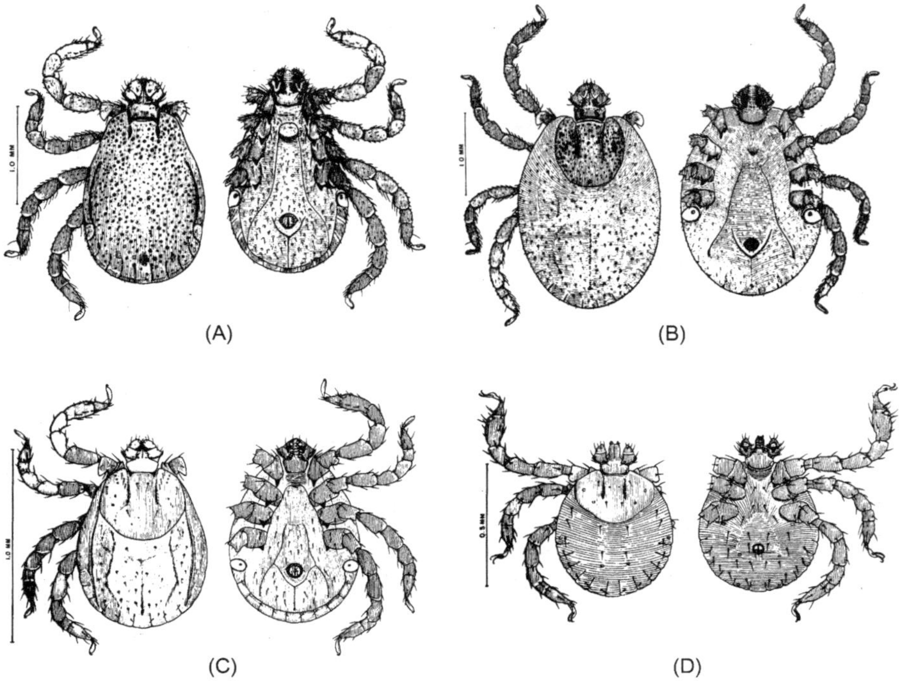

Figure 2.94 *Haemaphysalis kutchensis*. (A) Male, dorsal and ventral side; (B) female, dorsal and ventral side; (C) nymph, dorsal and ventral side; (D) larva, dorsal and ventral side. *Source*: Hoogstraal and Trapido (1963), Journal of Parasitology (vol 49(3)), Allen press Publishing Services.

Female (Figures 2.94 and 2.95)[55]

Mean body length is approximately 2.2 mm and width 1.3 mm.

Capitulum: Basis capitulum is slightly over twice as wide as long; cornua are widely triangular and only about one-third as long as base; porose areas are ovate, small, deep, separated from each other by a distance approximately twice the diameter of one porose area. Palpi are with basal margin of segment II, not so acutely angled as in male; thus, lateral margin of segment II is longer than in male and segments II and III are sub-equal in length. Hypostome is essentially as in male except for a few more small denticles in each file basally.

Scutum: Scutum is with length and width sub-equal; lateral margins are mildly convex on anterior two-thirds of length, more abruptly narrowing posteriorly in un-engorged specimens but less so in engorged specimens; posterior margin is broadly rounded. Cervical grooves are shallow, straight, parallel, extending approximately to mid-length of scutum. Punctations are large, moderately deep, and closely spaced on anterior half of lateral fields, irregularly sized and spaced.

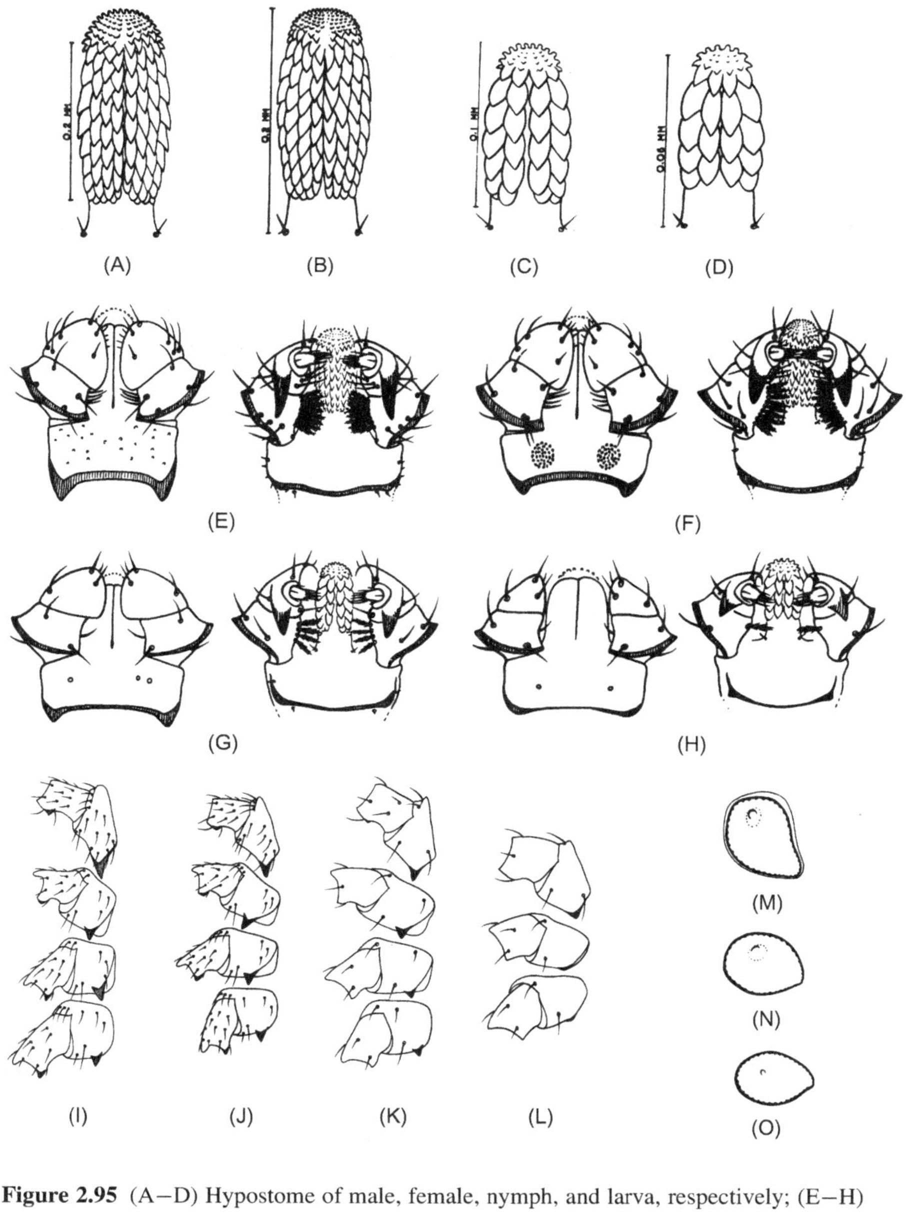

Figure 2.95 (A–D) Hypostome of male, female, nymph, and larva, respectively; (E–H) capitulum of male, female, nymph, and larva, respectively; (I–L) coxa of male, female, nymph, and larva, respectively; (M–O) spiracular plates of male, female, and nymph, respectively.
Source: NIV. unpublished drawings.

Legs: Legs are essentially as in male. Coxal spurs are quite similar to those of male except that spur of IV is somewhat smaller than that of III and those of II and III are sub-equal in length and width.

Nymph (Figures 2.94 and 2.95)[55]

Length measures approximately 1.0 mm, and width is around 0.65 mm.

Capitulum: Basis capitulum is approximately twice as wide as long; cornua are moderately wide and tapered. Palpi are with outline quite similar to that of female except for slightly greater ratio of width to length. Segment III is with ventral spur and dorsal margins similar to those of female. Hypostome is with 2/2 dental formula in files of approximately seven denticles; a moderate corona is present apically.

Scutum: Scutum is slightly broader than long; cervical grooves are as in female; setae number four in each scapular area, two anteriorly in central field.

Legs: Coxa are with spurs essentially as in female except that the spur of IV is smaller and directed medially.

Larva (Figures 2.94 and 2.95)[55]

Larva measures approximately 0.6 mm in length and 0.4 mm in width.

Capitulum: Basis capitulum is similar to that of nymph; cornua are not apparent in unmounted specimens but suggested in mounted specimens. Palpi are with outline similar to that of female and nymph except that inner margin dorsally is essentially straight and the contiguous margins of segments II and III are equal in length. Hypostome is with 2/2 dental formula in files of five denticles.

Scutum: Scutum is 1.5 times as wide as long; outline is as illustrated, with two pairs of setae and three pairs of sensilli.

Legs: Coxa I is with very small, triangular inner spur on basal margin; II and III each has an indefinite ridge-like spur on basal margin.

Related Species[55]

This species has some features in common with *H. howletti* Warburton, but it is easily distinguished because of virtual absence of palpal basolateral salience (marked in *howletti*); dental formula which is 4/4 (5/5 to 6/6 in *howletti*); different size and distribution of denticles; scutal outline of each sex which is shorter and wider and punctations which are less dense than in *howletti*; and coxal spurs which are somewhat larger than in *howletti*. The difference in length of the contiguous margins of palpal segments II and III dorsally is absent in *howletti*. From *H. campanulata*, this species is separated by lesser palpal width basally, longer ventral spur of palpal segment III, straight (not curved) basal margin of male palpal segment II, difference in length of contiguous margins of palpal segments II and III, smaller porose areas, more elongate tarsus IV lacking dorsal hump, larger coxal spurs, more elongate extension of spiracular plate (male), shorter and broader scutum, etc. This species differs from *H. bispinosa* and all others with a dorsobasal spur on palpal segment III by the absence of this spur. Of the known Indian *Haemaphysalis* larvae, that of *H. kutchensis* is closest in appearance to *H. turturis* and *H. intermedia*. From the former, it may be distinguished by the absence of cornua (present in *turturis*) and from the latter by the longer ventral spur of palpal segment III and the less acute angle at which the ventral basal margin and internal margin of palpal segment II meet. Of the known Indian *Haemaphysalis* nymphs, that of this species most closely resembles *H. turturis* and *H. intermedia*. From *turturis* it may be differentiated by the small but distinct sharp spurs of coxa II, III, and IV (reduced to bladelike projections in *turturis*); absence of ventral trochantal spurs (present in *turturis*); and presence of three or four lanceolate, closely set setae on the internal ventral margin of palpal segment II (contrasted with two slender, more widely spaced setae in *turturis*). The nymphs of *kutchensis* and *intermedia* are almost indistinguishable, but on side-by-side comparison it may be seen that the spur of coxa IV is directed posteriorly in *intermedia* while in *kutchensis* this spur points posterointernally.

Host[55,120]

Immature stages: Red winged bush lark, Indian robin, common babbler, white throat bird, rose colored starling, house sparrow.

Adults: Domestic animals, hare, painted partridge, gray partridge, common pea fowl, coucal, cat, mongoose, striped hyena, Indian wolf, jackal, dogs, sheep, caracal lynx, jungle cat.

Distribution[55,120]

India (Kutch area of Gujarat, Himachal Pradesh, Maharashtra, Rajasthan, Uttar Pradesh), Sri Lanka.

Disease Relationship

Not known.

2.7.15 Haemaphysalis turturis

This species was first described by Nuttall and Warburton in 1915 from a single specimen of male recorded in Ceylon. This single specimen was collected on spotted dove bird (*Turtur suratensis* or *Streptopelia chinensis ceylonensis*) in 1906. The name of this species is derived from the host's name (Figure 2.96).[138,143]

Male (Figures 2.97 and 2.98)[104,127,138]

Mean body length is approximately 2.2 mm and width 1.6 mm.

Capitulum: Capitulum is approximately 0.5 mm in length. Basis capitulum is dorsally approximately 1.5 times as wide as long, and surface has scattered superficial punctations; cornua are triangular, fairly sharply pointed, approximately

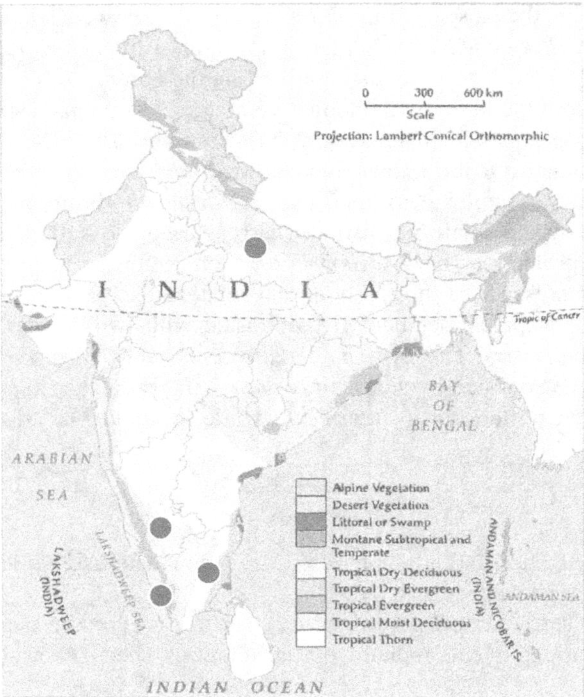

Figure 2.96 Distribution of *H. turturis* in India. (●) Place of first record. Map not to scale.

one-third the length of the basis capitulum. Palpi are conical, lacking basal salience; five narrowly lanceolate setae ventrally ("infrainternal setae") on segment II. Hypostome is with slightly convex lateral margins; dentition is 4/4, with denticles in files of six or seven.

Scutum: Scutum is 1.4 times as long as wide; margins are broadly rounded. Lateral grooves are almost obsolete, shallow, and not enclosing festoons. Cervical grooves are shallow, comma-shaped pits situated somewhat posterior of anterior margin. Punctations are shallow, small, numerous over entire surface as well as over festoons and periphery. Genital operculum is situated between coxa II. Festoons are clearly delimited, short, and broad.

Legs: Coxa, each has a triangular spur; spur on I is narrowly elongate, overlapping coxa II; spurs on II, III, and IV are wider, successively slightly shorter than the preceding; spurs on II and III are medial, and that on IV are closer to inner margin.

Female (Figures 2.97 and 2.98)[138]

Length of female is approximately 2.5 mm and width 1.8 mm.

Capitulum: Capitulum is somewhat larger than that of male and with a slight and recurved salience of the basolateral junctures of the palpi. Basis capituli are

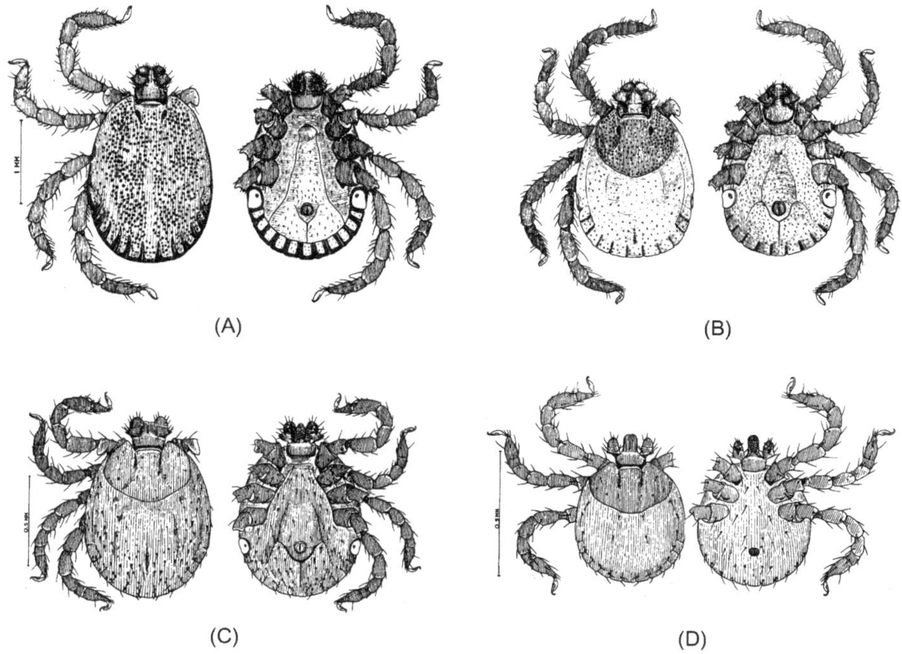

Figure 2.97 *Haemaphysalis turturis*. (A) Male, dorsal and ventral side; (B) female, dorsal and ventral side; (C) nymph, dorsal and ventral side; (D) larva, dorsal and ventral side. *Source*: NIV. unpublished drawings.

approximately 2.2 times as wide as long; porose areas are widely oval or sub-circular, moderate size, widely spaced.

Scutum: Scutum is sub-circular, width very slightly greater than length. Cervical grooves consist of pits only (as in male) or are also represented by almost obsolete, shallow extensions posterior of these pits. Punctations are similar to those of male. Integument is dorsally and ventrally as illustrated. Genital operculum is situated between coxa III or II.

Legs: Legs are essentially as in male except that spurs of coxa and trochanters tend to be somewhat smaller.

Nymph (Figures 2.97 and 2.98)[138]

Mean body length is approximately 1.1 mm and width 0.7 mm.

Capitulum: Basis capitulum is 2.1 times as wide as long; margins are straight; cornua are similar to those of female. Palpi are with margins similar to those of female, except that the dorsobasal margin of segment III is practically straight and lacks a spur; ventral spur of segment III is comparatively as long and as wide basally as in female. Hypostome is with convex lateral margins; dentition is 2/2, with files of six or seven denticles reaching almost to base of shaft.

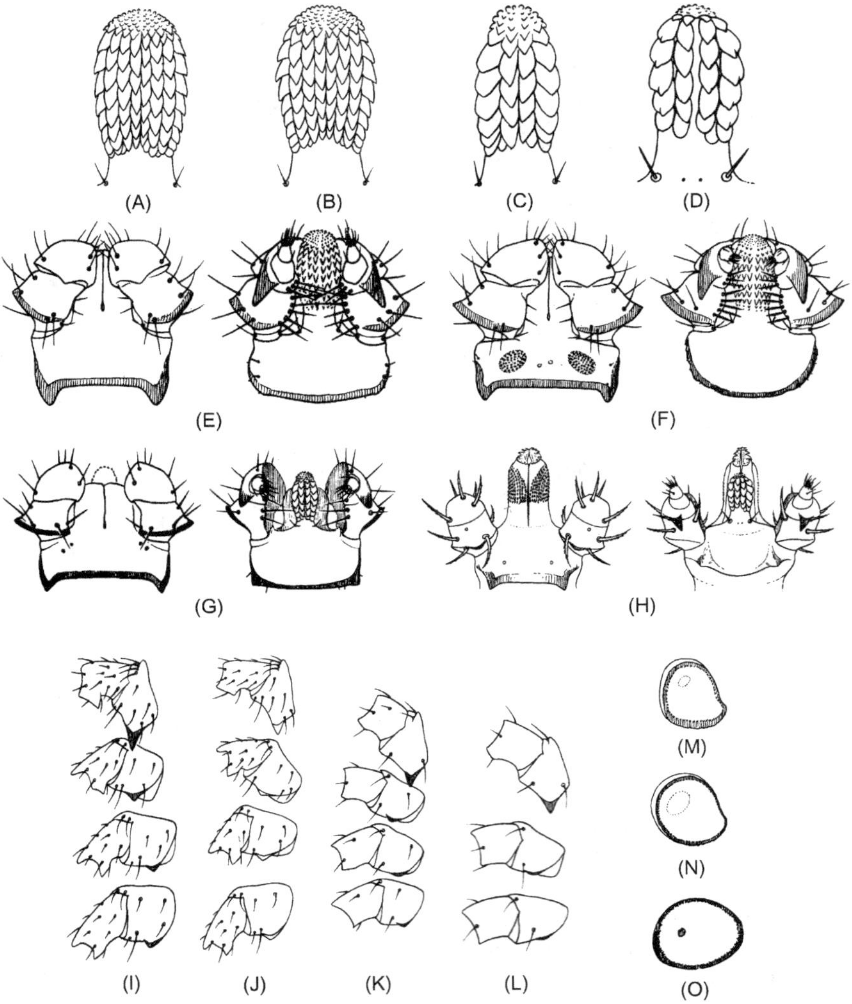

Figure 2.98 (A–D) Hypostome of male, female, nymph, and larva, respectively; (E–H) capitulum of male, female, nymph, and larva, respectively; (I–L) coxa of male, female, nymph, and larva, respectively; (M–O) spiracular plates of male, female, and nymph, respectively. *Source*: NIV. unpublished drawings.

Scutum: Scutum is approximately 1.3 times as wide as long. Cervical grooves are short, narrow, parallel, reaching almost to mid-length of scutum. Punctations are shallow, rare. Integument is dorsally and ventrally as illustrated.

Legs: Coxa I are with a fairly large, elongate triangular spur; II are with a shorter, wider spur just barely overlapping basal margin of coxa; III and IV are with small, more plate-like spurs that do not extend beyond margin.

Larva (Figures 2.97 and 2.98)[138]

Overall body length is approximately 0.51 mm.

Capitulum: Basis capituli are wide, short, with short, triangular cornua dorsally and shorter, wider cornua ventrally. Each palpus is bell shaped. Hypostome is with 2/2 dentition, lacking corona; a pair of prominent posthypostomal bristles is present.

Scutum: Scutum recalls that of nymph.

Legs: Coxa I is with a prominent triangular spur, II with a plate-like ridge, III with a slight suggestion of a ridge.

Related Species[56]

This species belongs to *turturis* group, which consists of *silvafelis*, *turturis*, and *paraturturis*. Among these species, adults of *H. paraturturis* are closest to those of *H. turturis*; they may be most readily separated by the number and the form of infrainternal setae. In *H. turturis*, infrainternal setae are slender and well spaced apart and four or five in numbers, whereas in *paraturturis* it is broad, closely set, and seven to nine in numbers.

The species *silvafelis* is very similar to *turturis*, but *turturis* may be separated from *silvafelis* by the absence of ventral trochanter spurs, seven broad infrainternal setae. The species *silvafelis* can be separated from *paraturturis* and *turturis* by the large, coarse punctations, densely scattered all over the scutum.

Host[114,117,119,139]

Immature stages: Asiatic jackal, toddy cat, wild boar, mouse deer, crested porcupine, black-naped hare, jungle cat, langur monkey, cattle, buffalo, rat (5, 6, 9), squirrel, shrew, leopard, wild dog, bonnet monkey.

Adults: Asiatic jackal, wild boar, mouse deer, sambar deer, crested porcupine, monkey, cattle, small mammals, bird species (spotted dove), wild goat, leopard, chital deer, barking deer, buffalo, jungle cat, leopard, wild hare, spotted deer, wild dog, bonnet monkey, langur monkey, and from flag dragging in Sagar district of Karnataka (India).

Distribution[113,114,117,143]

India (Karnataka, Kerala, Tamil Nadu, Uttar Pradesh), Sri Lanka.

Disease Relationship[23,131,134,135]

This species ranks second with regard to the abundance and number of KFD isolations recorded in the KFD area. As many as 431 isolations from nymphs and 265 isolations from adults with a single isolation from larva have been recorded during the period of 1961–1972. Experimental studies have also proved this species as an efficient vector of KFD under the laboratory conditions. In experimentally inoculated (parenteral inoculation) unfed *H. turturis*, the Ganjam virus was detected on days 15 and 35 post-inoculation. Two isolations of Kaisodi virus were recorded from adults of *H. turturis* during 1965–1970 from KFD area.

2.7.16 Haemaphysalis paraturturis

This species was first described by Hoogstraal et al. in 1963. This species has been collected from various carnivores as well as domestic cattle from India. The name has been coined to show its close morphologic resemblance to *turturis* (Figure 2.99).

Male (Figures 2.100 and 2.101)[56]

Mean body length measures approximately 1.9 mm and width 0.9 mm.

Capitulum: Basis capitulum is almost twice as wide as long; cornua are triangular, approximately half as long as the base of basis capituli; surface is with a few shallow punctations. Palpi are with a rudimentary, practically obsolete basolateral salience. Hypostome is extending to the apex of palpi; lateral margins are slightly convex. Dental formula is 4/4 in files of approximately seven denticles.

Scutum: Scutum typically is very slightly over half as wide as long but may be shorter and wider, widest at the level of coxa IV. Lateral grooves are narrow, deep, distinct, reaching the level of coxa III and enclosing one pair of festoons. Cervical grooves are deep, narrow, short, abruptly converging. Punctations are moderately

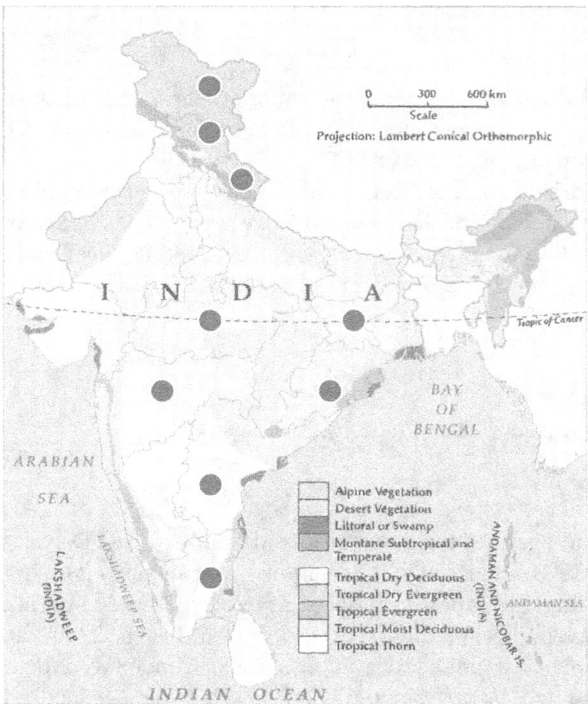

Figure 2.99 Distribution of *H. paraturturis* in India. (●) Place of first record. Map not to scale.

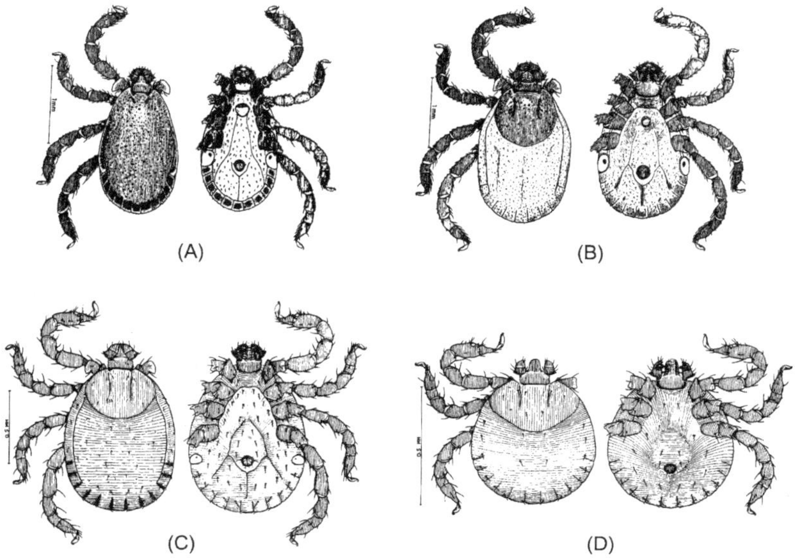

Figure 2.100 *Haemaphysalis paraturturis*. (A) Male, dorsal and ventral side; (B) female, dorsal and ventral side; (C) nymph, dorsal and ventral side; (D) larva, dorsal and ventral side. *Source*: Hoogstraal et al. (1963), Journal of Parasitology (vol 49(4)), Allen press Publishing Services.

numerous, moderately deep, irregularly sized, and distributed. Festoons number 11, clearly defined.

Legs: Legs are moderately long and robust. Coxa each is with a well-defined spur; I is with a large, elongate, moderately narrow, pointed spur; II and III each is with a short, more widely triangular, medially situated, sub-equal spur; IV is with a shorter, more narrowly triangular spur at juncture of inner and basal margins of coxa.

Female (Figures 2.100 and 2.101)[56]

Overall body length measures approximately 2.4 mm.

Capitulum: Basis capituli are shorter and wider than that of male; porose areas are sub-circular, widely spaced. Palpi are with a slightly more pronounced basolateral salience, and with a slight sub-apical depression of the lateral margin forming a very small, rounded apical nubbin. Infrainternal setae of segment II number seven to nine. Spur-like projection of dorsobasal margin of segment III is wider than in male; palpi are otherwise quite similar to those of male. Hypostome is similar to that of male.

Scutum: Scutum is with width and length sub-equal; lateral margins are widely convex; posterior margin is broadly rounded. Cervical grooves are deep, narrow, parallel, extending approximately to mid-length of scutum. Punctations are slightly larger and less numerous than those of male, very few posteriorly.

Legs: Legs are similar to those of male.

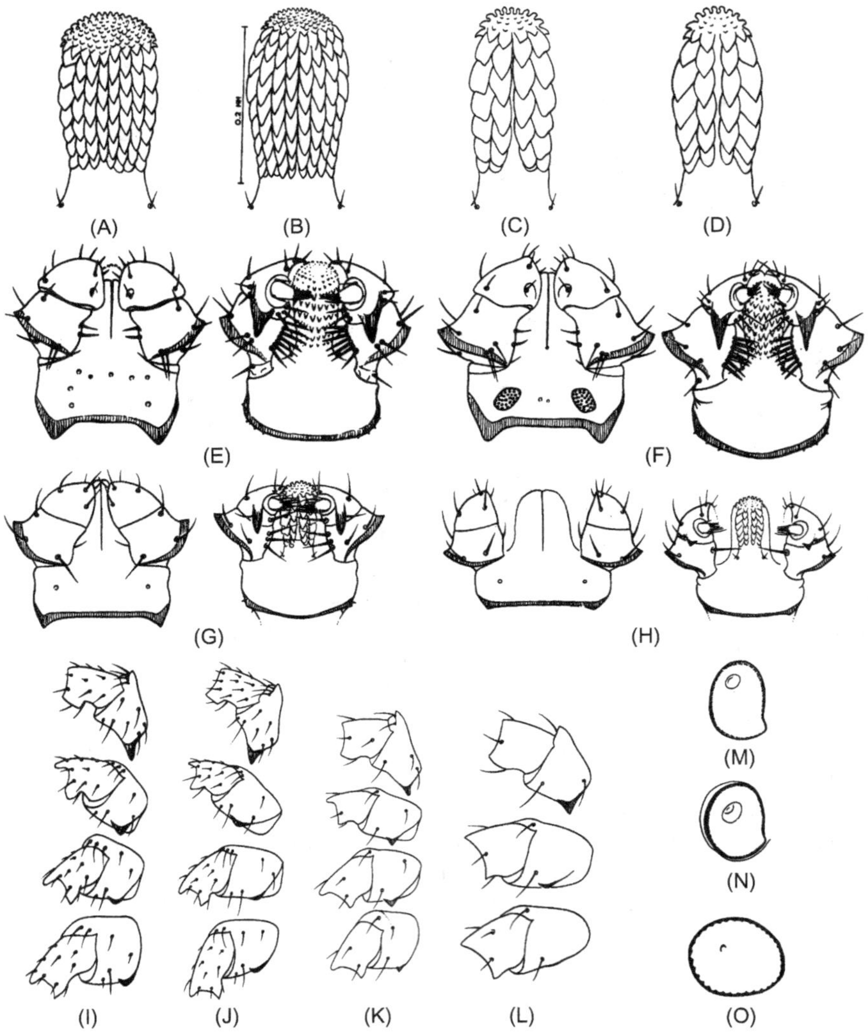

Figure 2.101 (A–D) Hypostome of male, female, nymph, and larva, respectively; (E–H) capitulum of male, female, nymph, and larva, respectively; (I–L) coxa of male, female, nymph, and larva, respectively; (M–O) spiracular plates of male, female, and nymph, respectively.
Source: NIV. unpublished drawings.

Nymph (Figures 2.100 and 2.101)[56]

The total body length of the nymph is about 1.1 mm.

Capitulum: Capitulum is 0.2 mm in length. Palpi are with length–width ratio slightly shorter than that of female and lacking spur on dorsobasal margin of segment III. Palps are compactly arranged. Infrainternal setae number three. Palpal segment II forms a good salience. Palpal segment III shows a moderate, elongately triangular, pointed spur, reaching to almost mid-length of palpal segment II. Hypostome is with 2/2 dental formula and showing six to seven denticles per file.

Scutum: The scutum is 1.4 times as wide as long. Punctations are very rare or equal to none. Festoons are well marked and deep. Cervical grooves are moderately deep and extend up to mid-length of the scutum.

Legs: Spurs of coxa are moderately long and pointed as compared to those in *H. turturis*, a closely related species.

Larva (Figures 2.100 and 2.101)

The total body length of larva is about 0.6 mm.

Capitulum: Capitulum is 0.3 mm long. Palpi are loosely arranged. A single infrainternal seta is present. Palpal segment II shows moderate to less salience. Palpal segment III shows a small spur just reaching to the upper margin of palpal segment II. Hypostome has a dental formula 2/2, with six denticles in each file.

Scutum: Scutum is 1.7 times as wide as long. No or very rare punctations are present on the scutum.

Legs: Coxa have moderate spurs; coxa I shows a moderately large spur while coxa II and III show small spurs.

Related Species[56]

Both sexes of *H. paraturturis* are quite similar to those of *H. turturis*. The male of this species is narrower and has distinct lateral grooves, which are absent in *H. turturis*. The infrainternal setae of palpal segment II number seven to nine and are closely spaced in *H. paraturturis*, and four or five widely spaced in *H. turturis*. The spiracular plates are much narrower in males of the *H. paraturturis*. The ventral spurs of the trochanters of both sexes are rudimentary in this species but distinct in *turturis*, and the scutal punctations of both sexes of this species are more irregular. *H. turturis* lacks the seta-bearing pit near the inner basal margin of palpal segment III, which is present in the *H. paraturturis* as well as in the related species *H. formosensis*, which is, however, a larger, more rugged species lacking any vestige of a basolateral salience of its palpi. *H. silvafelis* is also a member of the *turturis* group, but differs in reduction of the spur of the dorsobasal margin of palpal segment III, form of infrainternal setae of palpal segment II, scutal punctations, tarsal outlines, spiracular plate outlines, scutal outlines, etc.

Host[56,120,143]

Immature stages: Crow pheasants.

Adults: Carnivore, cattle, jungle cat, bullock, calf, bear, mongoose, birds, wild dogs, goat, wild boar, wolf.

Distribution[56,120,143]

India (Andhra Pradesh, Jharkhand (previously part of Bihar), Himachal Pradesh, Jammu and Kashmir, Maharashtra, Uttarakhand (previously part of Uttar Pradesh), Orissa, Madhya Pradesh, Tamil Nadu).

Disease Relationship

Not recorded.

2.7.17 Haemaphysalis silvafelis[54]

This species was described first by Hoogstraal and Trapido in 1963 collected from a red jungle cat (*Felis chaus*) in the Hyderabad area of Andhra Pradesh state of India, so the name may have been after the scientific name of the host. This species is closely related to *H. turturis* in many morphologic features but differs in some characteristics (Figure 2.102).

Male (Figures 2.103 and 2.104)[54]

Overall body length is approximately 1.7 mm and width 1.0 mm.

Capitulum: Basis capitulum is 1.7 times as wide as long; cornua are approximately half as long as the base of basis capitulum, acutely triangular; surface has several large, shallow punctations. Palpi are sub-conical, lacking a pronounced

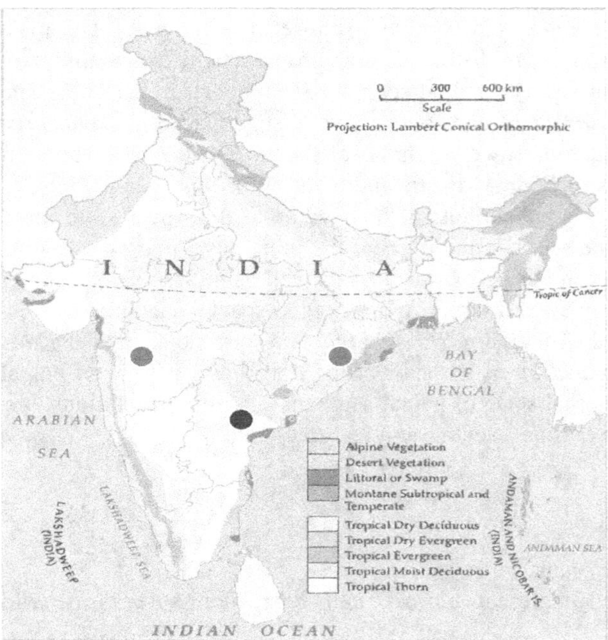

Figure 2.102 Distribution of *H. silvafelis* in India. (●) Place of first record. Map not to scale.

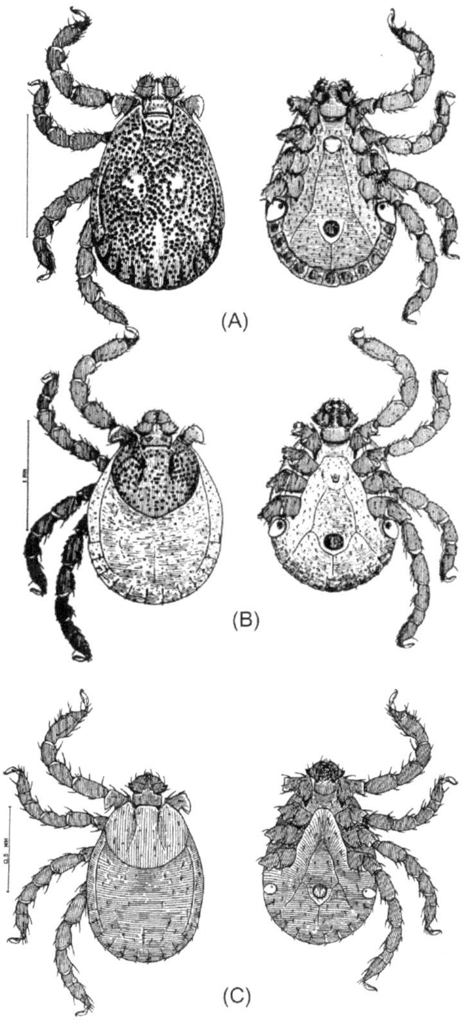

Figure 2.103 *Haemaphysalis silvafelis*. (A) Male, dorsal and ventral view; (B) female, dorsal and ventral view; (C) nymph, dorsal and ventral view.
Source: Hoogstraal and Tapido (1963), Journal of Parasitology (vol 49(2)), Allen press Publishing Services.

basolateral salience; palpal outline is approximately twice as wide as long. Hypostome is with slightly convex lateral margins. Dental formula is 4/4, with files of approximately five larger anterior and three smaller posterior denticles.

Scutum: Scutum is approximately two-thirds as wide as long, widest at the level of coxa IV. Lateral grooves are reaching the level of posterior margin of coxa 11, enclosing one pair of festoons. Cervical grooves are deep, short, slightly converging. Punctations are numerous, mostly large, moderately deep, many contiguous, scattered rather evenly over entire surface including festoons and lateral ridges. Festoon number 11, clearly defined.

Legs: Coxa are each with a well-defined spur; I with a large, elongate, moderately narrow, acutely pointed spur; IV with a similar but somewhat smaller spur.

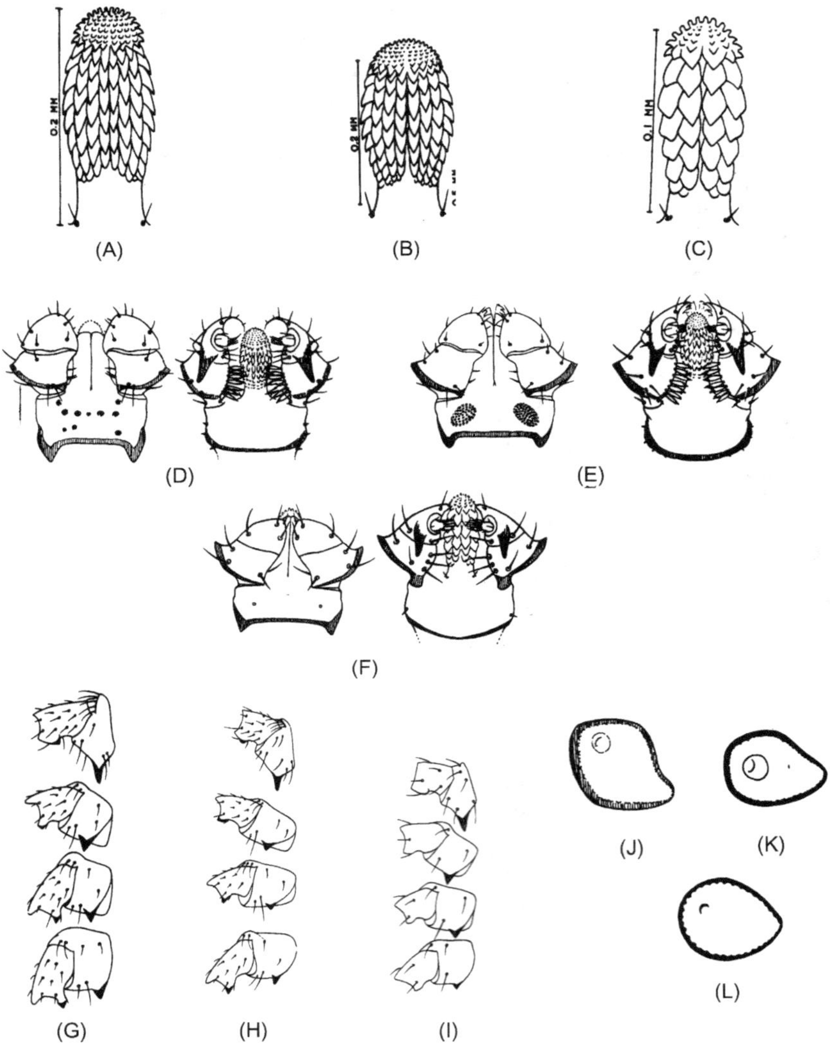

Figure 2.104 (A–C) Hypostome of male, female, and nymph, respectively; (D–F) capitulum of male, female, and nymph, respectively; (G–I) coxa of male, female, and nymph, respectively; (J–L) spiracular plates of male, female, and nymph, respectively. *Source*: NIV. unpublished drawings.

Female (Figures 2.103 and 2.104)[54]

Total body length measures approximately 1.7 mm.

Capitulum: Capitulum is slightly different from that of male. Basis capituli are shorter and wider than that of male; porose areas are sub-ovate, widely spaced. Palpi are with a slight basolateral salience formed by wider basal margins and more acutely recurved angle of juncture with lateral margin; palpi are otherwise quite similar to those of male. Hypostome is with one or two more denticles in each file than that of male, but otherwise similar.

Scutum: Scutum is with width very slightly exceeding length (sub-equal); lateral margins are widely convex, posterior margin is abruptly rounded and with a very small, broadly rounded median projection. Cervical grooves are deep, converging anteriorly and diverging posteriorly, extending approximately to mid-length of scutum. Punctations are less numerous but otherwise similar to those of male.

Legs: Coxal spurs and trochanters are similar to those of male.

Nymph (Figures 2.103 and 2.104)

Total body length is about 1.12 mm.

Capitulum: The capitulum is 0.2 mm long with a broad salience of palpal segment II. Palpal segment II also shows an externobasal pointed margin-like spur. Palpal segment III bears a very long, triangular, blunt spur extending to more than half of palpal segment II. Infrainternal setae number three. Hypostome is having 2/2 dental formula, with six to seven denticles in each file.

Scutum: The scutum is 1.3 times as wide as long. Punctations are very few, rare, small, and shallow. Cervical grooves are very long and extend to two-thirds of scutum length and are shallow.

Legs: Coxa are showing pointed spurs as compared to *H. turturis* and *H. paraturturis*, the closely related species. Coxa I shows an enlarged, elongately triangular pointed spur, reaching to almost upper margin of coxa II. Coxa II to IV bear short and triangular spurs.

Related Species[54]

The species is morphologically related to *turturis*. The male of *silvafelis* could be separated from *turturis* chiefly by the presence of long lateral groves, larger scutal punctations, rudimentary spurs on trochanters ventrally, and reduction of elevated palpal spurs of dorsobasal margin of segment III to a mere curvature of the margin in this place. The infrainternal setae of the palpi of *H. turturis* are narrow and number four or five; those of *H. silvafelis* are broad and number seven.

Host[54,93,143]

Immature stages: Not available.

Adults: Jungle cat, small mammals, hare.

Distribution[54,93,143]

India (Andhra Pradesh, Maharashtra, Orissa).

Disease Relationship

Not recorded.

2.8 Subgenus *Aborphysalis*

General Characteristics

Aborphysalis species are *H. aborensis*, *H. kyasanurensis*, *H. formosensis*, *H. atherurus*, and *H. capricornis*. Adult *Aborphysalis* are ectoparasites of deer and pigs but more frequently of porcupines, suggesting their coevolution with porcupines. *Aborphysalis* palpi differ from those of most haemaphysalines except *Segalia* in being broadly expanded in either larvae and nymphs or only in nymphs and compact or campanulate in adults. In *Ornithophysalis* and *Rhipistoma* both adult and immature palpi are broad. Adult *Aborphysalis* either have very small capitular, coxal, and trochantal spurs as in *H. capricornis*, or moderate sized coxal spurs as in the male of *H. atherurus* or moderately large coxal spurs in both sexes of *H. aborensis*, *H. kyasanurensis*, *H. formosensis*; and the males also have trochantal spurs ventrally as in *H. formosensis*. All capitular spurs (cornua and palpal segment III ventrally) are much reduced and no other hair-hooking devices except the leg spurs are present on these ticks. This unusual combination is likely to be associated with the variety of quills, bristles, and soft and hard hairs in different areas of bodies of these host genera. The reversal of the unusual immature-adult palpal form in *Aborphysalis* is postulated to be a functional adaptation for co-existence with spiny and quilled porcupines.

In India only two species, i.e., *H. kyasanurensis* and *H. aborensis* belonging to this species have been recorded.

Keys to Identify Species of Subgenus Aborphysalis

Males

Palpi with posteroexternal juncture of segment II are forming a small but prominent projection; ventral spur of segment III is not extending beyond intersegmental suture; cornua are triangular, sharp apically; scutal punctations are obsolete or few and obscure; coxal spurs are sharp apically; trochanter I ventral spur is distinctly developed..[*H. kyasanurensis*]

Palpi with posteroexternal juncture of segment II are at right angle or rounded or forming a minute projection; ventral spur of segment III is extending slightly beyond intersegmental suture; cornua are sub-triangular, bluntly pointed apically; scutal punctations are moderately numerous; few centrally, coxal spurs are blunted apically; trochanter I ventral spur is obsolete or indistinctly developed.................. [*H. aborensis*]

Females

Cornua are pointed apically; palpal segment II is with posteroexternal juncture forming a prominent projection; coxa I spur is elongately triangular (extending well beyond anterior margin of coxa II); scutal punctations are usually fewer and not discrete..[*H. kyasanurensis*]

Cornua are rounded apically (and usually shorter); palpal segment II is with posteroexternal juncture at an angle or rounded or minutely salient; coxa I spur is peg-like, blunt apically (extending only slightly)...................[*H. aborensis*]

2.8.1 Haemaphysalis kyasanurensis

This species was first described by Trapido et al. in 1964 from specimens collected in the area of Kyasanur village of Shimoga district of Karnataka state of India, hence the name (Figure 2.105).[38,141]

Male (Figures 2.106 and 2.107)[141]

Mean length measures approximately 2.8 mm and mean width 1.7 mm.

Capitulum: Basis capituli are rectangular, approximately 1.7 times as wide as long; cornua are at least as long as their basal width and one-third as long as base of basis capituli. Surface has a few faint punctations. Palpi are compact, subquadrate, with a slight, abrupt basolateral salience; each palpus is 1.5–1.7 times as long as wide. Hypostome is short, stout; lateral margins are convex; dental formula is 4/4, with denticles in files of approximately eight (inner) to twelve (outer).

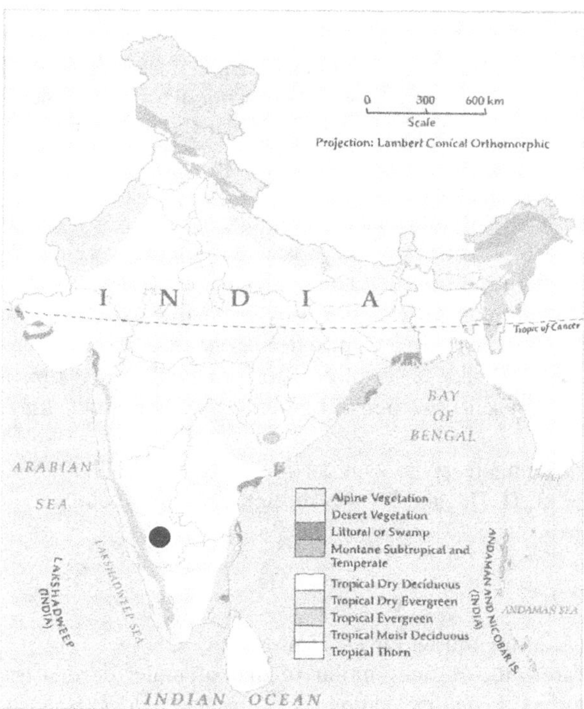

Figure 2.105 Distribution of *H. kyasanurensis* in India. (●) Place of first record. Map not to scale.

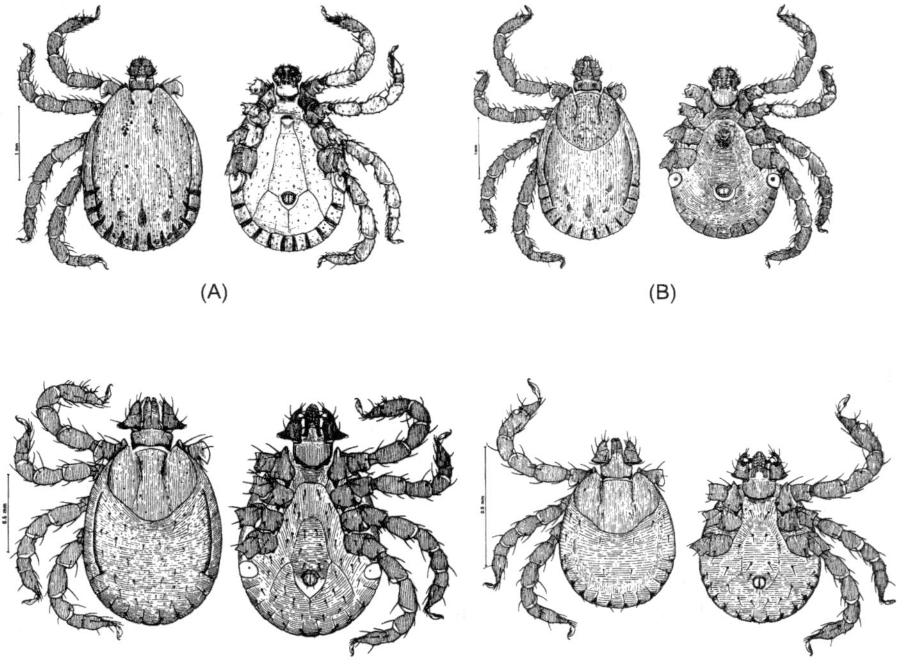

(A) (B) (C) (D)

Figure 2.106 *Haemaphysalis kyasanurensis.* (A) Male, dorsal and ventral side; (B) female, dorsal and ventral side; (C) nymph, dorsal and ventral side; (D) larva, dorsal and ventral side. *Source*: Trapido et al. (1964), Journal of Parasitology (vol 50(2)), Allen press Publishing Services.

Scutum: Scutum is flat or very slightly convex between deeply depressed lateral grooves and elevated lateral ridges; 1.5 times as long as wide; lateral margins are almost straight and parallel, or slightly convex. Lateral grooves are long, distinct, extending anteriorly to the level of coxa III, posteriorly enclosing two festoons. Cervical grooves are very narrow, converging from anterior emargination of scutum to short, deep, rounded cervical pits. Punctations are practically obsolete, a few shallow punctations may be present. Festoons number 11, exceptionally long, distinctly separated.

Legs: Coxa each is with an elongate spur; spur on I is longest, parallel sided, bluntly rounded apically; spurs on II, III, and IV are shorter, sub-equal, elongately triangular.

Female (Figures 2.106 and 2.107)[141]

Mean length is approximately 3.2 mm and width 1.8 mm.

Capitulum: Basis capituli are with outline similar to that of male; cornua are shorter than those of male; porose areas are shallow, sub-circular, widely spaced,

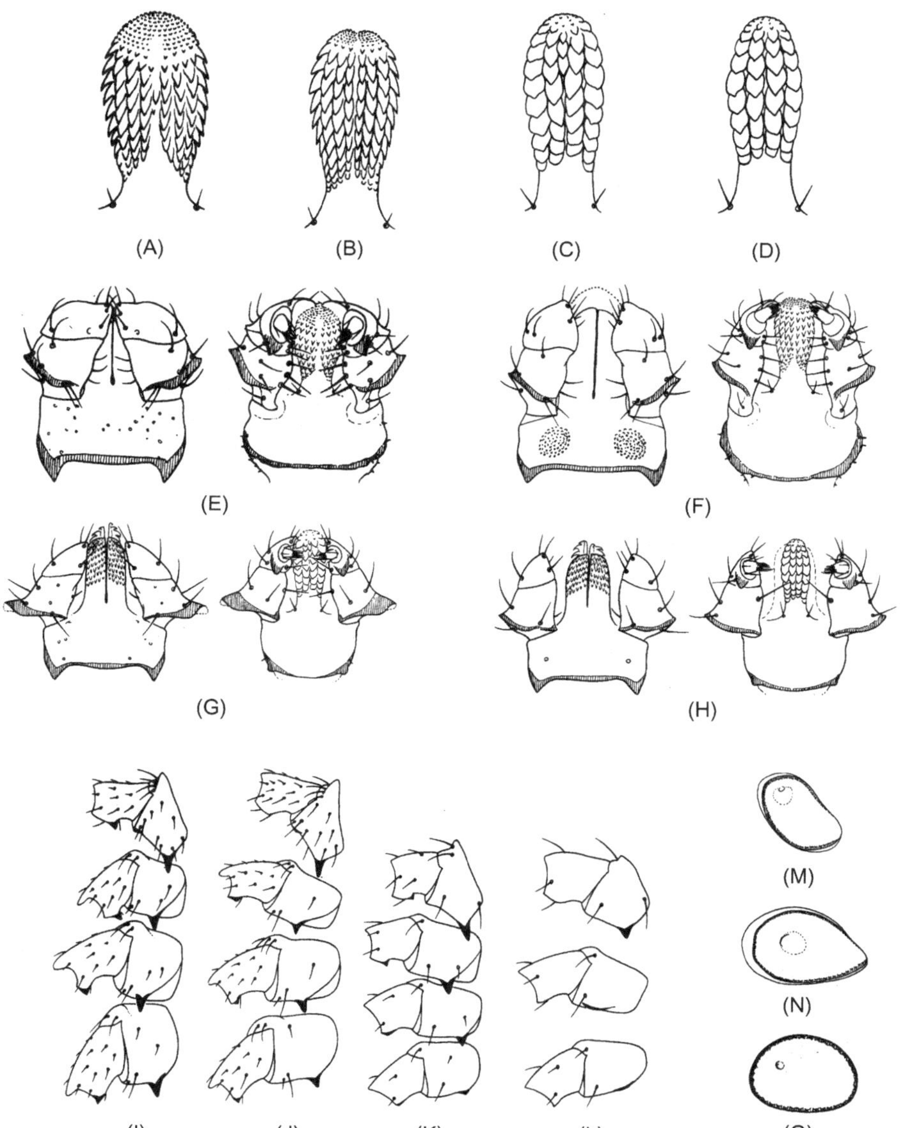

Figure 2.107 (A–D) Hypostome of male, female, nymph, and larva, respectively; (E–H) capitulum of male, female, nymph, and larva, respectively; (I–L) coxa of male, female, nymph, and larva, respectively; (M–O) spiracular plates of male, female, and nymph, respectively.
Source: NIV. unpublished drawings.

moderate size. Palpi are quite elongate, with a basolateral salience similar to that of male. Hypostome is elongate, slightly longer than palpi; lateral margins are slightly convex; dental formula is as in male but with one to four more denticles in each file.

Scutum: Scutum is glossy, sub-circular, length and width sub-equal, margins rounded. Cervical grooves are as in male except that shallow, diverging grooves extend from cervical pits to beyond mid-length of scutum. Punctations are shallow, indistinct, small and medium size, moderately numerous, widely spaced.

Nymph (Figures 2.106 and 2.107)[141]

Mean body length is approximately 1.4 mm.

Capitulum: Basis capitulum is dorsally approximately twice as wide as long; dorsal cornua are as in males; ventral cornua are short, bluntly rounded in each basolateral corner; a short seta is present just anterior of each basolateral juncture, and a pair of comparatively large posthypostomal bristles is also present. Palpi are with a wide basolateral salience. Hypostome is slightly longer than palpi, elongate; lateral margins are slightly convex; dental formula is 2/2, with denticles broad in files of nine to eleven.

Scutum: Scutum is slightly shorter than wide; margins are rounded to mid-length. Cervical grooves are deep and parallel or slightly converging, shallow and diverging on posterior half of scutum. Punctations are rare, obscure.

Legs: Coxa I is with a large, elongately triangular basal spur; II, III, and IV each is with a deltoid spur successively decreasing in length.

Larva (Figures 2.106 and 2.107)[141]

Mean body length is approximately 0.70 mm and width 0.52 mm.

Capitulum: Basis capitulum is quite similar to that of nymph, no ventral setae noted. Palpi are 1.6 times as long as wide. Hypostome is similar to that of nymph.

Scutum: Scutum is 1.4 times as wide as long; lateral margins are anteriorly gradually rounded. Cervical grooves are straight, parallel, extending to scutal mid-length. Punctations number four, each with a small seta.

Legs: Coxa I is with a large basal spur similar to that of nymph; II is with a very short, wide spur; III is with a minute, barely discernible inner spur.

Species Relationship[141]

H. kyasanurensis of southwestern India and Ceylon is closely related to *H. formosensis* of Formosa and southeastern Asia, a seldom-collected species whose immature stages remain unknown. Both species are moderately large and share structural characters that easily separate the *formosensis* group from others in this genus. *H. kyasanurensis* differs from *H. formosensis* in having longer palpal segment II with a distinct, abrupt basolateral salience; typically three rather than four infrainternal setae; a shorter ventral spur of segment III; 4/4 rather than 6/6 dental formula; longer cornua; fewer scutal punctations; longer coxal spurs, more slender and pointed; longer and narrower tarsi, etc.

Life History[38]

The life cycle of this species was studied by Geevarghese et al. in 1973 under ambient temperature (17–28 °C). All the nonparasitic stages were held at 90–100% relative humidity and room temperature. The maximum longevity of unengorged ticks was found to be 6 months. Larvae were fed on approximately 2-week-old white leghorn chicks and on white-tailed rats (*Rattus blanfordi*). For feeding, the best results were obtained when the newly emerged adults were held for a period of about 3 weeks before releasing them on the host. When released on the host, most adults were attached within 24 h. Copulating pairs were observed between the third day of release and the last day of engorgement. The females started dropping seventh day onward and continued up to the tenth day. The first phase lasted until the last but one day of parasitic phase. During this phase, the rate of increase in size was very slow. The second phase, which terminated with the dropping, was of a short duration, not beyond the last night of feeding. This was the period of active engorgement, during which the tick imbibed the bulk of blood meal and attained full repletion. The average amount of blood imbibed by the tick was approximately 135 times their average unengorged body weight (1.8 mg). The oviposition commenced 7–18 days after their detachment from the host and continued for 16–29 days. Each tick laid between 1,194 and 3,057 eggs. The number of eggs laid by each individual was directly proportional to its fed weight. The spent females died 6–28 days after the completion of oviposition. The loss in the initial weight after the oviposition ranged between 58% and 71%. Hatching commenced 47–67 days after the initiation of oviposition. The process started with the eggs laid on the first day and ended with those laid on the last day. In a single brood observed, the maximum longevity of the larvae was 180 days. The time taken for completion of larval engorgement was 4–7 days. As in the case of female, the feeding larvae also showed two well-defined phases. The larvae stayed on the chicks in the first phase for a minimum period of about 48 h, with 2 days and 2 nights, before passing on to the second phase. The larvae, which failed to pass on to the second phase on the third night, were carried on for the subsequent night for engorgement and the following day for dropping. This periodicity continued till the last larva dropped off. Engorged larvae molted into nymphs 15–18 days after dropping. The maximum longevity of the unengorged nymphs was observed to be 103 days. Nymphs, when released on chicks for feeding, followed a similar pattern of events as larvae. Engorged nymphs molted into adults 18–25 days after dropping (Table 2.9).

Host[141]

Immature stages: Mouse deer, wild dog, crested porcupine, cattle, buffalo, handpicking in KFD area of Karnataka (India).

Adults: Wild boar, wild dog, crested porcupine, small mammals.

Table 2.9 Life Cycle of *Haemaphysalis kyasanurensis* in Laboratory

Events	Number of Days	Number of Observations
Pre-oviposition	7–18	21
Oviposition	16–29	21
Incubation	47–67	10 egg masses
Pre-feeding of larva	5–7	6 broods
Feeding of larva[a,b]	4–7	6 batches
Molting of larva	15–18	6 batches
Pre-feeding of nymph	5–7	3 batches
Feeding of nymph[a]	4–7	3 batches
Molting of nymph	18–25	3 batches
Pre-feeding of adult	20–25	21 pairs
Feeding of female[c]	7–10	21

[a] Fed on chick.
[b] Fed on rats.
[c] Fed on rabbit.

Distribution[141]

India (Karnataka), Sri Lanka.

Disease Relationship[10]

A number of KFD isolations, single isolation from larvae, 41 isolations from nymphs, and 4 isolations from adults during the period of 1961–1972, have been made from the immature as well as adult stages of this species.[10]

Ecology[118]

This is a species with apparently restricted distribution in the western parts of India. It is fairly common in the KFD area, the adults of which chiefly parasitizing wild animals such as *Hystrix indica*, *Canis aureus*, etc.

2.8.2 Haemaphysalis aborensis

This species was briefly described from a single female collected from Abor hills of Arunachal Pradesh, India by Warburton in 1913, hence the name. In this species, the adult palpi are compact while the immatures are salient, the characteristic feature of the subgenera *Aborphysalis* (Figure 2.108).

Male (Figures 2.109 and 2.110)[78]

Overall body length is 2.33–3.08 mm (average 2.79 mm), breadth 1.49–2.00 mm (average 1.80 mm).

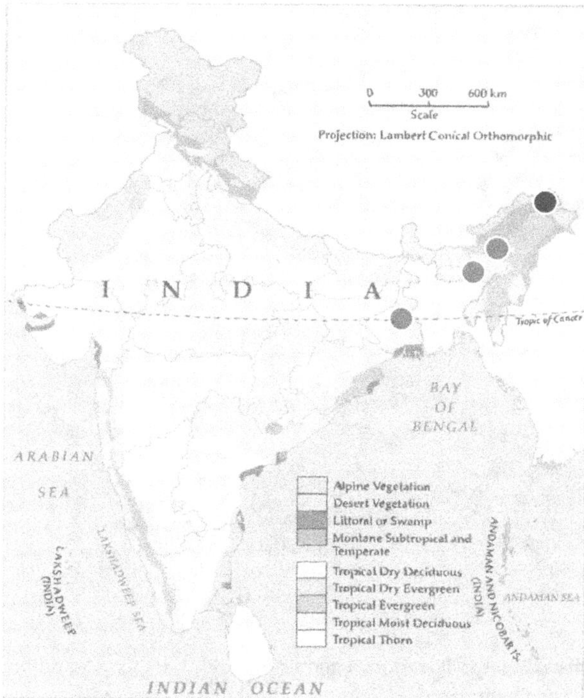

Figure 2.108 Distribution of *H. aborensis* in India. (●) Place of first record. Map not to scale.

Capitulum: Basis capitulum is dorsally approximately 1.9 times as broad as long; external margins are essentially straight; cornua are broadly triangular, approximately one-fourth as long as the base of basis capituli, bluntly pointed apically. Palpi are compact; setae number five dorsally, three or four ventrally. Hypostome is somewhat shorter than palpi; narrow posteriorly, small hooklets in five or six dense rows; dental formula is 5/5, with denticles in files of six to nine.

Scutum: Scutum is smooth, glossy, flat, and 1.26–1.31 times as long as broad. Cervical pits are small, sub-circular, deep, and sub-anterior. Cervical grooves are obsolete. Lateral grooves are narrow, distinct, extend to the level of coxa III, enclose first festoon; surface lateral of grooves is somewhat bulging. Punctations are shallow but distinct, small and medium size, concentrated chiefly posteriorly and anterolaterally, fewer or absent elsewhere. Festoons number 11. Spiracular plates are ovate, 1.5–1.8 times as long as broad.

Legs: Legs are long and narrow. All coxal spurs lengths are sub-equal.

Female (Figures 2.109 and 2.110)[78]

Mean length of body is 2.79–3.56 mm (average 3.32 mm), breadth 1.69–2.23 mm (average 2.04 mm).

Capitulum: Basis capituli are dorsally 2.5–2.8 times as broad as long; external margins are straight; cornua are short, broadly triangular, one-fourth to one-third as

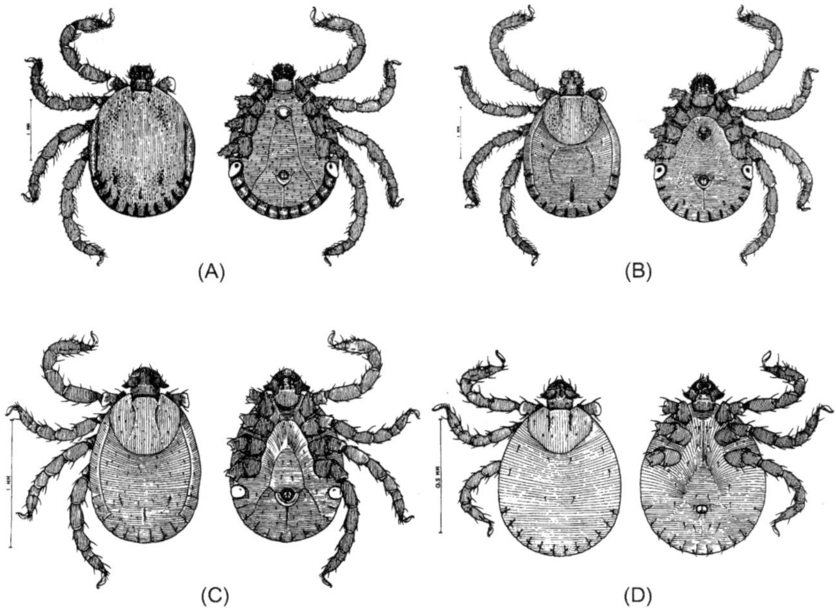

Figure 2.109 *Haemaphysalis aborensis*. (A) Male, dorsal and ventral side; (B) female, dorsal and ventral side; (C) nymph, dorsal and ventral side; (D) larva, dorsal and ventral side. *Source*: Hoogstraal et al. (1971), Journal of Parasitology (vol 57), Allen press Publishing Services.

long as the base of basis capituli; porose areas are small to moderately large, oval or sub-oval. Palpi are compact; dorsointernal and ventrointernal setae each number three or four. Hypostome is somewhat shorter than palpi, narrow posteriorly; dental formula is 5/5, with denticles in files of six to nine.

Scutum: Scutum is smooth, glossy, flat; cervical pits are small, sub-circular, deep, and sub-anterior. Cervical grooves are obsolete. Lateral grooves are narrow, distinct, extend to the level of coxa III, enclose first festoon; surface lateral of grooves is somewhat bulging. Punctations are shallow but distinct, small and medium size, concentrated chiefly posteriorly and anterolaterally, fewer or absent elsewhere. Festoons number 11. Legs are long and narrow.

Legs: All coxal spurs lengths are sub-equal.

Nymph (Figures 2.109 and 2.110)[78]

Overall body length is 1.21−1.54 mm (average 1.33 mm), breadth 0.71−0.82 mm (average 0.77 mm).

Capitulum: Basis capituli are dorsally approximately 1.6−1.8 times as broad as long; external margins are straight; cornua are elongately triangular, sharply pointed, approximately two-thirds as long as the base of basis capitulum. Palpi are broadly salient posteriorly; combined breadth is 1.5 times breadth of basis

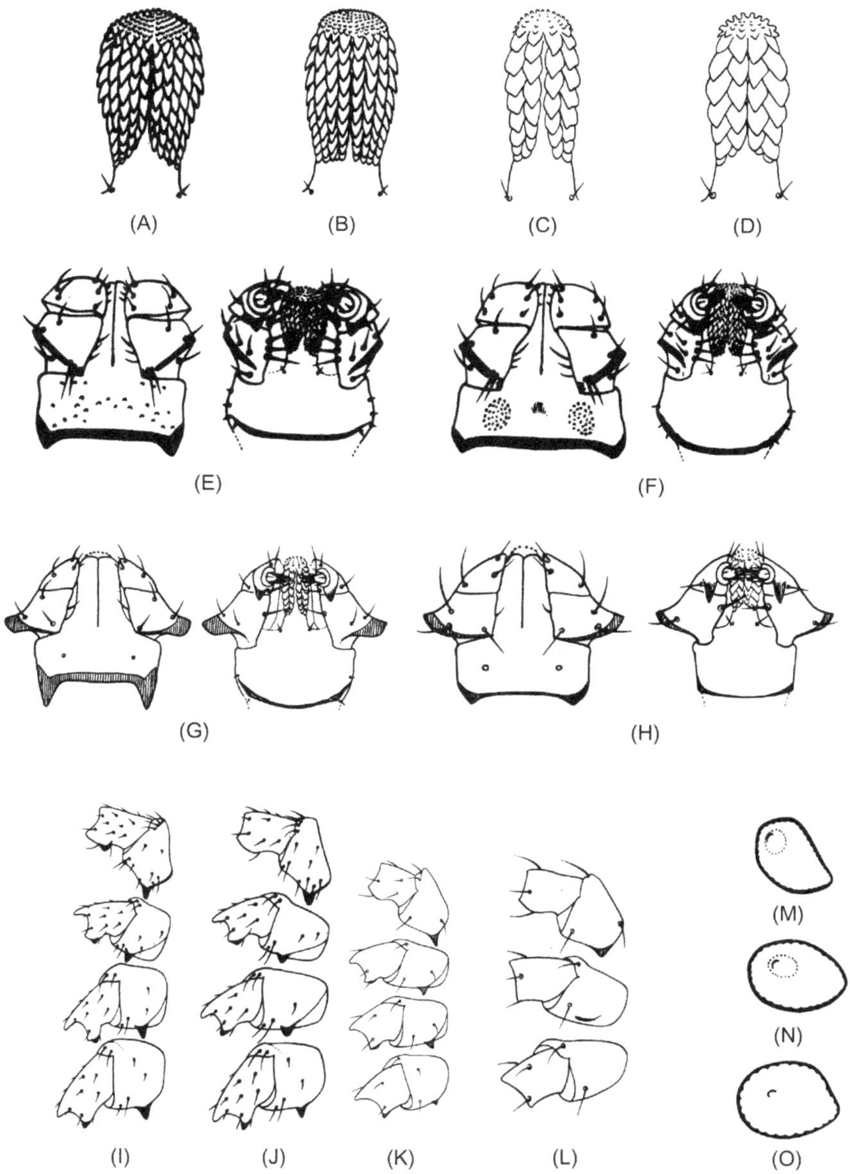

Figure 2.110 (A–D) Hypostome of male, female, nymph, and larva, respectively; (E–H) capitulum of male, female, nymph, and larva, respectively; (I–L) coxa of male, female, nymph, and larva, respectively; (M–O) spiracular plates of male, female, and nymph, respectively.
Source: NIV. unpublished drawings.

capitulum. Hypostome is as long as or longer than palpi; dental formula is 2/2, with denticles in files of seven or eight.

Scutum: Scutum is 1.1–1.2 times as broad as long. Cervical grooves are converging from sub-anterior pits, posteriorly diverging to beyond scutal mid-length. Punctations are few, inconspicuous. Spiracular plates are sub-circular.

Legs: Legs are moderately long and narrow.

Larva (Figures 2.109 and 2.110)[78]

Mean length is approximately 0.8 mm and breadth 0.5 mm.

Capitulum: Basis capitulum is dorsally approximately 2 times as broad as long; external margins are straight; cornua are broadly triangular, approximately one-fifth as long as the base of basis capitulum. Palpi are broadly salient posteriorly; combined breadth is approximately 1.45 times breadth of basis capitulum. Hypostome is slightly longer than palpi; dental formula is 2/2, with denticles in files of five to seven.

Scutum: Scutum is about 13 times as broad as long. Genital grooves are as arcs to scutal mid-length. Punctations are few, inconspicuous.

Ecology

This species is generally found on altitudes between 1,600 and 5,000 ft, from domestic animals or from forest vegetation or forest-dwelling vertebrates. One female crawling on a person was taken. The temperature in the collection localities ranges from 36 °C in summer to 7 °C in winter. The average annual rainfall of about 140 inches is mostly during the monsoon months of June to September, but the relative humidity remains high throughout the year owing to frequent heavy showers. The wet tropical evergreen forests in this area have dense shrub undergrowth.

Hosts[78]

Immature stages: Green backed tit birds, porcupines, domestic buffalo, domestic fowl, tree shrew, wild boar, hog badger.

Adults: Tiger, wild boar, barking deer, porcupine, zhum cattle (cow–yak hybrid), sambar deer, single adult crawling on human, flag dragging and handpicking in West Bengal.

Distribution[78]

India (Assam, Meghalaya, West Bengal, Arunachal Pradesh), Nepal, Burma, Thailand, Laos, Vietnam.

Species Relationship

H. aborensis is characterized by the compact palpi of adults and salient palpi of immatures which is the characteristic of subgenus *Aborphysalis*. Within the subgenus *Aborphysalis* this species is very closely related to *Haemaphysalis kyasanurensis* and *Haemaphysalis formosensis* by structural similarity. These two species with *aborensis* form an *aborensis* group in this subgenus. *H. formosensis* is not found in India. These three species differ from each other in degree of development of spurs and dental formulae. In females of *H. kyasanurensis*, the cornua are pointed apically whereas in *H. aborensis*, the cornua are shorter and rounded apically.

Disease Relationship

There have been no records of involvement of this species in disease transmission so far.

2.9 Subgenus *Segalia*

Santos Dias in 1963 proposed subgenus *Fonsecaia* for a single species, *H. montgomeryi*. Later, in 1968, Santos Dias replaced it with *Segalia*. *Segalia* consists of two groups of species, *H. parva* group and *H. obesa* group. As in *Aborphysalis*, immature *Segalia* palpi are broader than those of adults.

In *H. parva* group, immature palpal segment II forms large, unique posteroventral spurs and nymphal *H. montgomeryi* coxal spurs are unusually large. The adult palpi are campanulate in male and broadly campanulate in female *H. parva* and elongately campanulate in *H. montgomeryi*. The adult *H. parva* palpal segment III ventral spur and coxal spurs are only moderately large. However, *H. montgomeryi* palpal segment III spur is unusually elongate in male and even more in female. The male *H. montgomeryi* coxal and trochantal spurs are all extraordinarily large and the female coxal are moderately large.

H. parva group immatures are structurally similar to the most recently evolved *Rhipistoma* immatures. Adults are similar to those of the subgenus *Haemaphysalis* but are unspecialized in *H. parva* and specialized in *H. montgomeryi*.

The *H. obesa* group parallels in certain respect, the *H. parva* group. Immatures have broadly campanulate palpi with a moderately large ventral spur on segment III but lacking the segment II ventral spur characterizing the *H. parva* group and a moderately large spur on coxa I. Adult palpi are compact or compact elongate and have a moderately large ventral spur on segment III. Adult cornua are small or moderate sized; however, immature cornua are relatively large except larval *H. hirusta*.

Keys to Identify Species of Subgenus Segalia

Males

1. Palpi are compact; posterodorsal margin of palpal segment II is ridged while posteroventral margin bears a greatly pointed; triangular spur is on palpal segment II; posterodorsal margin of palpal segment III is plane while posteroventral margin of palpal segment III shows a very much pointed, large and elongate spur reaching to the base of palpal segment

II; infrainternal setae number seven; dental formula is 6/6 or 7/7; all coxa and trochanters show large, triangular spur; coxa I and IV show large, broadly triangular spur while coxa II and III show small but triangular and pointed spurs............................[*H. montgomeryi*]
Palpi are compact, so no salience; posterodorsal spur on palpal segment III is broadly triangular or ridge-like; posteroventral spur is conventionally triangular and overlaps half of palpal segment II; infrainternal setae number four; lanceolate, well spaced, and medium in length; dental formula is 5/5; all coxa and trochanters bear spur but are blunt and reduced in size from coxa I to IV..[*H. obese*]

Females

1. Posterodorsally palpal segment II shows broadly but conventionally triangular spur infrainternal setae are six and are long, lanceolate, and well spaced; dental formula is 7/7 or 8/8; porose areas are oval but elongate; well spaced but small, all coxa and trochanters bear small spurs but are triangular and pointed; coxa I and IV spurs are moderate and pointed but coxa II and III spurs are broadly triangular but small and pointed; trochanters show pointed spur but are small..[*H. montgomeryi*]
Posterodorsal margin of palpal segment II is ridged while posteroventral margin bears a moderate spur; conventionally triangular and reaching or overlapping one-third of palpal segment II; infrainternal setae number five or six, and are short but well spaced; porose areas are oval but inclined and large; coxa and trochanters bear spurs but coxa I spur is moderate and blunt while other coxa are ridged; trochanters are ridged..................
[*H. besa*]

2.9.1 Haemaphysalis montgomeryi

Nuttall first described this species in 1912 from the samples collected by R.E. Montgomery from Uttarakhand state (formerly a part of Uttar Pradesh state) of India in 1905. The name of the species is kept in honor of the collector of this species (Figure 2.111).

Male (Figures 2.112 and 2.113)[63]

Mean body length is 3.0 mm and average width is 1.6 mm.

Capitulum: Basis capitulum is dorsally approximately 1.3 times as wide as long; cornua are short, approximately one-third as long as the base of basis capitulum, widely triangular, apex sharply or bluntly pointed. Palpi are compact, lacking basolateral salience. Hypostome is as long as palpi; lateral outline is parallel; dental formula is 6/6 (rarely 6.5/6.5), with files of equal length consisting of approximately 15 narrowly elongate, closely spaced denticles.

Scutum: Scutum is bordered laterally by narrow strip of integument, approximately 1.5 times as long as wide; widest just anterior of the level of dorsal projection of spiracular plate. Scutal surface is glossy, convex within lateral ridges and anterior of festoons, depressed posteriorly. Lateral grooves are shallow, short, and usually enclose first festoon, and extend to or are very slightly beyond the level of

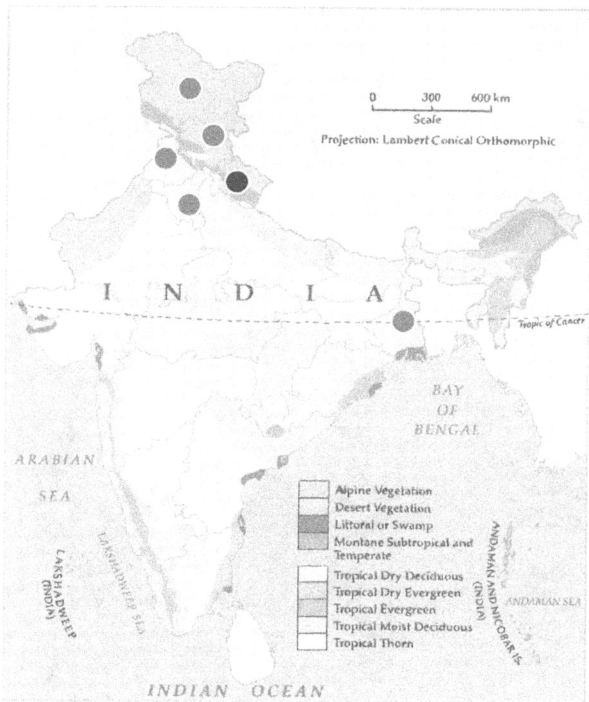

Figure 2.111 Distribution of *H. montgomeryi* in India. (●) Place of first record. Map not to scale.

widest area of scutum. Cervical grooves are short, deep, sub-parallel. Punctations are rare, widely scattered, shallow, obscure, mostly small, and some medium sized. Festoons number 11.

Legs: Legs are robust. Coxa each is with a prominent, elongately triangular, sharply pointed spur; spurs of II and III are sub-equal, slightly wider than others; spur of I is slightly longer than those of II and III; spur of IV is almost twice as long as those of II and III; each coxa is with numerous (12–20) long setae.

Female (Figures 2.112 and 2.113)[63]

Mean body length is approximately 3.2 mm, and width is 1.8 mm; length when engorged may attain 9.0 mm.

Capitulum: Capitulum resembles that of male. Basis capitulum is twice as wide as long; lateral margins are slightly diverging from cornua to apex; cornua are slightly shorter and often more blunt apically than those of male; porose areas are small, widely spaced, elongately oval. Palpi essentially are as in male though the dorsobasal margin of segment II is not bladelike; the basolateral juncture of segment II may form a minute salience, and the dorsobasal margin of segment III is less elevated than in male. Hypostome is slightly broader but otherwise similar to that of male; dental formula is 6/6, 6.5/6.5, or 7/7.

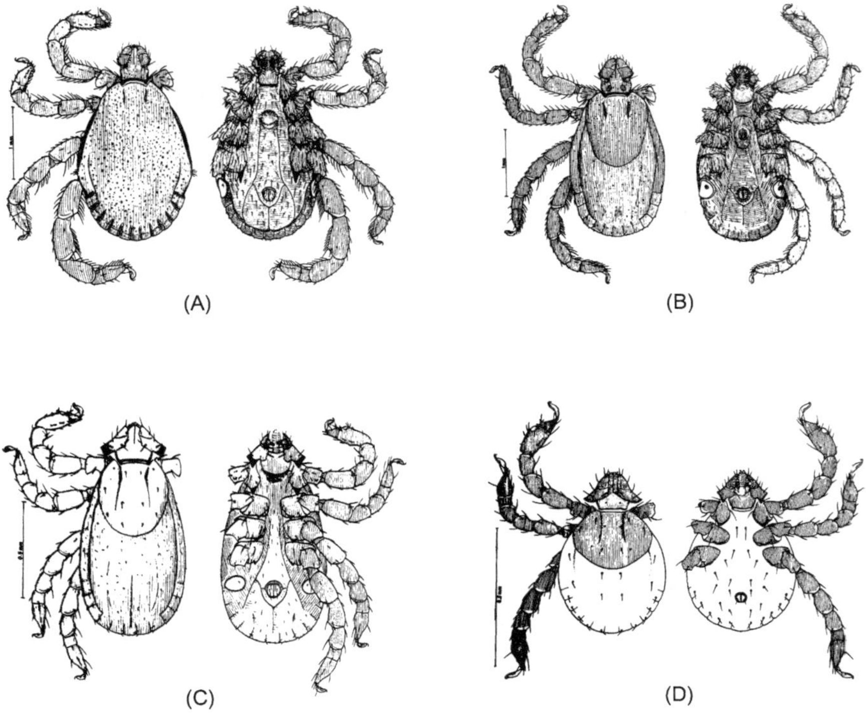

Figure 2.112 *Haemaphysalis montgomeryi.* (A) Male, dorsal and ventral side; (B) female, dorsal and ventral side; (C) nymph, dorsal and ventral side; (D) larva, dorsal and ventral side. *Source*: Hoogstraal et al. (1966), Journal of Parasitology (vol 52(1)), Allen press Publishing Services.

Scutum: Scutum is elongate, 1.25 times as long as wide; posterior margin is bluntly rounded; punctations are as in male. Genital operculum is highly distinctive, tongue-like, bordered laterally by labia.

Legs: Coxa are with spurs widely triangular, shorter than in male; spur of III is extending to or slightly beyond basal margin of coxa; spur of II, IV, and I each is slightly longer than the other in order listed; coxal setae are approximately as in male.

Nymph (Figures 2.112 and 2.113)[63]

This stage differs notably from the adult in that the palpi are widely salient. Length is approximately 1.1 mm, and width is 0.5 mm.

Capitulum: Basis capitulum is twice as wide as long; cornua are short, widely triangular, more or less sharply pointed apically, approximately one-third as long as the base of basis capituli. Palpi are widely salient (*leachi*-type). Hypostome is as long as palpi; dental formula is 2/2, with denticles large, in files of approximately seven.

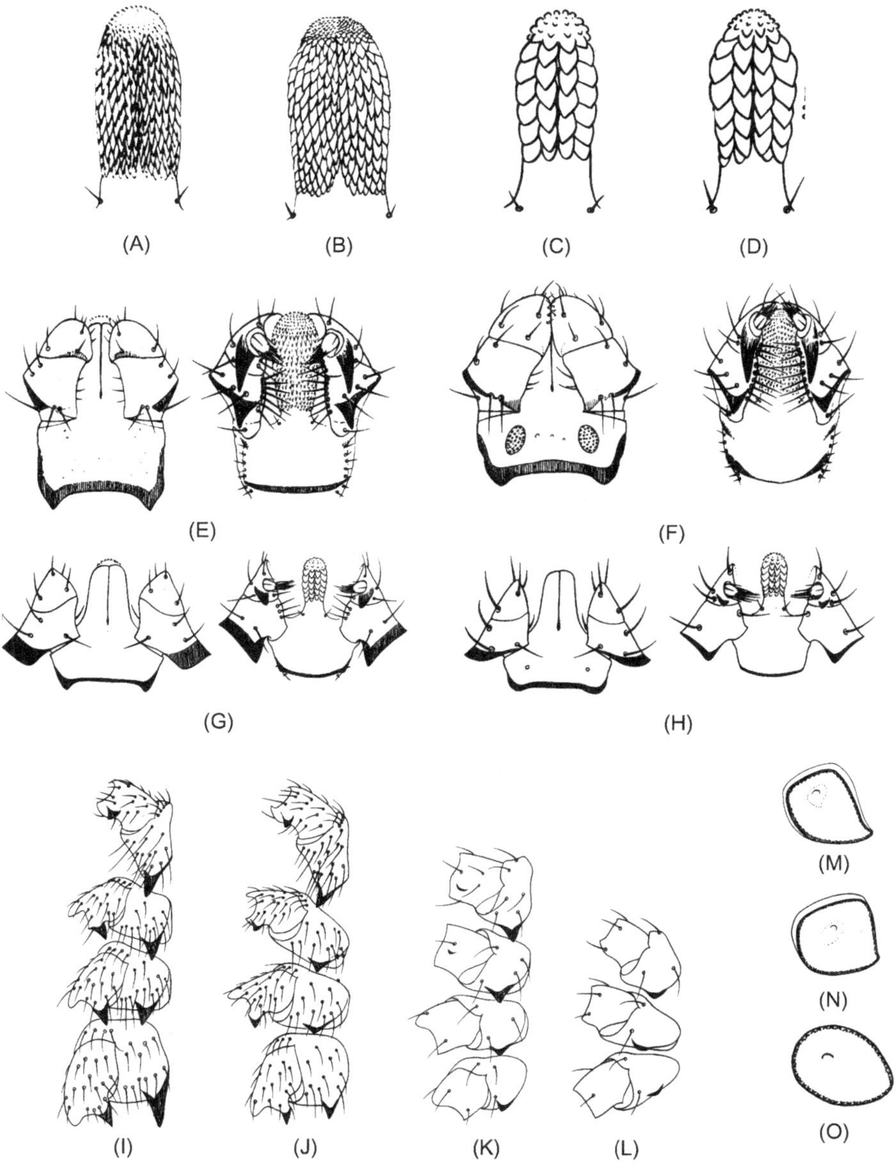

Figure 2.113 (A–D) Hypostome of male, female, nymph, and larva, respectively; (E–H) capitulum of male, female, nymph, and larva, respectively; (I–L) coxa of male, female, nymph, and larva, respectively; (M–O) spiracular plates of male, female, and nymph, respectively.
Source: NIV. unpublished drawings.

Scutum: Scutum is slightly longer than wide; lateral and posterior margins are gradually rounded. Cervical grooves are slightly converging anteriorly, slightly diverging and shallower posteriorly, extend somewhat beyond mid-length of scutum. Punctations are approximately 12 in number, each bearing a short seta. Spiracular plates are ovate.

Legs: Coxa are each with a short, widely triangular spur gradually increasing in length from IV to I.

Larva (Figures 2.112 and 2.113)[63]

Palpi of this stage are quite similar to those of nymph. Length is approximately 0.78 mm, and width is 0.38 mm.

Capitulum: Basis capitulum is slightly over twice as wide as long; cornua are reduced to small, rounded marginal bulges. Palpi are with outline and spurs similar to those of nymph. Number of setae on segment II is three dorsally, one ventrally; suprainternal and infrainternal setae each single; number of setae on segment III is three dorsally, three ventrally. Hypostome is essentially as in nymph.

Scutum: Scutum is 1.25 times as wide as long.

Legs: Coxa are each with a short, widely rounded spur extending slightly beyond coxal margin; spurs are gradually decreasing in length from I to III.

Hosts[63]

Immature stages: Natural host not reported.

Adults: Buffalo, cattle, goat, sheep, leopard, bullock, camel, cow, cow calf, rat, squirrel, dog, horse, mule, ponies, bull, serow, spotted nut cracker bird, marmots (rodent), one female removed while feeding on human in Nepal.

Distribution[63]

India (Himachal Pradesh, Jammu and Kashmir, Uttarakhand, Punjab, Haryana, West Bengal), Pakistan, Nepal.

Ecology[63]

H. montgomeryi is restricted ecologically to altitudes between 4,500 and 12,000 ft in the western Himalayas and its foot hills, from SWAT state of northern Pakistan to northern India farther to Kathmandu, valley of Nepal at altitudes. These areas extend from subtropical with mild winters to temperate with extreme winters. Data collected from Pakistan show feeding of both sexes between April (spring, earliest month of collecting) and July (summer), but suggest less female activity in August and almost none in September (latest month of collecting effort) though males still occur on animals. Nemenz (1962) reported two engorged females under stones in May in the Karakoram Range of Pakistan. Kashmir (India) data are chiefly from April and May (spring), when both sexes were common on animals. In Nepal, collections of *H. montgomeryi* were made in June, July, August, September, and

November (summer and fall). The pattern from June and July was like that in Pakistan but sufficient number of females was present in August and September.

Related Species

The species *montgomeryi* is closely related with *H. parva* morphologically. The species *H. parva* has not been recorded from India. Immature of *parva* and *montgomeryi* palpi II form large unique posteroventral spurs. The adult palpi are campanulate in male and broadly campanulate in female of *parva* while elongately campanulate in *montgomeryi*. The adult *parva* palpal segment III, ventral spur and coxal spurs are moderately large; however, in *montgomeryi* segment III ventral spur is unusually elongate, more in female. Coxal and trochantal spurs are extraordinarily large in male and moderate in female. Host adaptive structures are mild or reduced in adult *parva* and extremely developed in adult *montgomeryi*. Such structures are well developed in immatures of both species. Close phylogenetic relationship of these two species is revealed by similarity of distinctive female external genitalia.

2.9.2 *Haemaphysalis obesa*

This species was described for the first time by Larrousse in 1925 from Vietnam with a presumed host to be a barking deer (Figure 2.114).

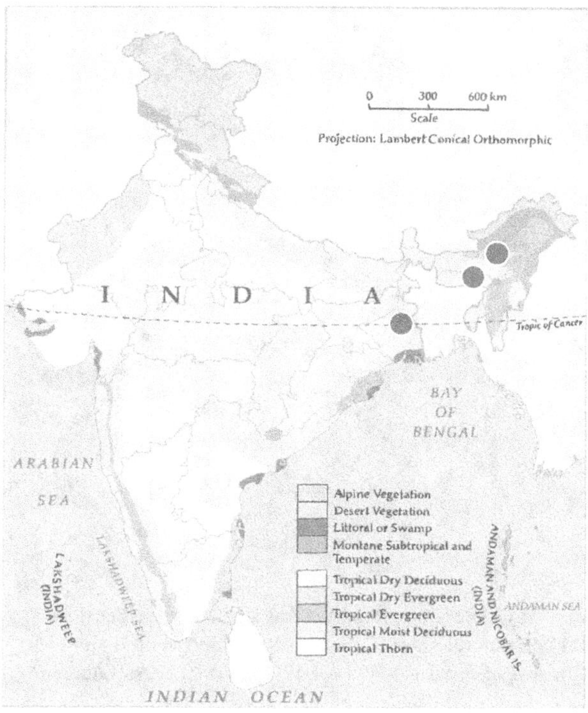

Figure 2.114 Distribution of *H. obesa* in India. (●) Place of first record.
Map not to scale.

Male (Figures 2.115 and 2.116)[63]

Mean body length is approximately 3.2 mm and width is 2.1 mm.

Capitulum: Basis capitulum is dorsally approximately 1.6 times as wide as long; cornua are approximately one-fourth as long as base, widely triangular, apex bluntly pointed. Palpi are compact, lacking basolateral salience. Hypostome is as long as palpi; outline is sub-parallel laterally; dental formula is 5/5, with denticles in files of 10–12 of which anterior and posterior two or three are smallest.

Scutum: Scutum is bordered laterally. Lateral grooves are indistinct, indicated as short, shallow depressions marking juncture between sloping scutal surface and lateral ridges, extending from anterior festoon to the level of coxa II or III. Cervical grooves are short, deep, sub-parallel. Punctations are shallow, more or less discrete, moderately numerous, irregularly arranged and distributed. Festoons number 11.

Legs: Coxa are each with a prominent spur; spur of I is peg-like, narrowly elongate, bluntly pointed apically, overlapping apical margin of coxa II; spurs of II and III are sub-equal, short, widely triangular, median, extending to but not overlapping following coxa; spur of IV is also triangular but shorter, situated close to inner margin of coxa.

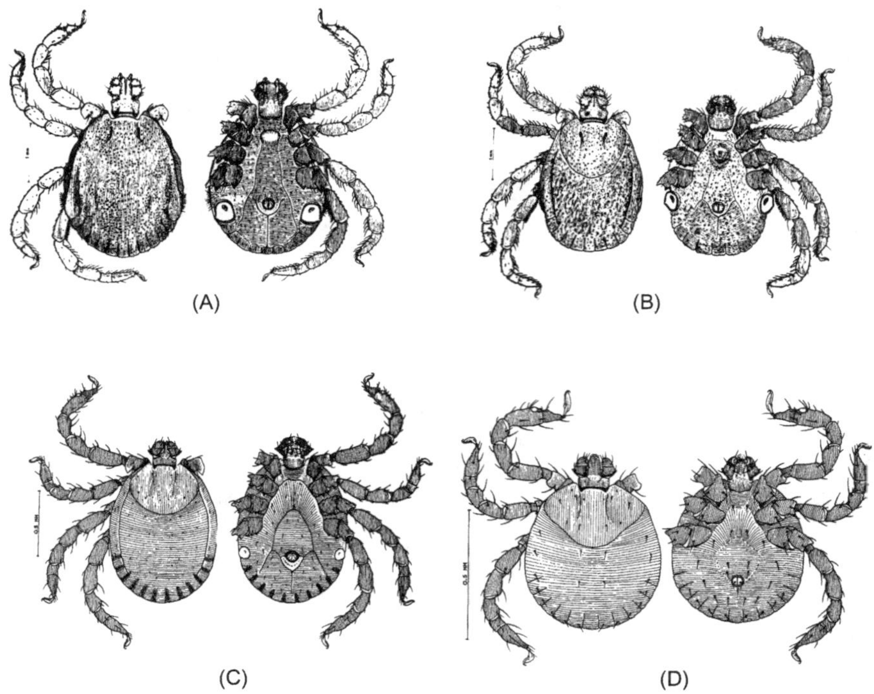

Figure 2.115 *Haemaphysalis obese*. (A) Male, dorsal and ventral side; (B) female, dorsal and ventral side; (C) nymph, dorsal and ventral side; (D) larva, dorsal and ventral side. *Source*: Hoogstraal et al. (1966), Journal of Parasitology (vol 52(1)), Allen press Publishing Services.

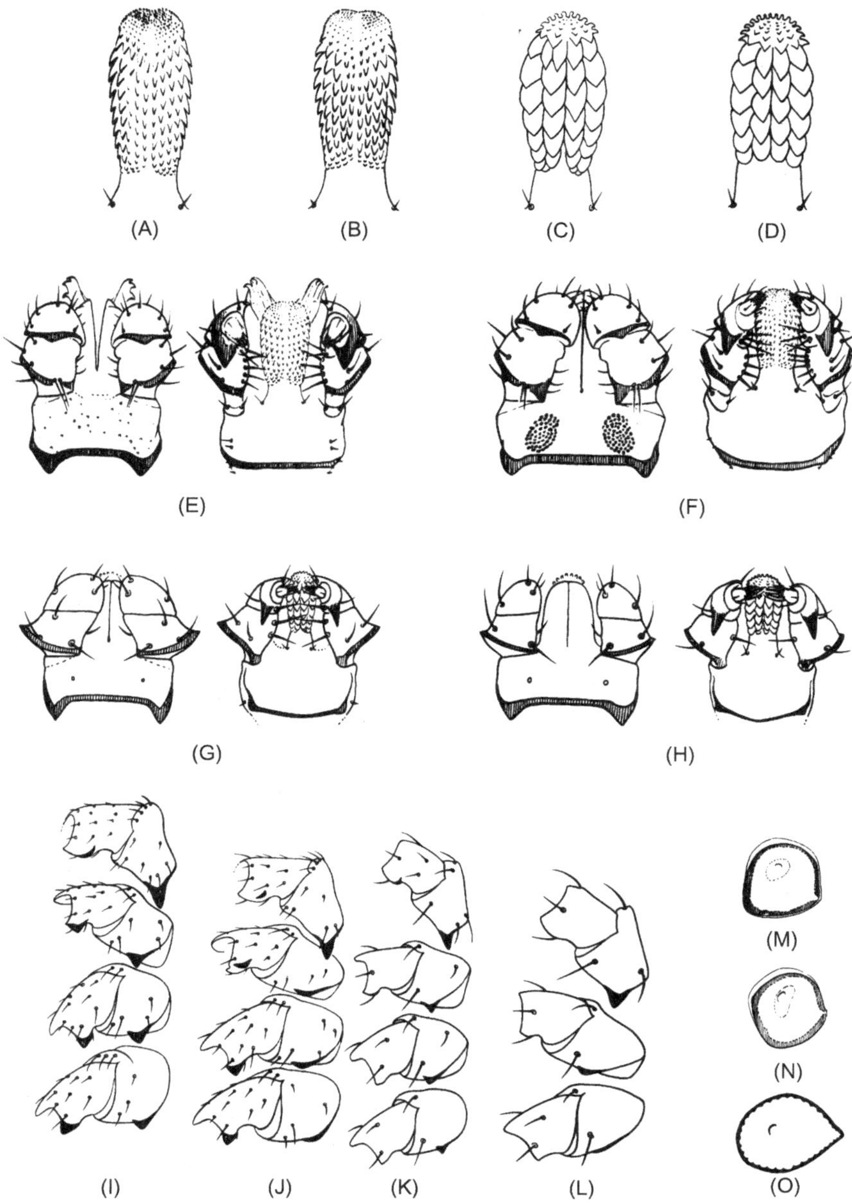

Figure 2.116 (A–D) Hypostome of male, female, nymph, and larva, respectively; (E–H) capitulum of male, female, nymph, and larva, respectively; (I–L) coxa of male, female, nymph, and larva, respectively; (M–O) spiracular plates of male, female, and nymph, respectively.
Source: NIV. unpublished drawings.

Female (Figures 2.115 and 2.116)[63]

Mean length is approximately 3.4 mm, and width is 2.0 mm.

Capitulum: Basis capituli are slightly wider than in male; porose areas are discrete, small, widely spaced, sub-circular. Palpi and hypostome are essentially as in male.

Scutum: Scutum is shield shaped, outline widest at apical one-fourth of length. Cervical grooves consist of small, deep sub-marginal pits anteriorly and shallow, parallel grooves extending to mid-length of scutum. Punctations are as in male anteriorly. Genital operculum is distinctive, triangular with truncate apex, and shallow, depressed sub-circular area centrally.

Legs: Coxa are with spurs similar to those of male, except that spur on IV is shorter.

Nymph (Figures 2.115 and 2.116)[76]

Overall body length is approximately 1.5 mm, breadth 0.8 mm.

Capitulum: Basis capitulum is dorsally 2 times as broad as long; margins are straight; cornua are broadly triangular, approximately one-half as long as the base of basis capitulum. Palpi are broadly campanulate; combined breadth is approximately 1.38 times breadth of basis capitulum. Hypostome is longer than palpi, approximately 2.3 times as long as broad; dental formula is 2/2 (one or two extra denticles may be inserted anteriorly and posteriorly), with denticles in files of seven or eight.

Scutum: Scutum is approximately 1.3 times as broad as long. Cervical grooves are narrow, linear, parallel, and extending to scutal mid-length. Punctations are few, obscure.

Legs: Legs are long. Coxa I spur is triangular, reaching anterior margin of coxa II; II to IV spurs each is broadly triangular, extending well beyond posterior margin, successively decreasing in size.

Larva (Figures 2.115 and 2.116)[76]

Total body length is approximately 0.7 mm, breadth 0.5 mm.

Capitulum: Basis capitulum is dorsally and ventrally essentially as in nymph except for more convex posteroventral ridge and absence of adjacent setae. Palpi outlines essentially as in nymph except for slightly narrower posterior margins of segment II. Hypostome is essentially as in nymph except for denticles in files of about six.

Scutum: Scutum is about 1.5 times as broad as long; anterior emargination and margin are as in nymph except for more rounded scapulae; margin is concave from mid-length, posteriorly narrow, bluntly rounded.

Legs: Legs are long. Coxa I spur is broadly triangular, extending almost up to anterior margin of segment II; coxa II and III spurs each is as a broadly rounded ridge.

Related Species[76]

Haemaphysalis obesa of Northeast India and Southeast Asia is closely related to *H. hirusta* found in Java. These two species are a part of the *H. obesa* group. The larval and nymphal stages of *hirusta* and *obesa* show a close relationship with each other. In the *obesa* nymph, the cornua and coxa I spur are approximately one-half the size of those in *hirusta*. The larval *obesa* is more easily separated from the larval *hirusta* in that its cornua are well developed which is almost obsolete in *hirusta* and the ventral spur of palpal segment III is approximately 3 times as long as in *hirusta*.

Life History of obesa[89]

Life cycle of this species has been studied by Ilkal et al. in 1981 under room temperature (17–26 °C and RH 90–95%). Most of the adults released on the host (cow calf) were found attached within 24 h of their release. They commenced feeding by the third day of their release. Copulating pairs were observed during feeding period. The engorged females started detaching from the sixth day onward and continued up to the tenth day. Two well-defined phases were observed at the time of feeding. The first or the "preparatory phase" started soon after attachment and lasted until last but one day of parasitic stage. The second phase or the "repletion" was of a shorter duration and lasted till the detachment of the fully engorged tick. The weight of the fed females ranged from 236.8 to 575.6 mg. The average amount of blood ingested by the females was approximately 220 times their unfed body weight (1.66 mg). The oviposition commenced 4–8 days after detachment from the host. The process lasted from 15 to 21 days. The number of eggs laid ranged from 3,740 to 5,245. The maximum number of eggs laid was on the third or fourth day of commencement of oviposition. The average weight of an egg ranged from 0.0479 to 0.0791 mg. The incubation period for hatching of the eggs ranged from 45 to 58 days at 20 °C. The eggs hatched in the same sequence as they were laid. The minimum pre-feeding period for larvae was 2 days, but the optimum number of fed larvae was recovered when they were released on the host after the 15th day of hatching. The time taken by the larvae to complete their engorgement ranged from 2 to 6 days. The largest number of engorged larvae was obtained on the third day after release. As in the case of females, the larvae also showed two phases while feeding. The period for the fed larvae to molt into nymphs ranged from 8 to 12 days after completion of engorgement. The nymphs were ready to feed a few hours after emergence. The feeding period (on chicks) for the nymphs ranged from 3 to 6 days. In feeding nymphs also, two phases, i.e., preparatory phase and repletion phase, were observed without any sharp demarcation. The fed nymphs moved for 3–5 days before they entered the molting phase. It took 12–14 days for the nymphs to molt into adults after completion of engorgement phase (Table 2.10).

Ecology[76]

In India, the species is known from West Bengal, Assam, and Meghalaya.

Table 2.10 Duration of Various Events in the Life Cycle of *Haemaphysalis obesa*

Events	Duration	Host
Female feeding	6–10	Calves
Pre-oviposition	4–8	–
Oviposition	15–21	–
Egg hatching (from commencement of oviposition)	45–58	–
Pre-larval feeding (optimum period)	15–20	–
Larval feeding	2–6	Chicks
Larva–nymph molt	8–12	–
Pre-nymphal feeding (optimum period)	20–30	–
Nymphal feeding	3–6	Chicks
Nymph–adult molt	12–14	–

From West Bengal, specimens were obtained from Himalayan foothills from Jalpaiguri district. The climate of the forest in this area is about 36 °C in summer and monsoon while 7 °C in winter. Relative humidity is high with frequent rain showers. *Michelia champaca* is one of the dominant vegetation in this forest (at 490 ft altitude). From Assam, it was collected from the capped monkey *Presbytis pileatus* subspecies, at around 1,560 ft altitude on the Garo hills of the Rongrengiri area.

Host[76]

Immature stages: Crawling on human, bear.

Adults: Cattle, capped monkey, human, Tibetan black deer, barking deer, wild boar, Malayan tapir, bear.

Distribution[76]

India (West Bengal, Meghalaya, Assam), Thailand, Malaysia, Republic of Vietnam.

Disease Relationship

Not known.

2.10 Subgenus *Rhipistoma*

This is probably the most recent haemaphysaline subgenera according to evolutionary plan. Adults of 13 of 27 *Rhipistoma* species are exclusive parasites of carnivores. All *Rhipistoma* immature stage and adult palpi are broadly expanded posteriorly; many are modified by spurs or grooves as hair-hooking devices. The basis capitulum has small to large cornua. Coxal and trochantal spurs are unspecialized, generally small or even obsolete.

Keys to Identify Species of Subgenus Rhipistoma

Male

1. Posterodorsal spur of palpal segment II elongately triangular, pointed apex; lateral grooves extending anteriorly to the level of coxa III, posteriorly enclosing first festoon...................[*H. canestrinii*]
Posterodorsal spur of palpal segment II is broadly triangular, blunt apex; lateral grooves are extending anteriorly to the level of coxa III, posteriorly to anterior margin of first festoon...................[*H. indica*]

Females

1. Cornua with pointed apex; posterodorsal margin of palpal segment II with one large, elongately triangular, pointed apex spur and one rudimentary spur, separated by indistinct groove; infrainternal setae are seven in number...................[*H. canestrinii*]
Cornua are with broadly rounded apex; posterodorsal margin of palpal segment II is with one broadly triangular, blunt apex spur; infrainternal setae number four...................[*H. indica*]

2.10.1 Haemaphysalis canestrinii

This species was previously confused with *Haemaphysalis leachi*, later on Supino described it as a separate species called *Haemaphysalis canestrinii* in 1897 (Figure 2.117).[74]

Adults and larvae were not available in our collection, hence no illustration.

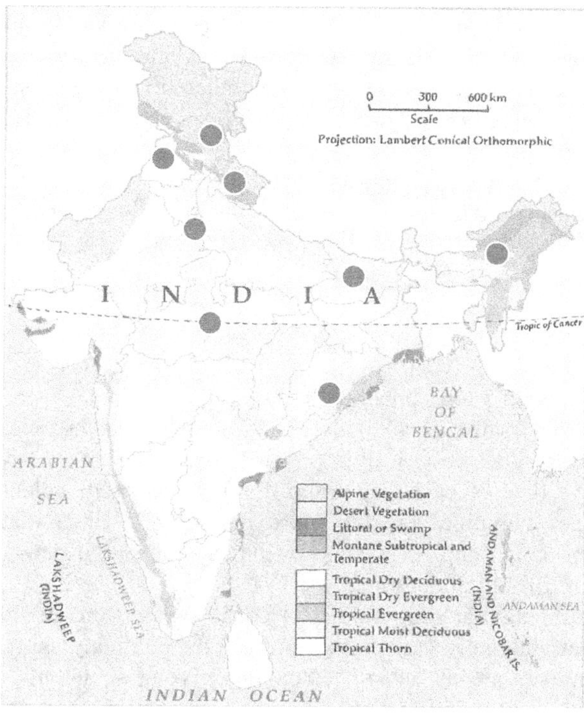

Figure 2.117 Distribution of *H. canestrinii* in India. (●) Place of first record. Map not to scale.

Male[74]

Mean body length is approximately 1.95–2. 63 mm, breadth 0.98–1.13 mm.

Capitulum: Basis capituli are dorsally approximately 1.45 times as broad as long; cornua are elongately triangular, sharply pointed, approximately five-eighths as long as the base of basis capituli. Palpi are broadly salient (*leachi*-type), combined breadth 2.1–2.3 times breadth of basis capituli. Hypostome is as long as or slightly longer than palpi; external profile is quite straight; dental formula is 4/4, with denticles in sub-equal length files of eight to ten (mostly nine).

Scutum: Scutum is 2.0–2.2 times as long as broad; margins are gradually rounded. Lateral grooves are deep, distinct, extend to anterior one-fourth of scutal length, enclose first festoon. Cervical grooves are narrow. Punctations are moderately numerous, irregularly scattered and sized. Festoons number 11.

Legs: Coxa are each with spur short, small, broadly triangular, sub-equal, extending slightly beyond posterior margin.

Female[74]

Overall body length is 2.55–2.85 mm, breadth 1.20–1.28 mm.

Capitulum: Basis capitulum is dorsally approximately 1.9 times as broad as long; external margins are converging posteriorly; cornua are broadly triangular, approximately one-fourth as long as the base of basis capitulum; porose areas are elongate and oval, slightly tilted, moderate size, widely spaced. Palpi are similar to those of male except as follows; segment II spur from posteroventral margin is slightly shorter and broader; ventrointernal setae usually number seven. Segment III is slightly longer, three setae from internoventral margin posteriorly. Hypostome is essentially as in male.

Scutum: Scutum is about 1.25 times as long as broad. Cervical grooves and punctations are essentially as in male. Genital operculum is sub-quadrate; margins are straight; junctures are rounded.

Nymph (Figure 2.118)[74]

Mean body length is approximately 1.3 mm, breadth 0.8 mm.

Capitulum: Basis capituli are dorsally approximately 2.8 times as broad as long, external margins somewhat constricted posteriorly; cornua are triangular, approximately one-third as long as the base of basis capitulum. Palpi are broadly salient posteriorly (*leachi/doenitzi*-type); combined breadth approximately 1.75 times that of basis capitulum. Hypostome is as long as or slightly longer than palpi; dental formula is 2/2, with denticles in files of about nine.

Scutum: Scutum length and breadth are approximately equal; greatest breadth is near mid-length, margin thence gradually converging, posteriorly relatively short, broadly rounded. Cervical grooves are as an arc, extending beyond scutal mid-length. Punctations are few and small.

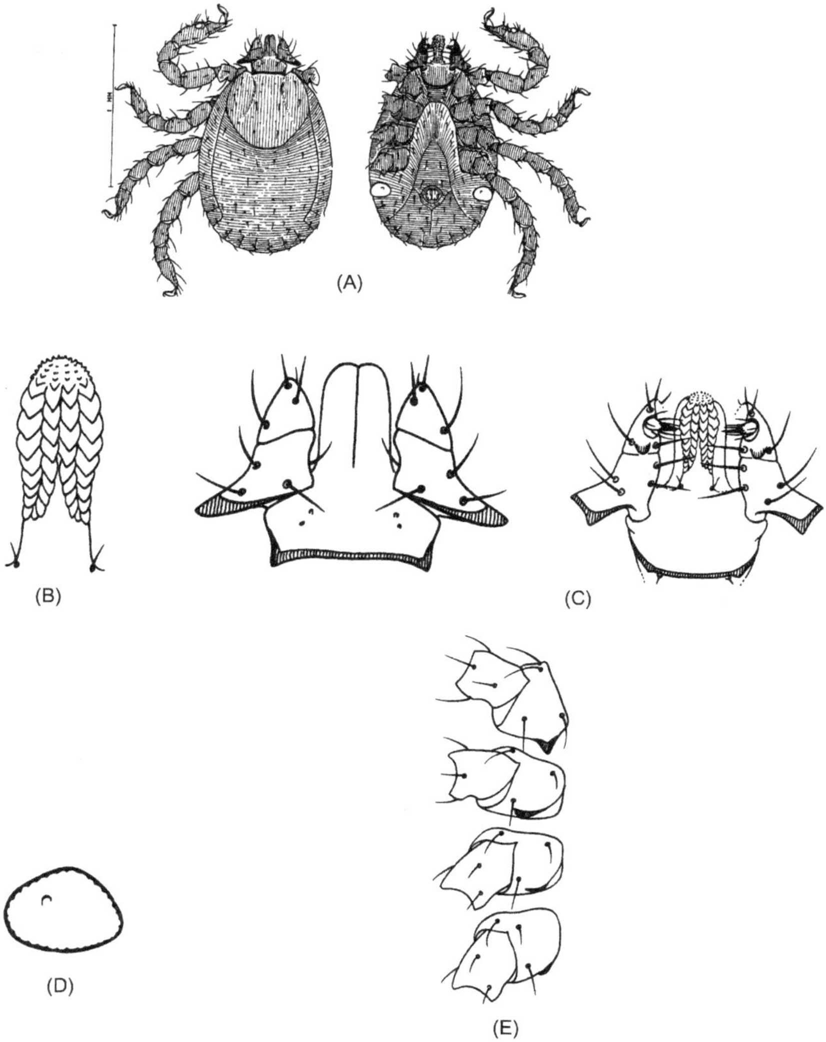

Figure 2.118 *Haemaphysalis canestrinii.* (A) Dorsal and ventral view of nymph, (B) hypostome, (C) dorsal and ventral view of capitulum, (D) spiracular plate, (E) legs. *Source*: Hoogstraal (1971), Journal of Parasitology (vol 57(1)), Allen press Publishing Services.

Legs: Legs are robust. Coxa I spur is broadly triangular, extending well beyond coxal margin; II to IV are each with spur reduced to a broadly rounded ridge, extending to or slightly beyond coxal margin.

Related Species[74]

H. (*R.*) *canestrinii* females are easily distinguished from the closely related species, *H.* (*R.*) *leachi*, by the presence of a prominent, posteriorly directed spur from the posteroventral margin of palpal segment II, which is absent in *leachi* females. The only readily apparent difference among males is that the lateral grooves enclose a single festoon in *canestrinii* but two or three festoons in *leachi*.

Ecology[74]

This species occurs from the Northwestern region of Asia, Pakistan through India and Nepal to South Eastern Asian region, to Thailand, Vietnam, and Burma. It occurs in altitude ranges of 300 ft in Burma to 2,600 ft altitude in Madhya Pradesh region of India and in Vietnam.

Hosts[74]

Immature stages: Human, tree shrew, rats, immature jungle cat/civet, five striped palm squirrel, mouse.

Adults: Red jungle fowl, Indian civets, tiger, leopard, Asiatic jackal, Bengal foxes, fishing cat, palm civet, hog badger, ferret badger, small Indian mongoose, jungle cat, hare, small Taiwan civet, large Indian civet, immature jungle cat.

Distribution[74]

India (Assam, Madhya Pradesh, Uttarakhand (formerly a part of Uttar Pradesh), Orissa, Bihar, Chandigarh, Delhi, Himachal Pradesh, Punjab), Thailand, Vietnam, Taiwan, Burma, Pakistan, Nepal.

2.10.2 Haemaphysalis indica

This species was first described from India by Warburton in 1910 as a variety of *Haemaphysalis leachi* but later on was given a separate species status (Figure 2.119).[71]

Male (Figures 2.120 and 2.121)[71]

Overall body length is 1.58–1.88 mm, breadth 0.75–1.05 mm.

Capitulum: Basis capituli are dorsally approximately twice as broad (anteriorly) as long; cornua are short, broadly triangular, more or less pointed apically, one-fourth or one-third as long as the base of the basis capituli. Palpi are broadly salient (*leachi*-type); combined breadth is approximately 1.6 times breadth of basis capituli. Hypostome is as long as palpi; dental formula is 4/4 (maybe irregularly 3/3 posteriorly), with denticles in irregular files of six to eight.

Scutum: Scutum is typically 1.5–1.7 times as long as broad (variable); margins are broadly rounded. Lateral grooves are deep, distinct, extend to anterior one-fourth of scutal length, enclose first festoon. Cervical grooves are short, shallow. Punctations are very few, range from small and discrete to large and shallow, irregularly scattered. Festoons number 11.

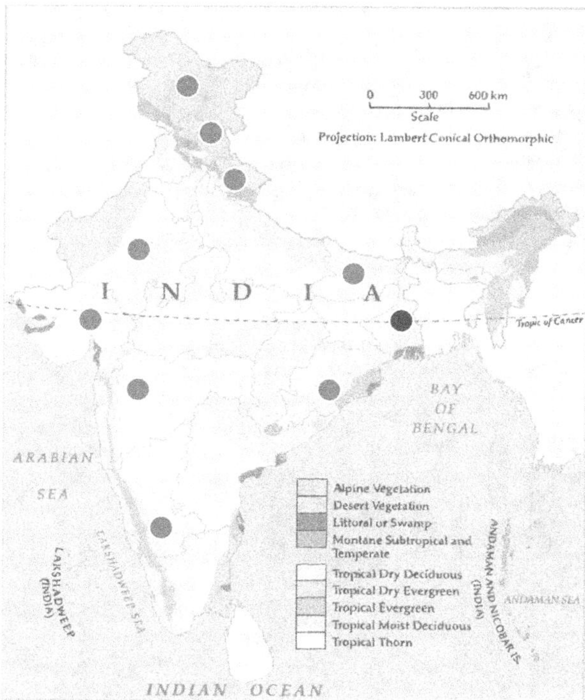

Figure 2.119 Distribution of *H. indica* in India. (●) Place of first record.
Map not to scale.

Legs: Coxa are each with spur broadly triangular, relatively small; spur of I is extending well beyond coxal margin; spurs of II to IV are about one-half as large as that of I.

Female (Figures 2.120 and 2.121)[71]

Mean body length is approximately 1.65–2.63 mm, breadth 0.90–1.50 mm.

Capitulum: Basis capitulum is dorsally approximately 2.5 times as broad as long; cornua outline is as in male, approximately one-third as long as the base of basis capituli; porose areas are shallow, elongated, irregularly oval, tilted, moderate size, widely spaced. Palpi are similar to those of male except as follows. Combined breadth is approximately 1.5 times breadth of basis capitulum. Hypostome is approximately twice as long as broad; denticles are in files of seven to nine, otherwise quite similar to that of male.

Scutum: Scutum is approximately 1.2 times as long as broad; outline is the broadest at about mid-length. Cervical grooves are sub-parallel, and extend to or are beyond scutal mid-length. Punctations are few, mostly rugose, size variable, shallow, irregular. Genital operculum is sub-quadrate; margins are straight; marginal junctures are rounded.

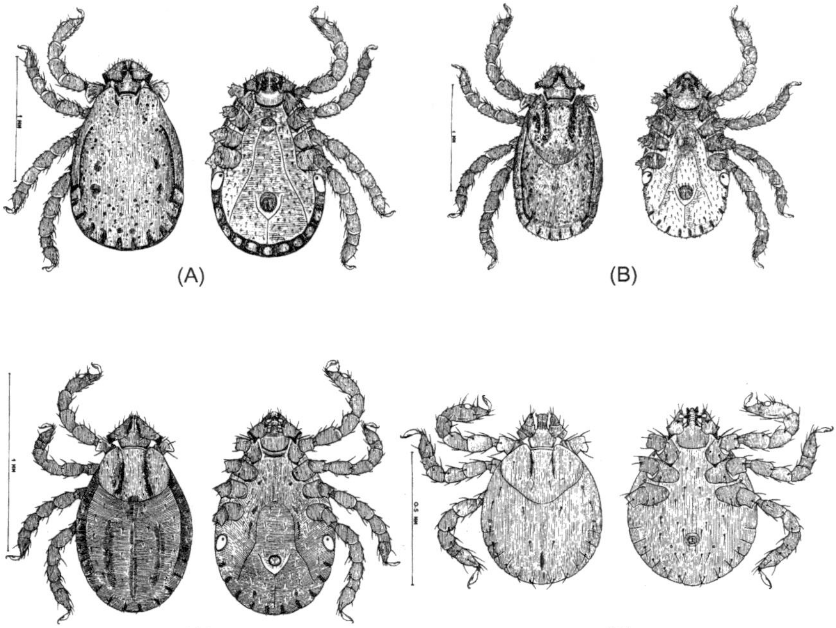

Figure 2.120 *Haemaphysalis indica*. (A) Male, dorsal and ventral side; (B) female, dorsal and ventral side; (C) nymph, dorsal and ventral side; (D) larva, dorsal and ventral side. *Source*: Hoogstraal (1970), Journal of Parasitology (vol 56(5)), Allen press Publishing Services.

Legs: Legs are longer and thinner than in male, otherwise similar except as follows: Coxa with spurs are relatively somewhat shorter; pulvilli are often shorter.

Nymph (Figures 2.120 and 2.121)[71]

Overall body length is approximately 1.1 mm, breadth approximately 0.6 mm.

Capitulum: Basis capituli are dorsally approximately 3.1 times as broad as long; cornua are very small, triangular, approximately one-fifth (or less) as long as the base of basis capitulum. Palpi are broadly salient posteriorly (*leachi*-type); combined breadth is approximately 1.6 times that of basis capitulum. Hypostome is as long as or slightly longer than palpi; dental formula is 2/2, with denticles in files of six or seven.

Scutum: Scutum is 1.3 times as broad as long; greatest breadth is near mid-length; margins are broadly rounded. Cervical grooves are as an arc, extending to posterior margin of scutum. Punctations are few, small, and setiferous.

Legs: Coxa I to III are each with spur broadly triangular, extending somewhat beyond coxal margin; IV with spur reduced to small, broadly rounded ridge on margin.

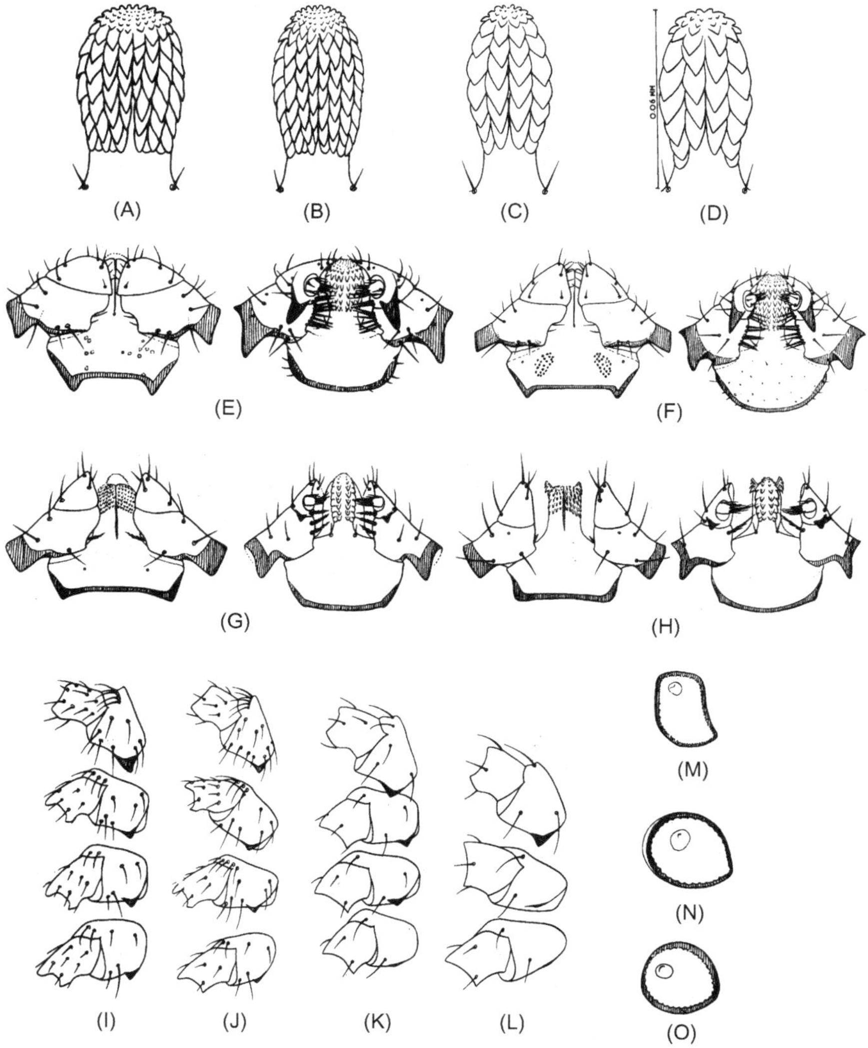

Figure 2.121 (A–D) Hypostome of male, female, nymph, and larva, respectively; (E–H) capitulum of male, female, nymph, and larva, respectively; (I–L) coxa of male, female, nymph, and larva, respectively; (M–O) spiracular plates of male, female, and nymph, respectively. *Source*: NIV. unpublished drawings.

Larva (Figures 2.120 and 2.121)[71]

Length is approximately 0.6 mm, breadth 0.4 mm.

Capitulum: Basis capituli are dorsally approximately 3 times as broad as long; cornua are obsolete. Palpi are approximately as in nymph. Hypostome is as long as or slightly longer than palpi; dental formula is as in nymph, with denticles in files of five to seven.

Scutum: Scutum is about 1.2 times as broad as long, broadest near mid-length, margin thence abruptly curving to relatively narrow, broadly rounded posterior margin. Cervical grooves are parallel, extending to or beyond scutal mid-length.

Legs: Coxa I is with spur broadly triangular, extending well beyond posterior margin; II and III are each with a faint ridge or ridge obsolete.

Related Species[71]

Sharif recognized two varieties of *Haemaphysalis leachi* in India, the typical form and the variety *indica*. The variety *indica* has now been raised to a species level and is reported from India only. The name *leachi* has been used for the ticks from Africa, Southern Asia, Australia, and New Zealand. In the male of *indica* the dorsal retroverted spurs on the lateral salience of palpal article II are blunt or obsolete, whereas in the *leachi* the dorsal retroverted spurs on the lateral salience of palpal article II are well developed and pointed. In the African ticks of *leachi*, the punctations are numerous and the cornua are more pointed than *indica*.

Haemaphysalis indica is related morphologically to *H. canestrinii*. The adults of *indica* are smaller (total length: male, 1.58–1.88 mm; female, 1.65–2.63 mm) than those of *canestrinii* (male, 1.80–2.70 mm; female, 2.25–2.85 mm), have a much broader basis capituli (1.45 mm (male of *indica*) vs. 2.00 mm (male of *canestrinii*) times as broad as long), few scutal punctations (moderately numerous in *canestrinii*), a definite dorsal projection on the male spiracular plate (obsolete or almost so in *canestrinii*), an ovate female spiracular plate (sub-quadrate in *canestrinii*), etc. The posterodorsal margin of palpal segment II differs distinctly; in the male *indica* the spur is shorter, broader, and more externally situated than in *canestrinii*; the broadly triangular spur-like angle of the female *indica* is replaced by a definite, elongately triangular spur in *canestrinii*.[84]

Host Preference and Ecology[71]

Small carnivores, especially mongooses, are important hosts of the adult stage. Other small- and medium-sized carnivores, such as the rattle, jackal, fox, and domestic dog also are parasitized by adults. Among medium- and large-size carnivores, immature stages were collected on the civet and leopard. It may be assumed that these animal species are also parasitized by the adult *H. indica*.

It is interesting to note that adults and immature stages have been taken together on various mongooses, jackal, and fox. One collection of nymphs was from a shrew. Thus, while the nymph and larva of *H. indica* may behave as burrow-infesting parasites, they do not appear to be so strictly confined to this niche as immature stages of the African *H. leachi*.

Host[71]

Immature stages: Fox, Indian gray mongoose, Indian small mongoose, Asiatic jackal, leopard, honey badger, house shrew, Indian hares, jungle cat, and large Indian civet.

Adults: Mice, mongoose, shrew, small mammals, forest wagtail bird, fox, jackal, domestic dog, honey badger, and chital deer.

Distribution[71]

India (Bihar, Gujarat, Himachal Pradesh, Jammu and Kashmir, Karnataka, Maharashtra, Orissa, Rajasthan, Uttarakhand (formerly a part of Uttar Pradesh), West Bengal).

Disease Relationship

Not known.

2.11 Proposed New Species

2.11.1 Haemaphysalis kumaonensis *(subgenus:* Herpetobia*)*

Male (Figures 2.122 and 2.123)

The body is elongate and total length of the body is 2.96 mm.

Capitulum: Capitulum is 0.5 mm long from the base of cornua. Cornua are broadly triangular, blunt, or rounded. Basis capitulum is rectangular in shape with few varied punctations. Palpi are compactly arranged with very little salience on palpal segment II. Palpal segment II bears seven long, lanceolate, and closely placed infrainternal setae. On dorsobasal side of palpal segment a ridge-like outgrowth appears. Palpal segment III shows a small broadly triangular but blunt and inwardly directed spur on ventral side, overlapping little on palpal segment II. Hypostome is 0.21 mm long, globular with 6/6 dental formula, and having eight to nine denticles per file.

Scutum: Scutum is 1.6 times as long as broad. Punctations are present all over the scutum and are varying in size and shape. Festoons are deeply marked but are not as long as broad. Cervical grooves are very short and convex but well marked. Lateral grooves include first festoon on both the sides and reach up to the base of coxa II.

Legs: Coxa are increasing in size from I to IV. Each coxa bears a hook-like spur. Coxa I spur is broadly triangular, moderately large with a hook-like structure. Coxa II and III spurs are small, triangular, and hook-like. Coxa IV spur is elongately triangular with pointed but hook-like or curved in shape. Trochanters show ridge-like outgrowths.

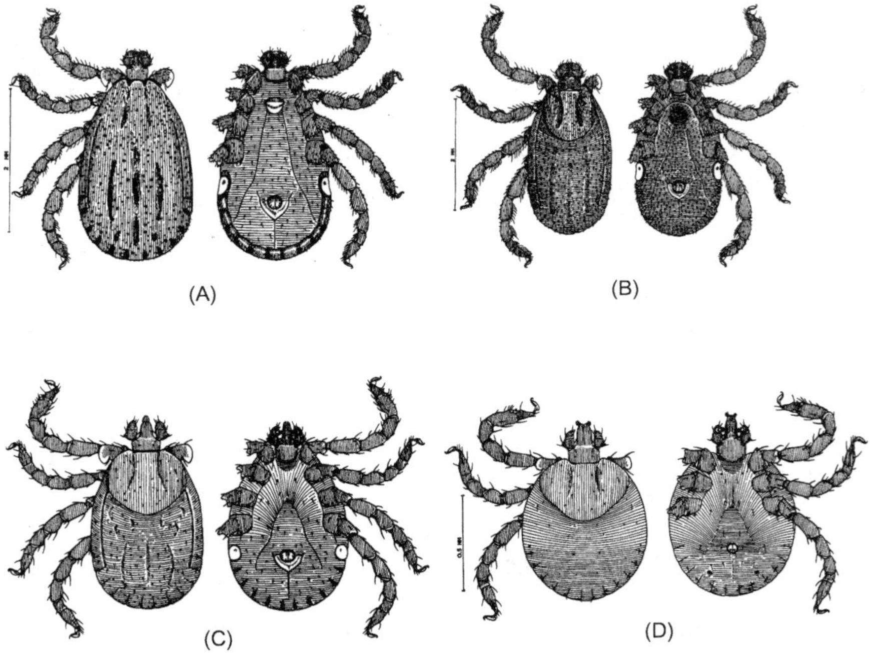

Figure 2.122 *Haemaphysalis kumaonensis*. (A) Male, dorsal and ventral side; (B) female, dorsal and ventral side; (C) nymph, dorsal and ventral side; (D) larva, dorsal and ventral side.

Female (Figures 2.122 and 2.123)

The total body length is 3.07 mm, showing elongated body structure.

Capitulum: Capitulum is 0.6 mm long with very little or low salience on palpal segment II. Basis capitulum is rectangular in shape with very small or no cornua. Porose areas are oval and large. Basis capitulum shows no punctations. Palps are compactly arranged. Palpal segment II ventrally shows seven long, lanceolate, and closely placed infrainternal setae. Dorsally palpal segment II shows ridge-like structure at the base. The space/suture between palpal articles II and III dorsally is signifying the gripping or holding development of evolution. Palpal segment III on ventral side shows broadly triangular, blunt, and inwardly directed spur, just reaching or crossing the base of palpal segment III. Hypostome is 0.3 mm long; dental formula is 6/6, with nine to ten denticles per file.

Scutum: Scutum is oval in shape. Cervical grooves are well marked, convex, and more than two-thirds of scutum length. Punctations are scattered all over the scutum with varying sizes and shapes. Festoons are not so deeply marked and broader than long. Chitinous body other than scutum displays lots of hairs.

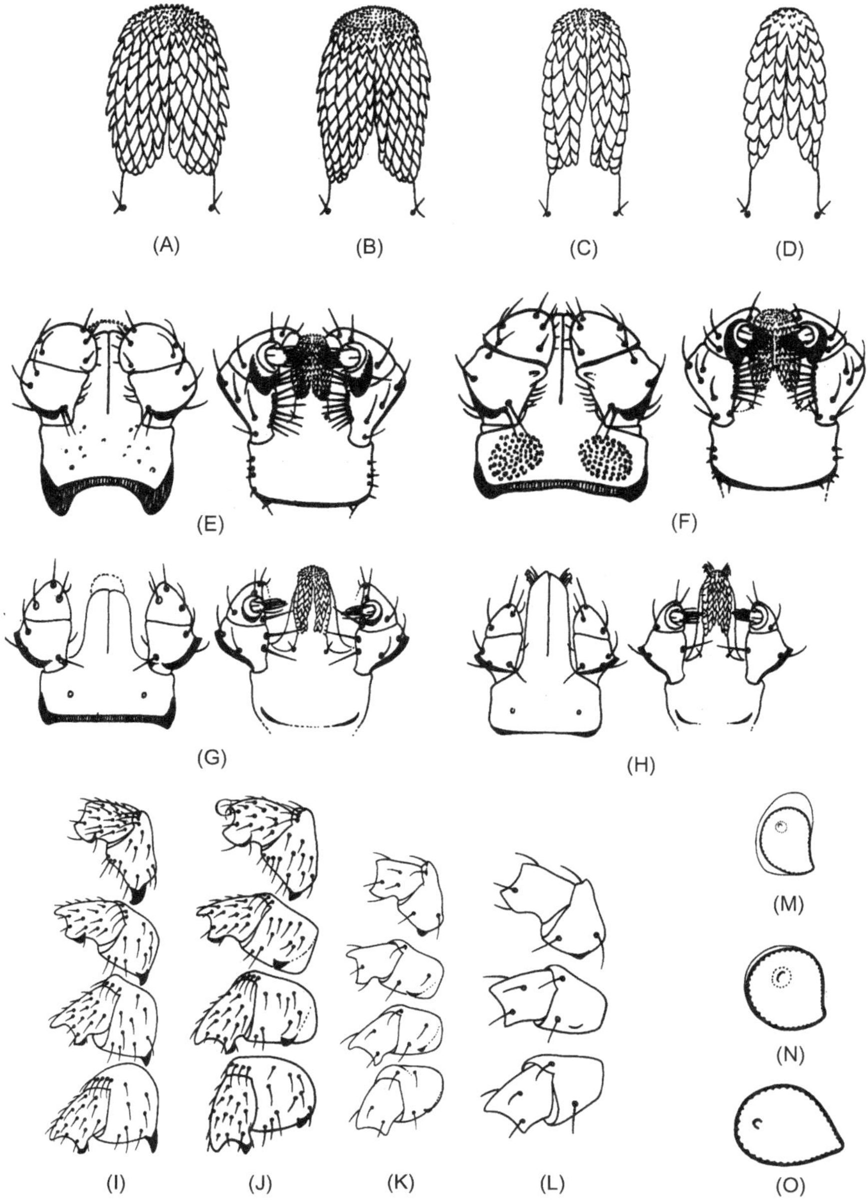

Figure 2.123 (A–D) Hypostome of male, female, nymph, and larva, respectively; (E–H) capitulum of male, female, nymph, and larva, respectively; (I–L) coxa of male, female, nymph, and larva, respectively; (M–O) spiracular plates of male, female, and nymph, respectively.

Legs: Coxa are increasing in size from I to IV with small hook-like spurs. Trochanters also show ridge-like outgrowths, bigger on trochanter IV.

Nymph (Figures 2.122 and 2.123)

The total body length is 1.5 mm; the body structure is more or less ovoid.

Capitulum: Capitulum is 0.29 mm long with no or very little cornua. Basis capitulum is rectangular in shape and not showing any punctations. Palpi are loosely arranged with moderate salience on palp II due to arrangements. Palp II dorsally on the basal side bears a ridge-like outgrowth. On the ventral side palp II shows two infrainternal setae, long and lanceolate and well spaced. Palp III on ventral side shows a small outgrowth or spur. Hypostome is with 4/4 dental formula and bearing nine to ten denticles per file.

Scutum: Scutum is wider and flat. The cervical grooves are deep, less convex, and are two-thirds of total scutum length. Punctations are very rare. Festoons are moderately marked and approximately equal in length and breadth. Body is little less hairy as compared to adults.

Legs: Coxa increase in size from I to IV with small ridge-like spurs.

Larva (Figures 2.122 and 2.123)

The body is oval in shape and is 0.9 mm long.

Capitulum: Capitulum is loosely arranged, with an extraordinarily long hypostome, almost double the total length of palps. The length of capitulum from tip of hypostome to base is 0.23 mm. Basis capitulum is rectangular in shape with no punctations and cornua. Palpi show no salience. Palpal segment II is dorsobasally ridged, showing single infrainternal seta on ventral side. Palpal segment III shows no spur on dorsal or ventral side. Hypostome is quite long measuring about 0.1 mm in length with 3/3 dental formula and seven to eight denticles per file.

Scutum: Scutum is more or less triangular in structure. Cervical grooves are moderately marked, reaching up to half of the scutum or little more. Festoons are very superficially marked and are as wide as long. Body is less hairy.

Legs: No coxa show any spur but a small ridge-like outgrowth on coxa I is present.

Species Relationship

This species differs from other *Herpetobian* species by having the following points:

1. A widely triangular dorsobasal spur on palpal segment II
2. Seven infrainternal setae in both sexes
3. Leathery body in female
4. Bigger and circular porose area
5. Broader hypostome
6. Surface of basis capitulum has more punctuations

Part 3

Discussion and Summary

3 Geographic Distribution and Ecologic Preference

Haemaphysaline ticks are found in different ecozones such as Indomalaya, Palearctic, and Southeast Asian regions, indicating their ecologic adaptations.[68,85,136] Of 42 species of Indian *Haemaphysalis* ticks, 16 species are confined to southern peninsular India and the remaining are distributed in northeastern and Himalayan regions. It is interesting to note that some of the species belonging to the latter regions such as *H. sulcata*, *H. doenitzi*, *H. campanulata*, and *H. bispinosa* extend their distribution to Palearctic and Australian ecozones. Certain species are restricted to only certain states, such as *H. kyasanurensis* and *H. megalaimae* in Karnataka state, *H. sambar* in Kerala state, and *H. sundrai* in Uttarakhand state, whereas species like *H. bispinosa* are widely distributed, occurring practically in almost all Indian states. Ecologic preferences of different subgenera of *Haemaphysalis* ticks are varied. Most of the structurally advanced (SA) populations occur below 1,500 m altitude in the Himalayas or elsewhere. All the species belonging to subgenus *Herpetobia* and other structurally primitive (SP) species are confined to sub-tropical or temperate steppes, semi-deserts, or mountains. These are the species that are supposed to have failed to find suitable host and remained in isolated location as relics. Species belonging to contemporary subgenus *Alloceraea* are notably absent in humid tropical regions, whereas many SA haemaphysaline are common in tropical Asian forests where two, three, or more species can be found infesting a single mammal. The single species belonging to *Alloceraea* in India, *H. aponommoides*, is confined to Himalayan and southern Chinese highlands (2,000–4,900 m altitude). In the cold high Himalayas, adults of *H. aponommoides* are active through much of the year. The cold season adult activity pattern is an adaptation to prevent the desiccation during the dry hot summer. The unique leathery *Alloceraea* integument probably is an adaptation for water conservation. The relict type ticks in the subgenus *Allophysalis* (*H. garhwalensis* and *H. warburtoni*) perhaps avoid competition from other ticks by being in the rocky biotopes in the 1,600–2,000 m altitude range of Asian mountains including the Himalayas.

3.1 Host Preferences[84,103,110,112,114,115]

Most of SA haemaphysalines, both adults and immatures, parasitize domestic and wild mammals and birds. Generally small mammals are infested by immature

Haemaphysalis **Ticks of India. DOI: 10.1016/B978-0-12-387811-3.00003-6**

stages that also parasitize birds. During a major ectoparasitic study in the KFD area involving 8,474 birds belonging to 184 species, the genus *Haemaphysalis* accounted for 99.5% of all the ticks collected and *H. spinigera* was the most common species of *Haemaphysalis*, closely followed by *H. turturis* and nine other species, including *H. megalaimae*. During a major study in India on birds of the Shimoga district, there was no evidence of birds transporting ticks from other countries.

Allophysalis adult parasitizes members of the rich Asian-mountain artiodactyl fauna and less often, domestic artiodactyls, marmots, and humans. Adult activity is recorded chiefly during spring and fall, and that of immatures from late spring to fall. Adults are recorded from domestic goats, sheep, cattle, and yaks, and from the wild goral, serow, thar, musk deer, and ibex. Immatures parasitize rodents, chiefly *Alticola*, *Cricetulus*, and *Marmota*, and also hares and the rock-dwelling pika *Ochotona* (Lagomorpha). Large ground-feeding birds such as the monal pheasant are also important hosts of immature *H.* (*A.*) *warburtoni*.

Immatures of each contemporary *Alloceraea* species commonly feed on shrews and rodents; bird hosts have also been recorded. Adults chiefly parasitize Artiodactyla such as the yak, a variety of deer, wild and domestic sheep and goats, and occasionally carnivores (bear, wild cat, etc.). Adults also infest domestic cattle, buffaloes, and horses. Few other *Haemaphysalis* species are associated with domestic herbivores; the fact that all SP adults (except *Sharifiella*) feed on domestic animals is apparently significant in the survival of this relict group. Lizards are important hosts of immature *H.* (*A.*) *inermis* in Palearctic lowlands. However, reptiles are rare or absent in the highland and more northern habitats where other *Alloceraea* species such as *H. aponommoides* have survived.

Aboimisalis species differs distinctly biologically and ecologically. Immature and adult *H.* (*A.*) *cornupunctata* are from wild or domestic artiodactyls. Immatures and adults are sometimes taken from one artiodactyl host, but immatures are not found on insectivores or rodents examined in the same collecting localities in Nepal, India, Pakistan, and Afghanistan. Parasitism of the same kind of host by immatures and adults, seen among SP subgenera only in *Sharifiella*, occurs in few species of SA subgenera. All *H.* (*A.*) *cornupunctata* records are from the spring and fall seasons. Ecologically, *H.* (*A.*) *cornupunctata* is similar to most other SP ticks but quite different from the three other *Aboimisalis* species.

Immatures of members of *Herpetobia*, *H. nepalensis*, and *H. sundrai* infest the same hosts as adults. Larval and nymphal *H. kashmirensis* chiefly infests the lizard *Agama tuberculata*. Hosts of immature *H. sulcata* include numerous lizards, snakes, tortoises, birds nesting in, on, or near the ground.

3.2 Host-Related Structural Adaptations and Evolution[85]

It is postulated that *Haemaphysalis* species co-evolved with birds and mammals throughout much of the world. During this process, these ticks have adapted structurally. Haemaphysaline host-adapted structures are more pronounced in males than in females, presumably because females are somewhat larger and possibly stronger, and do not wander on the host in search of several mating partners as do males. These adaptations are less pronounced in nymphs and larvae, presumably because these tiny ticks can slip between the host hairs. Among different species constituting SP subgenera, a stepwise progression is exhibited from species to species, from larvae to nymphs to adults as well as males and females of individual species. This progression provides a clear picture of structural evolution through history. This progression can be seen when one is passing through different species of SP, SI, and SA haemaphysaline ticks. Among the SP species, adults of Mesozoic-type relics, chiefly parasitizes Artiodactyla, are seen mostly in Himalayan and outlaying Palearctic and Oriental alpine and subalpine zones. These tick species either lack pronounced coxal spurs (as did Mesozoic reptile parasites) or have exceptionally large coxal spurs (hair-hooking devices). On the capitulum, specialized hair-hooking devices developed only after the narrowly elongate SP palpi changed to compact or posteriorly broadened structures. The primeval SPs are the largest in the genus. The dominating trend toward size reduction operates simultaneously with palpal broadening. Mammal-parasitizing haemaphysalines have a variety of hair-hooking devices—spurs on the coxal and trochanter spurs, basis capituli spurs (cornua), and palpal spurs and marginal indentations or slits—to assist the small tick in reaching a feeding site or a mate on the hairy host. These devices are especially luxuriant in smaller sized (physically weaker) species and in males, which are smaller than females and are promiscuous mate-seekers, moving about on the host more than females do. In bird-parasitizing haemaphysalines, there is little or no development of these devices except (atypically) in minute-sized species. Tenrecs-parasitizing species that feed among the spines and harsh hairs of the body have luxuriant capitular, coxal, and trochanter spurs, but those species that feed in the glabrous ears are virtually spurless (like bird-infesting species).

Ornithophysalis, with its broad palpi, appears to have evolved abruptly from SP subgenera with compact palpi when birds and mammals replaced reptiles as the world's dominant vertebrates. Significantly, of the 20 contemporary *Ornithophysalis* species, six parasitize only birds, five parasitize both birds and various mammals, three parasitize birds and marsupials or only marsupials, two parasitize only tenrecs, and four parasitize Oriental or Australian rodents. The only haemaphysalines which specifically parasitize the phylogenetically ancient tenrecs (Insectivora) are the SP species *H.* (*Sharifiella*) *theilerae* and the three SA species constituting the subgenus *Elongiphysalis*, which represents a highly specialized branch from *Ornithophysalis*. Except for the few species of the subgenus *Haemaphysalis* that parasitize birds as well as mammals, all other SA haemaphysaline adults are specific to mammals and never feed on birds, although immatures of a few SA species may infest birds. The four contemporary *Ornithophysalis* species specific to Oriental and Australian rodents suggest an early association between these ticks and mammals in the Oriental and Australian regions and, together with other evidence, with the origin of the genus *Haemaphysalis* in the Oriental region. The Australian and Oriental marsupial- and rodent-infesting species probably evolved from species like the contemporary bird-parasitizing *H.* (*O.*) *doenitzi* of the Oriental region and eastern New Guinea and Australia.

3.3 Impact of Deforestation, Urbanization, and Faunal Changes[20,45,83]

It is presumed that all *Haemaphysalis* ticks were once distributed in forest biotopes infesting wild animals. As original forest gave way to secondary forests and commercial tree plantations, the composition of the vertebrate fauna also changed. Larger mammals such as the deer, goral, monkey, bear, jackal, porcupine, and hare became uncommon and were replaced by domestic cattle, buffalo, sheep, and goats. Various squirrels, mice, rats, and shrews, and ground-feeding birds such as the red jungle fowl, common peafowl, and crow pheasant remain as potential hosts for immature stages of ticks. Some species like *H. bispinosa* evolved into more parasitic life on domestic animals by changing their diurnal rhythm of drop-off to that of nocturnal. Mass breeding of *H. bispinosa* in cattle sheds has been recorded in many localities in India. Some competitive species like *H. intermedia* has adapted to feeding on domestic goats, less competitive species like *H. cuspidata* and *H. aculeata*, are scarce now, consequent to depletion in the wild animal populations as a result of deforestation. Studies on the impact of ecologic changes due to urbanization on this prevalence of ixodid ticks in Pune and suburban areas of Maharashtra state of India have been reported recently. The study indicated that the infestation rates of *H. intermedia* and *H. bispinosa* on domestic animals has come down due to ecologic changes.

3.4 Seasonal Prevalence[46,110]

Seasonality of *Haemaphysalis* ticks have been extensively studied in the KFD area. Most of the *Haemaphysalis* species have a definite stage-wise seasonal activity. In southern India, *Haemaphysalis* species show definite seasonality for adults and immatures, mostly correlated with rainfall. In general, adult population is high during the rainy season, followed by larval and nymphal peaks during the post-monsoon period. The adults of all the *Haemaphysalis* including *H. spinigera* in the KFD area in Karnataka state become active after a few monsoon rains in June. The adult population reaches its peak during July and August and gradually declines in September. Larval population, which builds up in the monsoon months, remains dormant under the forest litter and becomes suddenly active when the litter dries up during the post-monsoon months (October and November), reaches the peak in November and December, and gradually declines thereafter. Nymphal activity follows the larval activity about a month later. Adult population builds up during the period of nymphal activity but remains dormant until the commencement of monsoon showers. The life span of each stage is restricted to one season and the life cycle is 1 year except in a few species, namely *H. kinneari*, where overlapping of two adult populations has been observed.

There are no systematic studies on the seasonality of ticks which are prevalent in northeastern and Himalayan regions. There are records of immature and adult activity of *H. nepalensis* of Himalaya from each month of the year. Adults of *Haemaphysalis* species such as *sulcata*, *warburtoni*, *kashmirensis*, and *garhwalensis* are chiefly recorded in spring and early fall and that of immature stages from late spring to fall (Figures 3.1 and 3.2).

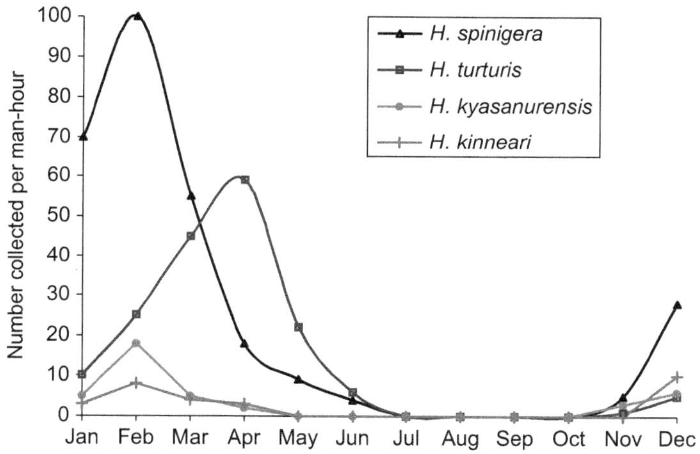

Figure 3.1 Seasonal prevalence of *Haemaphysalis* nymphs (averages for the year 1961–64).

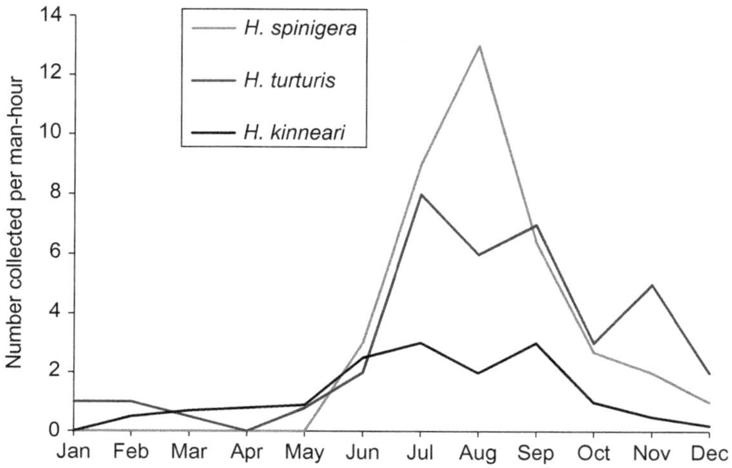

Figure 3.2 Seasonal prevalence of *Haemaphysalis* adults (averages for the year 1961–64).

3.5 Biology[11,12,14,17,38,87,88]

All *Haemaphysalis* species so far known have a three-host life cycle. The life histories of at least eight Indian *Haemaphysalis* species have been studied under the laboratory conditions, the life cycle followed a similar trend in all the species. Studies under laboratory conditions have shown that there was no difference in the life cycles of haemaphysaline ticks collected from southern and northern parts of India. In all except one species, *H. obesa*, eggs hatched under the ambient temperature and completed the developmental stages. These ticks fed readily on different laboratory animals such as guinea pigs, rabbits, and cow calves; and no difference in the host preference was observed. Duration of various developmental stages varied under different temperature conditions. Feeding duration of different stages was more or less same in different Indian species studied under the laboratory conditions, sometimes varying by 1 or 2 days. Fed weight of female adult fed on domestic cattle was generally higher than those fed on smaller laboratory animals like rabbit and guinea pigs. There was a direct correlation between fed weight and the number of eggs laid by the females. The weight of haemaphysaline ticks fed on different laboratory animals varied, the same being more in ticks fed on bigger mammals such as cow calves compared to smaller ones such as guinea pigs. The circadian periodicity of dropping of the engorged stages was closely associated with the diurnal activity of the host; and this activity insures a wider distribution of the species in the appropriate habit. However, in *H. bispinosa*, it has been observed that the dropping was during nighttime in the cattle shed when the animal was resting. The reproductive efficacy was more or less same as that observed in *H. bispinosa* and *H. campanulata* but was less than that of *H. spinigera*, *H. cuspidata*, and *H. intermedia*.

There is no report of biology of Indian species belonging to primitive species. However, reports from elsewhere indicated that female of *H. inermis* and *H. kitaokai* belonging to the primitive subgenus, deposit fewer than 1,000 eggs or less than 20% of the total egg production of most of other haemaphysalines. Each egg and the resulting larva were unusually larger than those of other subgenera. Immature stages of primitive species feed fully in 90 min to 6 h, while those of advanced species require more than 2 days for completion of feeding.

3.6 Distribution of *Haemaphysalis* Ticks in Different Biotopes in the KFD Area[16,46]

Different species of free-living *Haemaphysalis* ticks have varied preference for different forest biotopes. The KFD area has a number of diverse biotopes such as forest, cultivated valleys, and grasslands. Each biotope is distinguished into a number of associations. These associations are interspersed to form a mosaic. The forest biotope, which only provides the necessary physical and biotic environment, forms the main habitat of the tick fauna. This is divisible into three types: semi-evergreen,

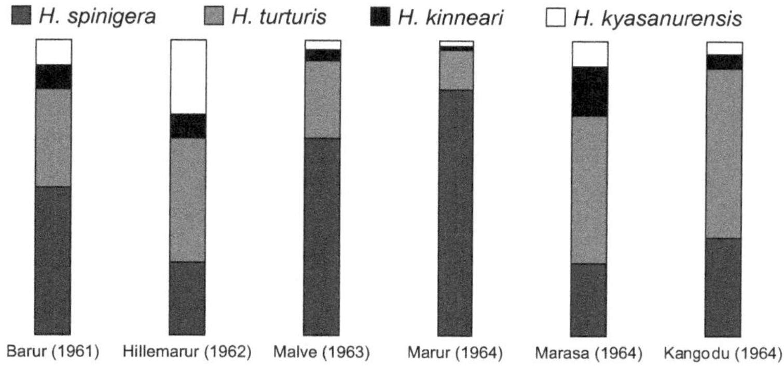

Figure 3.3 Vertical distribution of ixodid ticks on vegetation in the KFD area.

semi-deciduous, and deciduous. These three types are found in most of the localities adjacent to each other forming a mosaic. Delineation of the types is not clear cut, because they merge with each other and produce a transitional strip of ecotones between them. Along the border between the forests and the grasslands there are usually ecotones of scrub or thickets of various kinds. The grassland is usually dotted with small patches of vegetation, which most of the time form thick, impenetrable clumps. During a study undertaken in the KFD area, it was observed that tick fauna varies in such different ecologic situations (Figure 3.3). *H. turturis* and *H. kinneari* predominate in semi-evergreen and semi-deciduous forest. The relative prevalence of nymphs of *H. spinigera* and *H. turturis* varied in different localities depending upon the animal fauna and ecology (Figure 3.4) so much so that an experienced insect collector can recognize *H. turturis* area and *H. spinigera* area separately. Tick species of the KFD area, excepting *H. spinigera*, generally feed upon wild mammals and usually do not bite man. The tick *H. spinigera* feeds mainly upon cattle and therefore is preponderant in deciduous forest and plantation of teak and eucalyptus where cattle graze. During the study, it was observed that the heights of vegetation, at which the adults of various species settled, corresponded to the height ranges of the preferred hosts. Tick species such as *H. wellingtoni*, which generally feeds on ground birds, selected lowest height on vegetation, whereas *H. shimoga*, which feeds on buffalos and Gaur sambar deer were observed at higher heights (Figure 3.5).

3.7 Control of *Haemaphysalis* Ticks in the KFD Area[34,42,100,110,123]

In the KFD area, Drumund et al. have observed that lindane was more effective and toxic than carbaryl, which was more toxic than malathion. It appeared that one or two treatments of lindane at dosages of 0.1 to 1 lb/acre was highly effective in killing tick vectors of the KFD and particularly useful in areas where monkey

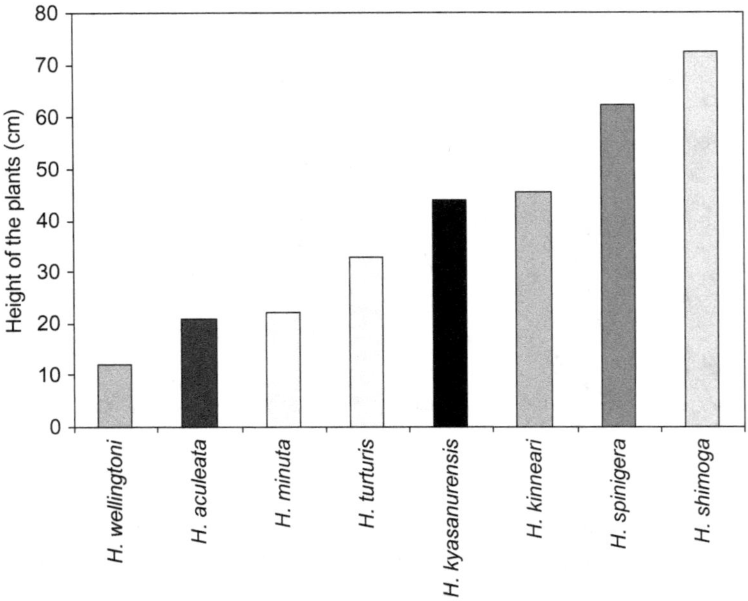

Figure 3.4 Adults of different species of ticks collected from different vegetation types at the KFD area between April and December 1971.

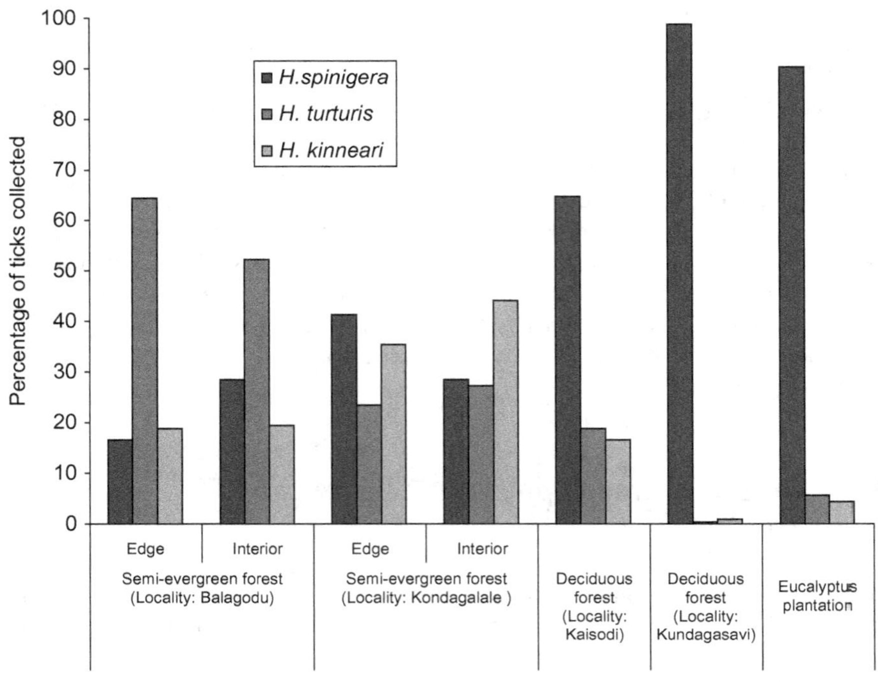

Figure 3.5 Relative proportion of nymphs taken in five localities during January to May.

deaths occur. Spraying of lindane of 1.12 kg/ha resulted in significant reduction of *Haemaphysalis* population for at least 12 weeks. As cattle constitute the major host for *Haemaphysalis* adult ticks, the effect of insecticide on cattle was also studied. Cattle were spread and dusted with different concentration of insecticides, that is 0.5% malathion, 0.25% and 0.5% of carbaryl, and 0.01% and 0.03% lindane, resulting in significant control of ticks on cattle.

Kulkarni and Naik studied the repellent effects of six chemicals, namely dibutylphthalate (DBP), dimethylphthalate (DMP), *N,N*-diethyl-*m*-toluamide (DEET), two proprietary preparations, "Insect repellent" (DEET + DMP) and "Mylol" (DBP + DMP), and pyrethrum were evaluated against larvae, nymphs, and adults of three species of ixodid ticks, namely *H. spinigera*, *H. turturis*, *H. kinneari*, and two species of argasid ticks, namely *Argas persicus* and *A. arboreus*. In addition, effect against adults of *H. intermedia*, *H. obesa*, and *H. wellingtoni* was also tested. DEET showed highest repellency against all stages of tick species except nymphs and adults of argasid ticks. The overall relative repellency of the six chemicals was in the following order: DEET > pyrethrum > DEET + DMP > Mylol > DMP > DBP.

3.8 Phylogeny of Indian *Haemaphysalis* Ticks[62,96]

Structurally primitive *Haemaphysalis*
Structurally intermediate *Haemaphysalis*
Structurally advanced *Haemaphysalis*

In all, 168 species have been included in this genus from all over the world. They are all mostly parasitic on birds and mammals, constitute the most useful assemblage in the superfamily Ixodoidea for displaying numerous interrelated structural–biological clues to affinities between tick species and groups, and to historical and contemporary host associations. A number of structurally primitive species, each with specialized biologic properties, provide indicators of the early history and contemporary survival and adaptation of *Haemaphysalis* species. *Haemaphysalis* ticks differ from those of other genera in that male capitular spurs and spur-like angles, and also coxal spurs, are almost invariably more luxuriant than those of females.

The dominating phylogenetic trend throughout the genus *Haemaphysalis* is the gradual reduction in the lateral salience of basis capituli and length of palpi which are characteristic of more primitive genera such as *Aponomma* and *Amblyomma*. These structures' evolution is more toward a rectangular basis capitulum with posterior cornua, compact to broadly basosalient palpi and various combinations of coxal and trochantal spurs and capitular spurs, angles, or emarginations. These adaptations apparently enabled the ticks to attach firmly on the host. These hair-hooking devices assist the small tick in penetrating a maze of stiff hairs and spines to reach a feeding site on the host integument. A trend toward smaller bodies and capituli is also strong. In small-sized ticks, posterior cornua on a rectangular basis

Table 3.1 Classification of Indian Haemaphysaline Subgenera

Sr. No.	Structurally Primitive (SP)	Structurally Intermediate (SI)	Structurally Advanced (SA)
1	*Alloceraea* (1 species)	*Herpetobia* (6 species)	*Ornithophysalis* (5 species)
2	*Allophysalis* (2 species)		*Haemaphysalis* (4 species)
3	*Aboimisalis* (1 species)		*Kaiseriana* (17 species)
4			*Aborphysalis* (2 species)
5			*Segalia* (2 species)
6			*Rhipistoma* (2 species)

capituli and compact or short, broad palpi function more effectively to force a passage though feathers or fur than do a broad basis capituli and elongate palpi. Large primitive ticks parasitizing glabrous reptiles were not faced with this problem.

On the basis of structural development as one of the criteria laid by Hoogstraal and Kim for the classification of different subgenera, 42 Indian *Haemaphysalis* ticks can be divided into 10 subgenera as given in Table 3.1.

Of the 42 Indian *Haemaphysalis* ticks recorded in 10 subgenera, species belonging to three subgenera, that is *Alloceraea* [1 species], *Allophysalis* [2 species], and *Aboimisalis* [1 species] are supposed to be structurally primitive (SP), six subgenera (*Ornithophysalis* [5 species], *Haemaphysalis* [4 species], *Kaiseriana* [17 species], *Aborphysalis* [2 species], *Segalia* [2 species], *Rhipistoma* [2 species]) are structurally advanced (SA) and includes 31 species. *Herpetobia* subgenus which includes six species, exhibit characters intermediate between the SA and SP species so they are considered as structurally intermediate haemaphysalids (SI) (Table 3.1).

3.8.1 *Structurally Primitive* Haemaphysalis

A total of four species are included in these subgenera. The most basic criterion of the four SP species is the presence, in each stage, or only in larvae and nymphs, of a lateral convexity of the basis capituli or of a projection from each side of the basis capituli. SP palpi are elongated and compact, but not basolaterally salient. SP capitular and leg-spur development is with few notable exceptions, exceedingly slight. In all other SA haemaphysaline species, the basis capitulum is rectangular, usually with posterior cornua; lateral projections are absent. SP palpi are elongated and compact, but not basolaterally salient. The three SP subgenera link *Haemaphysalis* and other non-rhipicephaline ixodids and differ distinctly from other haemaphysaline species in other subgenera.

There is only one species included under the subgenus *Alloceraea* in India, that is *H. aponommoides* Warburton, which is found in the highlands of Central and eastern Himalayan range of Southern China. *Alloceraea* males, nymphs, and larvae each also have a laterally convex or otherwise laterally projecting basis capituli, lacking posterior cornua, and elongated palpi lacking a ventral spur. These are the only males with a 2/2 dental formula in the entire genus *Haemaphysalis*. The various spurs and spur-like angles of the body appendages (capitulum and legs) that function as hair-hooking devices in so many other haemaphysalines are obsolete or only very slightly developed in each *Alloceraea* stage. The unique leathery *Alloceraea* integument is probably also specially adapted for water conservation.

3.8.2 *Structurally Intermediate* Haemaphysalis

The subgenus *Herpetobia* is supposed to be an intermediate between structurally SP and SA haemaphysaline. The subgenus *Herpetobia* is a relict, pivotal branch in *Haemaphysalis* phylogeny. All *Herpetobia* immatures and adults have an SA-pattern rectangular basis capituli. Their palpi, however, which are compact, but are slightly salient in certain nymphs, represent the forerunner of the broad palpi characterizing SA haemaphysalines. The species included under SI group are *H. sulcata*, *H. sundrai*, *H. kashmirensis*, and *H. himalaya*.

3.8.3 *Structurally Advanced* Haemaphysalis

There are 29 species belonging to six subgenera included under this SA group. The subgenera are *Aborphysalis*, *Haemaphysalis*, *Kaiseriana*, *Rhipistoma*, *Ornithophysalis*, and *Segalia*. The basis capituli of both immatures and adults in this group is rectangular and usually bears posterior cornua. The palpi of immatures and adults mostly show some basal broadening of segment II; some retain a compact form. The salience begins with a slight extension of the posterior breadth and reaches the broadly triangular outline characterizing most species of this genus. The dental formula is 2/2 in immatures and 4/4 in adults, rarely 3/3 or 4/4 in the former, and occasionally 3/3 or 5/5 to 7/7 in the latter. Adults vary in length between 2.2 and 3.5 mm; a few are even smaller; but very few exceed 3.5 mm. Most of the species occur in humid, wooded zones. In the subgenus *Kaiseriana* adults are structurally specialized, some extremely, for parasitizing Artiodactyla; most of the adults feed on Carnivora, but other hosts are exceptional. Immatures parasitize small mammals, especially Rodentia, and those of a few species feed on Artiodactyla and Carnivora as well as Rodentia. The hallmark of *Kaiseriana* is a hair-hooking spur extending from the posterodorsal margin of the adult palpal segment III. In *cornigera* group males this spur is supplemented or replaced by a gap in the external surface of the palpus. The generally small, often frail *cornigera* group adults have moderately to very large cornua on the basis capituli, an extraordinary variety of hair-hooking devices on the greatly broadened palpi, and pronounced spurs on most or all coxae. The coxa IV spur is lanceolate in *spinigera;* double (scissor like) in *anomala*, *cornigera*, and *shimoga*.

The subgenus *Rhipistoma*, probably the most recent of haemaphysaline subgenera, in contrast to *Ornithophysalis*, is one of the earliest types of SA haemaphysaline. The subgenus *Haemaphysalis* apparently evolved from *Herpetobia* (or a related, now extinct ancestor) in the Indian subregion of the Oriental region and fanned southeastward into the Oriental region and northeastward into the Palearctic region. Wild and domestic Artiodactyla (various deer, wild pigs, serow, goral, blackbuck, etc.) are the chief hosts of most adults of the subgenus *Haemaphysalis*. Carnivora are secondary hosts of some of these 11 species.

In *Aborphysalis* palpi differ from those of most haemaphysalines except *Segalia* is being broadly expanded either in larvae and nymphs or only in nymphs, and compact or campanulate in adults. All capitular spurs (cornua and palpal segment III ventral spur) are much reduced, and no other hair-hooking devices except the leg spurs are present on these ticks. The capitulum does not function for hair hooking, but the leg segments may do so with their sundry spurs, none of which are large. The reversal of the usual immature-adult palpal form in *Aborphysalis* is postulated to be a functional adaptation for coexistence with spiny and quilled porcupines.

Rhipistoma is probably the most recent haemaphysaline subgeneric assemblage. It shows signs of contemporary speciating associated with small- and medium-sized Carnivora in the Oriental, Ethiopian, and Malagasy regions. Only two species are found in India under this subgenus. All *Rhipistoma* immature-stage and adult palpi are broadly expanded posteriorly; spurs or grooves as hair-hooking devices modify many. The basis capituli has small to large cornua. The capitulum (basis capituli and palpi), assists *Rhipistoma* movement among the host pelage. Coxal and trochantal spurs are unspecialized, generally small, or even obsolete.

3.9 Tick-Borne Diseases in India[48,87,94,95,106,123]

Protozoan diseases
Bacterial diseases
Viral diseases
Rickettsial diseases
"Q" fever
Tick typhus
Other tick-borne maladies

Haemaphysalis ticks are efficient vectors of a number of protozoan, viral, rickettsial, and bacterial diseases of man and domestic animals. The immature stages of these ticks often feed on rodents, lagomorphs, birds, etc., which are resources of some of these disease-causing pathogens. The pathogens may be transmitted to man and domestic animals when the immature stages of these ticks feed on them.[3,48,95] A brief review of tick-borne diseases is given below.

Protozoan Diseases

Theileriosis and babesiosis are two common important protozoan diseases, affecting cattle population in India. *Theileria annulata* is the agent causing theileriosis in many countries including India. The vectors of *T. annulata* are various species and *Hyalomma*, which have been summarized by Geevarghese and Dhanda. The common protozoan parasite causing babesiosis in India is *Babesia bigemina*.

One important vector of *B. bigemina* is *Boophilus microplus*. With regard to mortality and morbidity, these two diseases are probably comparable to malaria in humans in India.[42,95]

Bacterial Diseases

The important bacterial tick-borne disease of man and animals in India is the relapsing fever caused by spirochaetes. Human infections with spirochetes are known in the northern colder parts of India, particularly in Kashmir. The disease is caused by *Borrelia persica* and is transmitted by the Argasid ticks, *Ornithodoros tholozani*. The natural reservoirs are rats, mice, and other rodents. The disease is clinically characterized by recurring bouts of fever separated by a febrile interval of 9–10 days. But the mortality rate is rather low. Army personnel working in the endemic areas are prone to the disease. Use of several synthetic insecticides like BHC, diazinon, dichlorovas, and Malathion is feasible and has been recommended. Tularemia, a plague-like disease, caused by *Francisella tularensis* and carried by *Dermacentor* sp. or *Amblyomma* sp. occurs in many parts of the world, including the USA, Canada, USSR, Japan, and southern Europe, but there are no reports of its occurrence in India so far.

Viral Diseases

There are a number of arboviruses which are known to be transmitted by Ixodid and Argasid ticks. Most of them have animal reservoirs either in domestic animals or in field animals in the natural environment. All over the world, over 80 such viruses have been recorded with tick involvement. However, in India the isolation of viruses from ticks and their hosts has been restricted only to a few areas.

Kyasanur Forest Disease (KFD)[5,19,20,46,144,145]

This is the most important tick-borne viral disease in India. It was first recognized in 1957 in the forests of the Shimoga district of Karnataka state as a fatal infection of monkeys and subsequently as a serious disease in man. It is still restricted to the same general area, but with some local extensions along the neighboring districts like north Kanara, South Kanara, Chikkamagaluru, and Udupi. The disease in man varies considerably from mild pyrexia for a few days to serious illness with hemorrhagic manifestation leading to death. The virus has been isolated in nature from man and monkeys and several species of ticks. The virus seems to be unusually hardy, capable of being isolated from blood or tissues several days after collection and transportation on ordinary ice.

The mammalian hosts other than man are also of many and varied types. Monkeys of two species present in the region, namely *Presbytis entellus* and *Macaca radiata*, are highly susceptible. Monkeys get infected by bites of larvae and nymphal stages of ticks, mostly of the genus *Haemaphysalis* sp. While the *P. entellus* monkey normally succumbs to the infections, the *M. radiata* monkey shows a certain degree of ability to overcome the disease and to survive. It is the infection and death of monkeys in the forest that generally heralds the epidemic season and the activity of the virus in the locality. The incidence in man coincides with the season of the prevalence of the nymphs of *Haemaphysalis*, that is generally December to March and occasionally up to June. KFD infects many species of rodents such as *Rattus r. wroughtoni*, *R. blanfordi*, and *Mus booduga* and the shrew *Suncus murinus*. It can also infect the porcupine, the bandicoot, and the flying squirrel as well as bats and many species of birds. Domestic cattle are rather poor hosts of the virus. It has perhaps the widest host range among arboviruses.

The vectors include several species of ticks, predominantly of the genus *Haemaphysalis*. *H. spinigera* is the most widespread vector with a wide host range. It is perhaps the sole vector in relation to man. *H. turturis* is an equally important vector with a major role in the extra human transmissions. Several other species have been found infected in nature or in the laboratory. Species of *Ixodes* and *Dermacentor* are also known to be infected in nature and laboratory transmission has been obtained with several of them. *Ornithodoros* sp. inhabiting the bat-roosting caves has also been found infected. KFD may also be transmitted by contact or through aerosols. Several laboratory infections have occurred among those handling virus-infected material. Though numerous aspects of the epidemiology of KFD have been studied and understood there are still a few nagging questions, in particular, the public health aspects related to the control of the disease and prevention of its further spread demand urgent attention.

The distribution of KFD virus in India has been the subject of much speculation. Neutralizing antibodies to the virus, which is the same or very similar, have been seen in horses, camels, and man in Kutch district of Gujarat state, but no disease of man or animals attributable to it has been recognized. Stray serologic positives in man have been seen in a few other parts of India, but their relation to KFD is not clear.

Ganjam Virus Disease[21,23,113]

Ganjam virus was first isolated in 1954 from *H. intermedia* infesting goats in Orissa state. But no relationship to a lumbar paralysis disease prevailing at the time in the exotic strains of goats was found. A little later the virus was isolated from *Culex vishnui* mosquitoes in Vellore in Tamil Nadu. The virus was also isolated from two persons, a young boy suffering from high fever in Vellore area and a laboratory worker who was handling the virus. Neutralizing antibodies of the virus were found in human sera in several parts of India. Recently, Ganjam virus was recorded to be antigenically identical with Nairobi sheep disease virus, a common fatal disease of sheep in quite large areas in Africa.

Wanowrie Virus Disease[42]

The Wanowrie virus has been isolated from ticks *Hyalomma marginatum isaaci* and *Culex fatigans* near Pune. It has so far not posed any problem in India, but the same virus was isolated from the brain of a sick woman in Sri Lanka. The illness ended in death. This certainly makes Wanowrie virus a pathogen of potential human importance.

Crimean/Congo Hemorrhagic Fever[122]

Report of serologic studies made in India in Tamil Nadu and Rajasthan indicate the occurrence of CHF virus in human and domestic animals. No disease resembling CHF has, however, been reported in the country. CHF is a well-known disease in Bulgaria, Turkey, Southern USSR (known since 1944), and Iran. A further interest in this virus, so far as the Indian sub-continent is concerned, arises from the discovery of this virus in Pakistan in *Hy. anatolicum.* Moreover, recent investigations have shown that the CHF virus is identical to the Congo virus which was first isolated in 1956 in Congo and subsequently in many parts of the African continent, both from humans and ticks particularly *Hy. marginatum* and *Hy. anatolicum.* Hazara virus, isolated in Pakistan from *Ixodes redicorzevi*, was seen also to be related to Congo virus. In view of the fact that the Congo/CHF virus produces a serious typhoid-like illness and is known to be transmitted by ticks, vigilance for this virus and its vectors in India need utmost attention.

Other Tick-Borne Viruses[42,109]

Other tick-borne viruses include Kaisodi from *H. spinigera* ticks and a bird (ground thrush), Wadmadani from *Hyalomma* sp., Nyamanani from paddy birds, Barur from ticks (*H. intermedia*) and rodents. Banja virus isolated from *H. intermedia* ticks of sheep collected in Orissa state is now known to cause human infection in Yugoslavia. Dhori virus isolated from *Hyalomma dromedarii* ticks of camels seems to infect man in the Western India, on the basis of serologic surveys. Chobar Gorge, a virus isolated from *Ornithodoros* sp. in Nepal seems to infect man in the Himalayan region on the basis of HI results.

Rickettsial Diseases[107,108]

The important tick-borne rickettsia diseases in India are "Q" fever and Indian Tick typhus.

"Q" Fever

The causative agent is *Coxiella burnetii.* The natural reservoirs include many kinds of animals. The clinical symptoms are remittent fever, chills, sweating, anorexia, etc. and can vary considerably in severity. Mortality is about 1%. The normal mode of transmission appears to be through close contact with the infected animals, milk, clothing, fecal matter, etc. However, over 40 species of both Ixodid and Argasid ticks are known to harbor the organisms, but no proved instance of transmission by bites of ticks is recorded, in nature although vector potentiality of many species of ticks have been proved experimentally. The incidence of "Q" fever in India is rather low and perhaps is not reported well.

Tick Typhus[101]

Also known as Boutonneuse fever, tick typhus occurs in many parts of India. The causative agent is *Rickettsia conorii* and is transmitted by *Rhipicephalus sanguineus*, the domestic dog tick in India. But in the African continent, it is transmitted by other ticks such as *R. evertsi*, *H. leachi*, and *Amblyomma* sp. The clinical symptoms of the disease have been described as fever, headache, and malaise lasting about 12–14 days and mortality is low. A maculopapular rash occurs on the second to fifth day of fever.

Other Tick-Borne Maladies

Apart from being vectors of disease organisms, ticks also play an important role in ill health in several ways. Tick bites sometimes produce wheals and allergic urticaria in sensistized persons. In one instance, a fresh tick bite by *H. spinigera* was known to produce allergic reactions at the sites of all previous bites. Paralysis in cattle due to tick bites have been reported to be a serious problem in several countries. However, in India, tick paralysis either has not received the attention it deserves or is absent or rare. However, in laboratory studies in Pune (NIV), paralysis in rabbits has been induced from bites, but the paralysis disappeared on the removal of the tick from the body.

Appendix I: A New Species Described from Kerala[111]

Haemaphysalis knobigera sp. nov.

Measurements

Length: 2.00 mm (2.00–2.2 mm)
Width: 1.1 mm (1.00–1.2 mm)

Dorsum

Basis capitulum rectangular, breadth including cornua more than 1.2 times greater than length, lateral margins almost straight. Posterior margin slightly concave; surface with few punctations. Cornua moderately long; less than the size of basis capitulum. Palpi conical, lacking basolateral salience. Palpal segment II without lateral salience. Median ridge-like dorsal palpal spur (Figure 1A–C). Hypostome reaching apex of palpi; cheliceral teeth protrude well above the level of palpi.

Scutal outline elongate-oval; width approximately two-thirds of length, widest at level of coxa IV. Lateral margins gradually rounded, posterior margin bluntly rounded. Lateral grooves of medium length; cervical grooves formed by moderate pits and shallow indistinct grooves. Punctations moderate and numerous. Festoons 11, well separated.

Venter

Segment III ventrally with a spur reaching midlength of segment II. Infra-internal setae seven. Hypostomal dental formula 4/4 (Figure. 1D).

Coxa I with a small knob-like internally directed spur. Coxae II to IV with small spurs (Figure 1E). Trochanter without ventrolateral spur. Tarsi stout, tapering gradually to the distal end, claws and pulvilli long and strong (Figure 1G). Anal groove posterior to anus, well marked. Spiracle plate elongate, oval (Figure 1F). Genital groove prominent, not reaching up to festoons. Anal groove with lateral arms almost reaching genital grooves and basal arm almost reaching median festoon. Integument with few scattered small setae.

Distribution

Kerala, India.

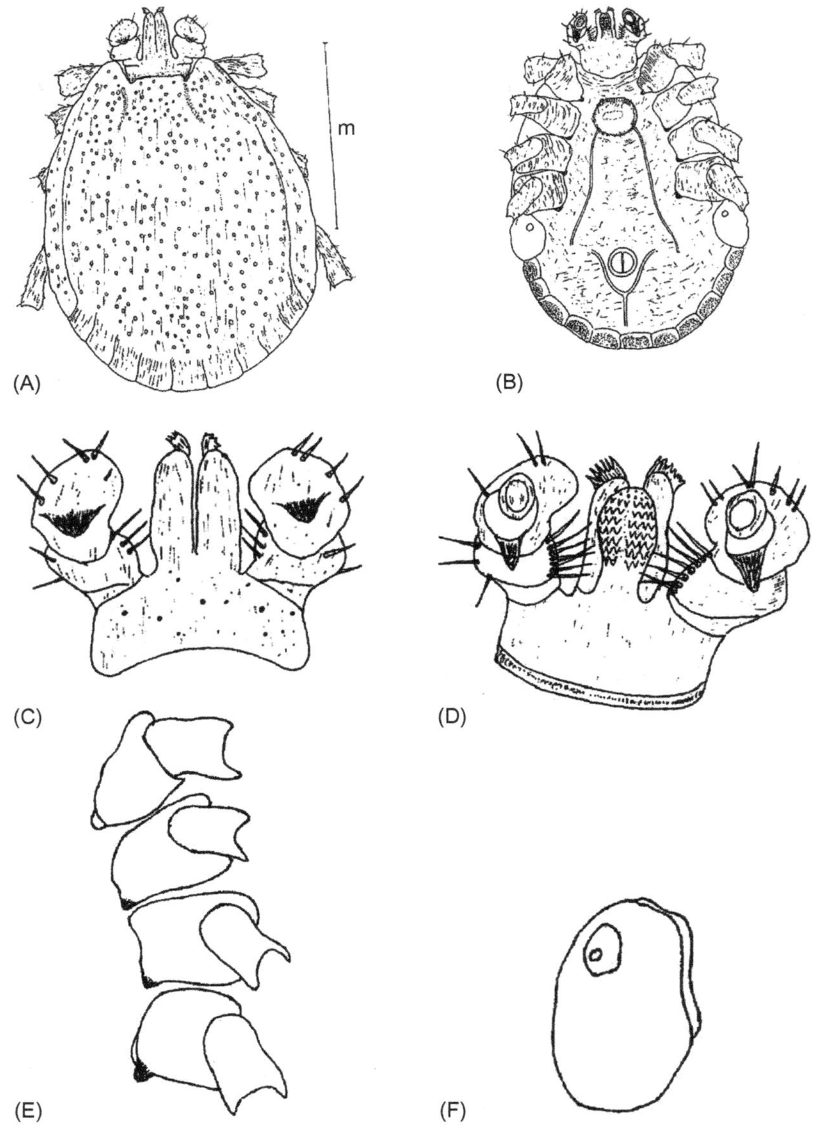

Figure 1 *Haemaphysalis knobigera* sp. nov. (A) Dorsal view; (B) ventral view; (C) capitulum—dorsal view; (D) capitulum—ventral view; (E) legs—coxae and trochanter I–IV; (F) spiracle.

Host

Cow and goat.

Material Examined

Holotype-a" collected from cow, Chelari cattle market, Malappuram district, coll. K. Prakasan on 2-4-2002, presently deposited in the Acarology Division, Department of Zoology, University of Calicut, Calicut, India.

Paratypes-14 <*f*<*f* and 12 9 9 from cow, 18<*f*<*f* from goat Chelari cattle market, Malappuram district, collected by K. Prakasan on 9-4-2002, deposited in Acarology Division, Department of Zoology, University of Calicut.

Remarks

The present specimen resembles *H. intermedia* in general characters. However, the species has been given the status of a new taxon by the possession of the following characters which are unique.

- Cornua less than the size of basis capitulum.
- Ventral palpal spur reaches near to mid-length of palpal segment II.
- Coxa 1 with small knob-like spur.

Appendix II: Scientific and Common Names of Host Species

Barking deer/Indian muntjac	*Muntiacus muntjac vaginalis* Boddaert, 1785
Formosan Reeves muntjac	*Muntiacus reevesi micrurus* Sclater, 1875
Badger, hog	*Arctonyx collaris* subsp.
Badger, stink	*Mydaus multiceps* subsp.
Badger, ferret	*Melogale personata* subsp.
Badger, honey (ratel)	*Mellivora capensis indica* Kerr, 1792
Black buck	*Antilope cervicapra* subsp.
Buffalo	*Bubalus b. bubalis* Linnaeus, 1758
Cat, jungle	*Felis chaus fulvidina* Thomas, 1929, *Felis chaus prateri* Pocock, 1939
Cat, fishing	*Prionailurus viverrinus* Bennet, 1833
Caracal lynx	*Caracal caracal schmitzi* Matschie, 1912
Deer, black Asiatic/Tibetan	*Ursus t. thibetanus* G. [Baron] Cuvier, 1823
Deer, chital	*Axis axis axis* Erxleben, 1777
Deer, mouse	*Tragulus/Moschiola memmina* Erxleben, 1777
Deer, Sambar	*Cervus/Rusa unicolor* Kerr, 1792
Donkey	*Equus asinus* Linnaeus, 1758
Foxes, Bengal	*Vulpes bengalensis* Shaw, 1800
Gaur/Indian bison	*Bos frontalis gaurus* C. H. Smith, 1827
Goat	*Capra hircus* Linnaeus, 1758
Goat, wild	*Hemitragus hylocrius* Ogilby, 1837
Himalayan thar	*Hermitragus jemlahicus* C. H. Smith, 1826
Goat antelope	*Nemorhaedus bubalinus* Hodg
Goral	*Nemorhaedus goral* subsp. Hardwicke, 1825
Hares	*Lepus (Indolagus) nigricollis ruficaudatus* Geoffroy, 1826
Hare, black-napped	*Lepus nigricolis* F. Cuvier, 1823
Himalayan black bear	*Ursus torquatus* Wagn
Indian civets	*Viverra zibetha picta* Wroughton, *Viverra zibetha zibetha* Linnaeus
Indian civet/Small Taiwan civet	*Viverrich/Viverricula indica pallida* Gray
Jackal, Asiatic	*Canis aureus*
Leopard	*Felis (Panthera) pardus*
Marmot rodents	*Marmota c. caudata*

Mice	*Mus booduga* Gray, *Mus musculus homourus* Hodgson, *Mus platythrix* Bennett
Mice, field	*Apodemus flavicollis*
Monkey, capped	*Presbytis pileatus* subsp.
Monkey, langur	*Presbytis entellus*
Monkey, bonnet	*Macaca radiata*
Mongoose (gray), Indian	*Herpested edwardsi*
Mongoose, small Indian	*Herpestes auropunctatus siamensis* Kloss, *H. a. pallipes* Blyth
Mountain voles	*Alticola roylei*
Porcupine, crested	*Hystrix indica*
Porcupine, Bengal	*Hystrix bengalensis*
Porcupine	*Hystrix hodgsoni subcristata*
Porcupine, Asiatic brush tailed	*Atherurus macrourus*
Rats	*Rattus rattus rufescens* Gray, *Rattus r. gangutrianus* Hinton, *Rattus (Millardia) meltada*, *Rattus rattus narbade* Hinton, *Bandicota bengalensis* Gray and Hardwicke, *Rattus blanfordi* Thomas, *Rattus cutchinus*, *Rattus rattoides* Hodgson
Serow	*Capricornis sumatraensis thar* Hodgson, *C. s. humei* Pocock
Sheep	*Ovis aries*
Shrew, tree	*Tupaia glis versurae* Thomas
Shrew	*Suncus murinus*
Squirrel	*Funambulus tristratis tristratis* Waterhouse
Palm squirrel, five striped	*Funambulus pennati* subsp.
Tapir, Malayan	*Tapirus indicus* Desmarest
Tiger	*Felis (Panthera) t. tigris*
Toddy cat/palm civet	*Paradoxurus hermaproditus*
Tortoise	*Geoemyda spinosa*
Wild boar	*Sus scrofa cristatus* (Thailand)
Wild dog, Indian	*Cuon alpinus dukhunensis*
Wild ox/Malaya archipelago	*Bos sondaicus*
Wolf	*Canis lupup pallipes* Sykes
Yak	*Bos grunniens* Linnaeus
Zebu cattle	*Bos indicus*

Bird[50,53]

Babbler, white headed	*Turdoides affinis* Jerdon
Babbler, common	*Turdoides c. caudatus* Dumont
Babbler, large scimitar	*Pomatorhinus h. hypoleucos*
Bulbul	*Pycnonotus jocosa jocosa* Linnaeus

Bulbul, red vented	*Pycnonotus cafer* Linnaeus
Bush lark	*Mirafra assamica* Horsfield
Bush lark, red winged	*Mirafa erythroptera* Blyth
Brown shrike	*Lanius cristatus* Linnaeus
Common wood shrike	*Tephrodornis pondicerianus* Gmelin
Crow pheasant bird	*Centropus sinensis intermedius*
Coucal	*Centropus sinensis sinensis* Stephens
Gray partridge	*Francolinus pondicewrianus* Gmelin
Green backed tit	*Parus monticolus* Vigors
House sparrow	*Passer domesticus indicus* Jardine and Selby
Indian pitta	*Pitta brachyuran* Linnaeus
Indian roller	*Coracias bengalensis*
Lesser coucal	*Centropus bengalensis* Gmelin
Fowl, jungle	*Gallus gallus gallus* Linnaeus
Fowl, jungle (Grey)	*Gallus sonneratti*
Fowl, painted spur	*Galloperdix lunulata* Valenciennes
Fowl, redspur	*Galloperdix spadicea*
Monal pheasant	*Lophophorus impejanus* Latham
Myna (common) bird	*Acridotheres tristis* Linnaeus
Myna, brahminy	*Sturnus pagodarum* Gmelin
Peafowl, common (Peacock)	*Pavo cristatus* Linnaeus
Pied harrier	*Circus melanoleucos* Forster
Partridge, painted	*Francolinus pictus* Jardine and Selby
Partridge, grey	*Francolinus pondicerianus* Gmelin
Robin, magpie	*Consychus saularis* Linnaeus
Robin, Indian	*Saxicoloides fulicata* Linnaeus
Rose colored starling	*Pastor roseus* Linnaeus
Spotted nutcracker	*Nucifraga caryocatactes multipunctata*
Tailor bird	*Orthotomus sutorinus* Pennant
Turtle dove	*Streptopelia turtur* Linnaeus
White throat (small) bird	*Sylvia minuta, Silvia communis* subsp.
Wood shrike, common	*Tephrodornis pondicerianus* Gmelin
Blyth's red warbler	*Acrocephalus dumetorum* Blyth
Malabar grey hornbill	*Tockus griseus*

References

[1] Anastos G. The scutate ticks or Ixodidae of Indonesia. Entomol Am 1950; (new series), 30: 1–144.

[2] Ansastos G. *The ticks or Ixodids of the U.S.S.R., A review of the literature*. U.S. Dept. of Health Education and Welfare; 1957. p. 397.

[3] Arthur DR. Ticks and Diseases. Oxford: Pergamon Press; 1962. p. 445.

[4] Balashov Yu S. Blood sucking ticks (Ixodoidea)- vectors of diseases of man and animals. Entomological Society of America; 1972. pp. 161–376 (miscellaneous publication).

[5] Banerjee K. Kyasanur forest disease. In: Monath TP, editor. The arboviruses: epidemiology and ecology. New York: CRC Press Inc.; 1990. p. 93–116.

[6] Barker SC., Murrell A. Systematics and evolution of ticks with a list of valid genus and species names, Parasitology, 2004, 129, S15–S36.

[7] Bhatt PN, Kulkarni KG, Boshell J, Rajagopalan PK, Patil AP, Goverdhan MK, Pavri KM. Kaisodi virus, a new agent isolated from *Haemaphysalis spinigera* in Mysore state, south India. I: Isolation of strains. Am J Trop Med Hyg 1966;15(6):958–60.

[8] Bhat HR. Localized mass breeding of *Haemaphysalis bispinosa* Neumann, 1897 (Acarina: Ixodidae) in Kyasanur Forest Disease area, Shimoga district, Mysore state, India. J Bombay Nat Hist Soc 1971;68(2):485–9.

[9] Bhat HR, Geevarghese G, Sreenivasan MA. Life history of *Haemaphysalis cornigera shimoga* Trapido and Hoogstraal, 1964 (Acarina: Ixodidae). Indian J Anim Sci 1974; 44(6):370–5.

[10] Bhat HR, Sreenivasan MA, Goverdhan MK, Naik SV. Transmission of Kyasanur Forest Disease virus by *Haemaphysalis kyasanurensis* Trapido, Hoogstraal and Rajagopalan, 1964 (Acarina: Ixodidae). Indian J Med Res 1975;63(6):879–87.

[11] Bhat HR, George JP, Sreenivasan MA. Life history of *Haemaphysalis cuspidata* Warburton, 1910 (Acarina: Ixodidae). Indian J Anim Sci 1975;45(8):546–50.

[12] Bhat HR, Jacob PG. Life history of *Haemaphysalis wellingtoni* Nuttall and Warburton, 1907 (Acarina: Ixodidae). Indian J Med Res 1977;65(1):1–7.

[13] Bhat HR, Naik SV. Transmission of Kyasanur Forest Disease virus by *Haemaphysalis wellingtoni* Nuttall Warburton, 1907 (Acarina: Ixodidae). Indian J Med Res 1978;67:697–703.

[14] Bhat HR. Life history of *Haemaphysalis spinigera* Neumann 1897 (Acarina: Ixodidae). Indian J Anim Sci 1979;49(7):517–22.

[15] Bhat HR, Sreenivasan MA. Further records of the ticks of some reptilian and mammalian hosts in the Kyasanur Forest Disease area, Karnataka, India. Indian J Parasitol 1981;5(2):207–10.

[16] Bhat HR, George JP, Geevarghese G. Pattern of vertical distribution of *Ixodid* ticks (Acarina: Ixodidae) on vegetation in Kyasanur Forest Disease area, Karnataka, India. Indian J Parasitol 1983;7(1):13–8.

[17] Bhat HR. Observations on the biology of *Haemaphysalis spinigera* Neumann, 1897 (Acarina: Ixodidae) under natural conditions in KFD area. J Bombay Nat Hist Soc 1985;82(3):548–62.

[18] Bhat HR, Sreenivasan MA, George JP. *Ixodid* ticks infesting goats in the Kyasanur Forest Disease area, Shimoga district, Karnataka. Indian J Parasitol 1986;10(1):39–45.

[19] Boshell JM, Rajagopalan PK, Patil AP, Pavri KM. Isolation of Kyasanur Forest Disease virus from *Ixodid* ticks: 1961–1964. Indian J Med Res 1968;56(Suppl. 4):541–68.

[20] Boshell JM. Kyasanur Forest Disease: ecologic consideration. Am J Trop Med Hyg 1969;18(1):67–80.

[21] Boshell JM, Desai PK, Dandawate CN, Goverdhan MK. Isolation of Ganjam virus from ticks *Haemaphysalis intermedia. Indian J Med Res* 1970;58(5):561–2.

[22] Dandawate CN, Shah KV. Ganjam virus: a new arbovirus isolated from ticks *Haemaphysalis intermedia* Warburton and Nuttall, 1909 in Orissa, India. Indian J Med Res 1969;57(5):799–804.

[23] Dandawate CN, Singh KRP, Dhanda V. Experimental infection of certain mosquitoes and ticks with Ganjam virus. Indian J Exp Biol 1981;19(12):1185–6.

[24] Dhanda V. Description of immature stages of *Haemaphysalis howletti* (Ixodoidea, Ixodidae) and redescription of adults. J Parasitol 1964;50(2):459–65.

[25] Dhanda V. *Haemaphysalis nepalensis* Hoogstraal, 1962 (Ixodoidea: Ixodidae), systematic position based on description of the nymph, and new host and locality records. J Parasitol 1964;50(6):783–5.

[26] Dhanda V, Ramachandra Rao T. A report on a collection of *Ixodid* ticks made in the North East frontier agency India. Indian J Med Res 1964;52(11):1139–53.

[27] Dhanda V, Bhat HR. *Haemaphysalis* (*Haemaphysalis*) *indoflava* sp. n. (Acarina: Ixodidae) from the Himalayan region of Uttar Pradesh and Madras, India. Indian J Parasitol 1968;54(5):1063–7.

[28] Dhanda V, Bhat HR. *Haemaphysalis* (*Allophysalis*) *garhwalensis* sp. n. (Acarina: Ixodidae) parasitizing sheep and goats in Garhwal region, Uttar Pradesh, India. J Parasitol 1969;54(4):674–8.

[29] Dhanda V, Kulkarni SM. Immature stages of *Haemaphysalis cornupunctata* Hoogstraal and Varma, 1962 (Acarina: Ixodidae) with new host and locality records, and notes on its ecology. Oriental Insects 1969;3(1):15–21.

[30] Dhanda V, Bhat HR. Immature stages of *Haemaphysalis indoflava* Dhanda and Bhat, 1968 (Acarina: Ixodidae), and notes on its ecology. Orient Insects 1970;4:253–8.

[31] Dhanda V, Hoogstraal H, Bhat HR. *Haemaphysalis* (*Kaiseriana*) *ramachandrai* sp. n. (Ixodoidea, Ixodidae), a parasite of man and domestic and wild mammals in northern India and Nepal. J Parasitol 1970;56(4):823–31.

[32] Dhanda V, Bhat HR. *Haemaphysalis* (*Herpetobia*) *sundrai* Sharif (Ixodoidea: Ixodidae) a tick parasitizing sheep in the western Himalayas, redescription of female, description of male, and ecological observations. J Parasitol 1971;57(3):646–50.

[33] Dhanda V, Bhat HR, Kulkarni SM. *Haemaphysalis* (*Allophysalis*) *garhwalensis* Dhanda and Bhat (Acarina: Ixodidae) description of immature stages, hosts, distribution and ecology. Indian J Parasitol 1977;1(1):9–14.

[34] Drumund RO, Rajagopalan PK, Sreenivasan MA, Menon PKB. Tests with *Ixodicides* for the control of the tick vectors of Kyasanur Forest Disease. J Med Entomol 1969;6 (3):245–51.

[35] Doube M, Kemp DH. Notes on the tick *Haemaphysalis* (*Ornithophysalis*) *doenitzi* Warburton and Nuttall (Acarina: Ixodidae) in Australia. J Aust Entomol Soc 18:169–70.

[36] Fernandes S. A review on Indian Ticks (Acari: Ixodoidea) with references to hosts and distribution. Dissertation submitted to Pune University, Pune (India), October 1984.

[37] Filippova NA. Fauna of Russia and neighboring countries; Ixodid ticks of subfamily Amblyomminae, 1997, Arachnoidea, vol. IV, issue 5. Russian Academy of Sciences, Zoological Institute, St. Petersburg: "NAUKA" Publishing house (in Russian).

[38] Geevarghese G, Bhat HR, Sreenivasan MA. Life history of *Haemaphysalis kyasanurensis* Trapido, Hoogstraal and Rajagopalan, 1964 (Acarina: Ixodidae) in the laboratory. Indian J Med Res 1973;61(5):639–44.

[39] Geevarghese G, Sreenivasan MA. Observations of *Hunterullus* sp. (Hymenoptera: Encyrtidae) parasitizing *Haemaphysalis bispinosa* Neumann, 1897 (Acarina: Ixodidae). Indian J Med Res 1973;61(7):992–7.

[40] Geevarghese G. A new species of *Chalcid* (Hymenoptera: Encyrtidae), parasitizing *Haemaphysalis bispinosa* from Karnataka, India. Orient Insects 1977;11(1):49–52.

[41] Geevarghese G, Dhanda V. Studies on Indian species of *Hunterellus* (Hymenoptera: Encyrtidae) parasitizing Ixodid ticks (Acarina: Ixodidae). Indian J Parasitol 1983; 7(1):93–5.

[42] Geevarghese G, Dhanda V. The Indian Hyalomma ticks (Ixodoidea: Ixodidae). New Delhi (India): ICAR Press; 1987. p. 119.

[43] Geevarghese G, Dhanda V. *Ixodid* ticks of Maharashtra state, India. Acarologia 1995;36(4):309–13.

[44] Geevarghese G, Fernandes S, Kulkarni SM. A checklist of Indian ticks (Acari: Ixodoidea). Indian J Anim Sci 1997;67(5):566–74.

[45] Geevarghese G, Kanojia PC. Impact of urbanization on the prevalence of tick fauna in Pune and its suburbs, Maharashtra state, India. Acarologia 2003;43(1):3–7.

[46] Geevarghese G, Mishra AC. Kyasanur Forest Disease—Current status, Published by National Institute of Virology (NIV) Commemorative Compendium, 2004, pp. 406–27.

[47] Ghalsasi GR, Dhanda V. Taxonomy and biology of *Haemaphysalis (Kaiseriana) spinigera* (Acarina: Ixodoidea). Orient Insects 1974;8(4):505–20.

[48] Goodman JL, Dennis DT, Sonenshine DE. Tick-borne diseases of humans. ASM Press; 2005.

[49] Hoogstraal H. African Ixodoidea. Vol. I. Ticks of Sudan. Dept. of Navy, USA: Bureau of Medicine and Surgery; 1956. pp. 1–110.

[50] Hoogstraal H, Kohls GM. The *Haemaphysalis* ticks (Acarina: Ixodidae) of birds, I. *H. ornithophila* N. Sp. from Burma and Thailand. J Parasitol 1959;45(4):417–20.

[51] Hoogstraal H, Varma MGR. *Haemaphysalis cornupunctata* sp. n. and *H. kashmirensis* sp. n. from Kashmir, with notes on *H. sundrai* Sharif and *H. sewelli* Sharif of India and Pakistan (Ixodoidea: Ixodidae). J Parasitol 1962;48(2):185–94.

[52] Hoogstraal H. *Haemaphysalis nepalensis* sp. n. from a Himalayan rodent and man and description of the male of *H. aponommoides* Warburton (N. Comb.) (Ixodoidea; Ixodidae). J Parasitol 1962;48(2):195–203.

[53] Hoogstraal H, Kaiser MN, Traylor MA, Guindy E, Gaber S. Ticks (Ixodidae) on birds migrating from Europe and Asia to Africa, 1959–1961. Bull WHO 1963;28:235–62.

[54] Hoogstraal H, Trapido H. *Haemaphysalis silvafelis* sp. n., a parasite of the jungle cat in southern India (Ixodoidea: Ixodidae). J Parasitol 1963;49(2):346–9.

[55] Hoogstraal H, Trapido H. *Haemaphysalis kutchensis* sp. n., a common larval and nymphal parasite of birds in northwestern India (Ixodoidea: Ixodidae). J Parasitol 1963;49(3):489–97.

[56] Hoogstraal H, Trapido H, Rebello MJ. *Haemaphysalis paraturturis* sp. n., a carnivore parasite of the *H. turturis* group in India (Ixodoidea, Ixodidae). J Parasitol 1963; 49(4):686–91.

[57] Hoogstraal H, Trapido H. Redescription of cotypes and all stages of *Haemaphysalis intermedia* Warburton and Nuttall, 1909 (=*H. parva* Newman, preoccupied) from Ceylon and India (Ixodoidea Ixodidae). J Parasitol 1963;49(5):838–46.

[58] Hoogstraal H, Trapido H, Kohls GM. Studies on Southeast Asian *Haemaphysalis* ticks (Ixodoidea, Ixodidae). H. *atherurus* sp. n. and redescription of type material of *H. birmaniae* Supino, 1897. J Parasitol 1965;51(1):114–25.

[59] Hoogstraal H, Trapido H, Kohls GM. Studies on Southeast Asian *Haemaphysalis* ticks (Ixodoidea, Ixodidae): the identity, distribution and hosts of *H.* (*Kaiseriana*) *hystricis* Supino. J Parasitol 1965;51(2):467–80.

[60] Hoogstraal H, McCarthy VC. Hosts and distribution of *Haemaphysalis kashmirensis* with descriptions of immature stages and definition of the subgenus *Herpetobia* Canestrinii (resurrected). J Parasitol 1965;51:674–9.

[61] Hoogstraal H, Kohls GM, Trapido H. *Haemaphysalis* (*Rhipistoma*) *eupleres* sp. n., from a Madagascar carnivore and definition of the subgenus *Rhipistoma* Koch (resurrected) (Ixodoidea; Ixodidae). J Parasitol 1965;51(6):997–1000.

[62] Hoogstraal H, 1965. Phylogeny of *Haemaphysalis* ticks. Proceedings of the 12th International Congress on Entomology (London, July 1964), pp. 760–61.

[63] Hoogstraal H, Trapido H, Kohls GM. Studies on Southeast Asian *Haemaphysalis* ticks (Ixodoidea: Ixodidae). Speciation in the *H.* (*Kaiseriana*) *obesa* group: *H. semermis* Neumann, *H. obesa* Larrousse, *H. roubaundi* Toumanoff, *H. montgomery* Nuttall, and *H. hirsuta* sp. n. J Parasitol 1966;52(1):169–91.

[64] Hoogstraal H. *Haemaphysalis* (*Allophysalis*) *pospelovashtromae* sp. n. from USSR and redescription of the type material of *Haemaphysalis* (*A.*) *warburtoni* Nuttall from China (Ixodoidea; Ixodidae). J Parasitol 1966;52(4):787–800.

[65] Hoogstraal H. *Haemaphysalis* (*Herpetobia*) *himalaya* sp. n. (Ixodoidea; Ixodidae), a parasite of the himalayan thar (Artiodactyla, Caprinae) in north-western India. J Parasitol 1966;52(4):805–9.

[66] Hoogstraal H, Trapido H. Redescription of the type material of *Haemaphysalis* (*Kaiseriana*) *bispinosa* Neumann (India), *H.* (*K.*) neumanni Doenitz (Japan), *H.* (*K.*) *lagrangei* Larrousse (Vietnam) and *H.* (*K.*) *yeni* Toumanhoff (Vietnam) (Ixodoidea, Ixodidae). J Parasitol 1966;52(6):1188–98.

[67] Hoogstraal H, Kohls GM, Trapido H. Studies on the Southeast Asian *Haemaphysalis* ticks (Ixodoidea: Ixodidae). *H.* (*Kaiseriana*) *anomala* Warburton: redescription, hosts and distribution. J Parasitol 1967;53(1):196–201.

[68] Hoogstraal H, Lim Boo-Liat. *Haemaphysalis* (*Kaiseriana*) *bispinosa* Neumann (Ixodoidea: Ixodidae), evidences for consideration as an introduced species in Malay Peninsula and Borneo. J Parasitol 1969;55(5):1075–7.

[69] Hoogstraal H, Dhanda V. *Haemaphysalis* (*H.*) *darjeeling* sp. n., a member of the *H.* (*H.*) *birmaniae* group (Ixodoidea: Ixodidae) parasitizing artiodactyl mammals in the Himalayan forests of India, and in Burma and Thailand. J Parasitol 1970;56(1):169–74.

[70] Hoogstraal H, Dhanda V, Bhat HR. *Haemaphysalis* (*Kaiseriana*) *davisi* sp. n. (Ixodoidea: Ixodidae), a parasite of domestic and wild mammals in northeastern India, Sikkim, and Burma. J Parasitol 1970;56(3):588–95.

[71] Hoogstraal H. Identity, hosts and hosts of *Haemaphysalis* (*R.*) *indica* Warburton (resurrected) (Ixodoidea: Ixodidae), a carnivore parasite of the Indian subregion. J Parasitol 1970;56(5):1013–22.

[72] Hoogstraal H, El Kammah KM. *Haemaphysalis* (*Herpetobia*) *himalaya* Hoogstraal (Ixodoidea: Ixodidae), description of immature stages, hosts and distribution. J Parasitol 1970;56(5):1023–7.

[73] Hoogstraal H. *Haemaphysalis (H.) birmaniae* Supino and *H. (H.) goral* sp. n. (Ixodoidea: Ixodidae), Asian parasite of Artiodactyl mammals. J Parasitol 1970;56(6):1227–38.
[74] Hoogstraal H. Identity, hosts and distribution of *Haemaphysalis (Rhipistoma) canestrinii* Supino (resurrected), the postulated Asian progenitor of the African *H. leachi* complex (Ixodoidea: Ixodidae). J Parasitol 1971;57(1):161–72.
[75] Hoogstraal H. *Haemaphysalis (H.) sambar* sp. n. (Ixodoidea: Ixodidae), a parasite of the Sambar deer in Southern India. J Parasitol 1971;57(1):173–6.
[76] Hoogstraal H, Saito Y, Dhanda V, Bhat HR. *Haemaphysalis (H.) obesa* Larrousse (Ixodoidea: Ixodidae) from northeast India and Southeast Asia: description of immature stages and biological observations. J Parasitol 1971;57(1):177–84.
[77] Hoogstraal H, Mitchell RM. *Haemaphysalis (Alloceraea) aponommoides* Warburton (Ixodoidea: Ixodidae), description of immature stages, hosts, distribution, and ecology in India, Nepal, Sikkim, and China. J Parasitol 1971;57:635–45.
[78] Hoogstraal H, Dhanda V, El Kammah KM. *Aborphysalis*, a new subgenus of Asian *Haemaphysalis* ticks; and identity, distribution, and hosts of *H. aborensis* Warburton (resurrected) (Ixodoidea, Ixodidae). J Parasitol 1971;57:748–60.
[79] Hoogstraal H. *Haemaphysalis (Allophysalis) warburtoni* Nuttall, description of immature stages, adult structural variation, and hosts and ecology in Nepal, with a redefinition of the subgenus *Allophysalis* Hoogstraal (Ixodoidea: Ixodidae). J Parasitol 1971;57:1083–95.
[80] Hoogstraal H, Dhanda V, Bhat HR. *Haemaphysalis (Kaiseriana) anomala* Warburton (Ixodoidea: Ixodidae) from India: description of immature stages and biological observations. J Parasitol 1972;58(3):605–10.
[81] Hoogstraal H, Wassef HY. The *Haemaphysalis* ticks (Ixodoidea: Ixodidae) of birds, 3. *H. (Ornithophysalis)* subgen. n. Definition, species, hosts, and distribution in the Oriental Palearctic, Malagasy, and Ethiopian faunal regions. J Parasitol 1973;59: 1099–117.
[82] Hoogstraal H. Biology of ticks, in J.K.H. Wilde (Eds.) Tick-borne diseases and their vectors (pp. 3–14). *Proceedings of International Conference*, Edinburgh, September–October 1976.
[83] Hoogstraal H. Changing patterns of tickborne diseases in modern society. Annu Rev Entomol 26: 75–99.
[84] Hoogstraal H, Aeschlimann A. Tick–host specificity. Bull Soc Entomol Suisse 1982;55:5–32.
[85] Hoogstraal H, Kim Ke Chung, Coevolution of parasitic arthropods and mammals, bublished by New York: John Wiley and Sons, Inc. in 1985.
[86] Horak IG, Camicas JL, Keirans JE. The Argasidae, Ixodidae and Nuttallielli (Acari: Ixodida): a world list of valid tick names. Exp Appl Acarol 2002;28:27–54.
[87] Horsefall WR. Medical entomology—arthropod and human diseases. New York: Ronald Press Company; 1962. pp. 467.
[88] Ilkal MA, Dhanda V. Laboratory studies on the life history of *Haemaphysalis intermedia* Warburton and Nuttall 1909 (Acarina: Ixodidae). Indian J Anim Sci 1981;50(2):198–202.
[89] Ilkal MA, Dhanda V, Goverdhan MK, Deshmukh PK. Life history of *Haemaphysalis obesa* Larrousse (Acarina: Ixodidae) and its susceptibility to Kyasanur Forest Disease virus. Indian J Parasitol 1981;5(1):27–32.
[90] Itagaki S, Noda R, Yamaguchi T. Studies on the Ticks of Domestic Animals in the Far East, Japan Society for the Promotion of Science, Tokyo, Japan, 1959.
[91] Jagannath MS, Alwar VS, Lalitha CM. Ixodid ticks of domestic stock in Tamil Nadu. Indian J Anim Sci 1973;43:119–24.

[92] Kaul HN, Mishra AC, Dhanda V, Kulkarni SM, Guttikar SN. Ectoparasitic arthropods of birds and mammals from Rajasthan state, India. Indian J Parasitol 1978;2(1): 19–25.

[93] Kaul HN, Dhanda V, Mishra AC. A survey of *Ixodid* ticks in Orissa state, India. Indian J Anim Sci 1979;49(9):707–12.

[94] Kaul HN, Shetty PS, Ghalsasi GR, Dhanda V. Survey of ticks (Acarina: Ixodidae) for Crimean hemorrhagic fever virus activity in Jammu and Kashmir state, India. Indian J Med Res 1990;91(1):5–8.

[95] Kettle DS. Medical and veterinary entomology. Wallingford, Oxon, U.K.: Cab International; pp. 1–658.

[96] Klompen JSH, Black IV WC, Keirans JE, Oliver Jr JH. Evolution of ticks. Ann Rev Entomol 1996;41:141–61.

[97] Kohls GM. *Haemaphysalis centropi*, a new species of ticks from birds in the far east. J Parasitol 1949;35(3):388–90.

[98] Kohls GM. Ticks (Ixodoidea) of the Philippines, National Institute of Health Bulletin No. 192, Federal Security Agency, Public Health Service, 1950.

[99] Kohls GM. Two new species of ticks from Ceylon (Acarina: Ixodidae). J Parasitol 1950;36(4):319–21.

[100] Kulkarni SM, Naik VM. Laboratory evaluation of six repellents against some Indian ticks. Indian J Med Res 1985;82(7):14–8.

[101] Magaw J. Indian tick typhus. *Indian Med Gazette* 1925;60:58–61.

[102] Miranpuri GS, Naithani RC. A checklist of Indian ticks (Ixodoidea: Acarina). Izatnagar: Indian Veterinary Research Institute; 1978. p. 50.

[103] Mishra AC, Dhanda V, Kulkarni SM. Ectoparasites of small mammals from Western Ghats of Poona district, Maharshtra. J Comm Dis 1977;9(1):40–8.

[104] Nuttall GHF, Warburton C. Ticks. A monograph of the Ixodoidea, part III, genus *Haemaphysalis*. London: Cambridge University Press; 1915. pp. 349–550

[105] Oliver Jr JH. Biology and systematics of ticks (Acari: Ixodida). Annu Rev Ecol Syst 1989;20:397–430.

[106] Padbidri VS, Gupta NP. Rickettsiosis in India: a review. Indian J Med Res 1978;71:104–7.

[107] Padbidri VS, Bhat HR, Rodriguez FM, Joshi MV, Shukla RN. Tick-borne Rickettsioses in Karnataka. Indian J Med Res 1982;75(4):507–14.

[108] Padbidri VS, Rodrigues JJ, Shetty PS, Joshi MV, Rao BL, Shukla RN. Tick-bornerickettsioses in Pune district, Maharashtra, India. Int J Zoon 1984;11(1):45–52.

[109] Pavri KM, Casals J. Kaisodi virus: a new agent isolated from *Haemaphysalis spinigera* in Mysore state, south India. II. Characterization and identification. Am J Trop Med Hyg 1966;15(6):961–3.

[110] Pomerantzev BI. Arachnoidea, Vol. 4, Ixodid Ticks (Ixodidae). Washington: American Institute of Biological Science; 1959.

[111] Prakashan K, Ramani N. Two new species of Ixodid ticks (Acarina: Ixodida) from Kerala, India. Intern J Zool Res 2007;3(4):169–77.

[112] Rajagopalan PK. *Haemaphysalis megalaimae* sp. n., a new tick from the Small Green Barbet, *Megalaima virdis* in India. J Parasitol 1963;49(2):340–5.

[113] Rajagopalan PK, Patil AP, Boshell J. Studies on *Ixodid* ticks populations on the forest floor in the Kyasanur Forest Disease area (1961–1964). Indian J Med Res 1966;56 (Suppl. 4):497–509.

[114] Rajagopalan PK, Patil AP, Boshell JM. *Ixodid* ticks on their mammalian hosts in the Kyasanur Forest Disease area of Mysore state, India, 1961–1964. Indian J Med Res 1968;56(4Suppl. 4):510–26.

[115] Rajagopalan PK, Anderson CR. Transmonsoonal persistence of Kyasanur Forest Disease virus in *Haemaphysalis nymphs* infected in nature. Indian J Med Res 1970;58 (9):1184–7.

[116] Rajagopalan PK, Sreenivasan MA, Paul SD. Isolation of Ganjam virus from the bird tick *Haemaphysalis wellingtoni* Nuttall and Warburton 1907. Indian J Med Res 1970;58(9):1195–6.

[117] Rajagopalan PK, Anderson CR. Further studies on ticks of wild monkeys of Kyasanur Forest Disease area, Shimoga district. Indian J Med Res 1971;59(6):847–60.

[118] Rajagopalan PK. *Ixodid* ticks (Acarina: Ixodidae) parasitizing wild birds in Kyasanur Forest Disease area of Shimoga district, Mysore state, India. J Bombay Nat Histol Soc 1972;69(9):55–78.

[119] Rajagopalan PK, Sreenivasan MA. *Ixodid* ticks on cattle and buffaloes in the Kyasanur Forest Disease area of Karnataka state. Indian J Med Res 1981;73(6): 880–9.

[120] Ramachandra Rao T, Dhanda V, Bhat HR, Kulkarni SM. A survey of haematophagous arthropods in western Himalayas, Sikkim and hill districts of West Bengal—a general account. Ind J Med Res 1973;61(10):1421–61.

[121] Rebello MJ, Reuben R. A report on ticks collected from birds and small mammals in North Arcot and Chittoor district, south India. J Bombay Nat Hist Soc 1966;63(2): 283–9.

[122] Rodrigues FM, Padbidri VS, Ghalsasi GR, Gupta NP, Mandke VB, Pinto BD, Hoon RS, Bapat MB, Mohan Rao CV. Prevalence of Crimean haemorrhagic fever - Congo virus in Jammu and Kashmir state. Indian J Med Res 1986;84:134–8.

[123] Roy, D.N. and Brown, W.A. 1970. Entomology (medical & veterinary) including insecticides & insect & rat control. Third edition. Bangalore Printing and Publishing, Bangalore.

[124] Santos Dias. About two entities of the kind *Haemaphysalis* C. L. Koch, 1844, Memorias e Estudos do Museu Zoologic da Universidade de Coimbra; 1956, pp. 1–9.

[125] Sartbaev SK. Materials on the biology of *Haemaphysalis warburtoni* Nutt., 1912, under Kyrgyz conditions (preliminary information) (In Russian), *Tr Inst Zool Parazit Akad Nauk Kirgiz SSR* 1912; 4:121–7 [in English: NAMRU2- T12].

[126] Sen Purnendu A. Check- and host—list of Ixodoidea (ticks) occurring in India. Ind J. Vet Sci Anim Husb 1938;8(15):133–47.

[127] Sharif M. A revision of the Indian Ixodidae with special reference to the collection. *Indian Museum, Records of Indian Museum*, Vol. 30, part 3, Zoological Survey of India Publications, 1928.

[128] Sharma BD. Ticks and diseases in the Kumaon Himalayas, Uttarakhanda Bharti, vol. 2, No. 1.

[129] Singh KRP. A note on tick paralysis in rabbits. Curr Sci 1963;32(3):116.

[130] Singh KRP, Pavri KM, Anderson CR. Experimental transovarial transmission of Kyasanur Forest Disease virus in *Haemaphysalis spinigera*. Nature 1963;199 (4892):513.

[131] Singh KRP, Pavri KM, Anderson CR. Transmission of Kyasanur Forest Disease virus by *Haemaphysalis turturis, Haemaphysalis papuana kinneari* and *Haemaphysalis minuta*. Indian J Med Res 1964;52(6):566–73.

[132] Singh KRP, Goverdhan MK, Ramachandra Rao T. Experimental transmission of Kyasanur Forest Disease virus to small mammals by *Ixodes petauristae, I. ceylonensis* and *Haemaphysalis spinigera*. Indian J Med Res 1968;56(Suppl. 4):594–609.

[133] Sonenshine DE. Biology of Ticks, Vol. 1 and 2. Oxford University Press; 1991.

[134] Sreenivasan MA, Rajagopalan PK, D'lima LV. Isolation of Kaisodi virus from Ixodid ticks in the Kyasanur Forest Disease area of Shimoga district, Mysore state, India. Indian J Med Res 1973;61(1):15–22.

[135] Sreenivasan MA, Bhat HR, Rajagopalan PK. Studies on the transmission of Kyasanur Forest Disease virus by partly fed ixodid ticks. Indian J Med Res 1979;69:708–13.

[136] Sreenivasan MA, Rajagopalan PK, Bhat HR. Spatial distribution of infected *Haemaphysalis* nymphs in the epizootic localities of Kyasanur Forest Disease. Indian J Med Res 1983;78(10):531–6.

[137] Sreenivasan MA, Rajagopalan PK, Bhat HR. Isolation of Kyasanur Forest Disease virus from Ixodid ticks collected between 1965 and 1972. Prog Acarol 1989;1(1): 37–44.

[138] Trapido H, Hoogstraal H, Varma MGR. Redescription of the holotype and description of all stages of *Haemaphysalis turturis* Nuttall and Warburton from Ceylon and India (Ixodoidea, Ixodidae). J Parasitol 1963;49(4):678–85.

[139] Trapido H, Goverdhan MK, Rajagopalan PK, Rebello MJ. Tick ectoparasitic on monkeys in the Kyasanur Forest Disease area of Shimoga district, Mysore state, India. Am J Trop Med Hyg 1964;13(2):763–72.

[140] Trapido H, Hoogstraal H, Varma MGR. Status and description of *Haemaphysalis p. papuana* Thorell (n. comb.) and of *H. papuana kinneari* Warburton (n. comb.) (Ixodoidea, Ixodidae) of Southern Asia and New Guinea. J Parasitol 1964;50(1): 172–88.

[141] Trapido H, Hoogstraal H, Rajagopalan PK. *Haemaphysalis kyasanurensis* sp. n., a member of the formosensis group in southern India and Ceylon (Ixodoidea, Ixodidae). J Parasitol 1964;50(2):295–302.

[142] Trapido H, Hoogstraal H. *Haemaphysalis cornigera shimoga* sub sp. n., from southern India (Ixodoidea, Ixodidae). J Parasitol 1964;50(2):303–10.

[143] Trapido H, Varma MGR, Rajagopalan PK, Singh KRP, Rebello MJ. A guide to the identification of all stages of the *Haemaphysalis* ticks of south India. Bull Entomol Res 1964;55(5):249–70.

[144] Varma MGR, Trapido H, Rajagopalan PK. Studies on ticks as possible vectors of Kyasanur Forest Disease. Proceedings of the Ninth Pacific Science Congress 1957;17:88–90.

[145] Varma MGR, Webb HE, Pavri KM. Studies on the transmission of Kyasanur Forest Disease virus by *Haemaphysalis spinigera* Newman. Trans R Soc Trop Med Hyg 1960;54(6):509–16.

[146] Warburton C. On some new and obscure species of the genus *Haemaphysalis* of the Ixodidae. Proceedings of Cambridge Philosophical Society, 1908, Vol. XIV, Part V, pp. 508–19.

[147] Yamaguti Tipton, Keegan Toshioka. Ticks; *Haemaphysalis campanulata* Ecology and Biology. Brigham Young Univ Sci Bull, Biol Ser 1971;15(1):43–9.

www.ingramcontent.com/pod-product-compliance
Lightning Source LLC
LaVergne TN
LVHW052304200926
848481LV00002B/460
* 9 7 8 0 3 2 3 1 6 5 2 7 3 *